D0821041

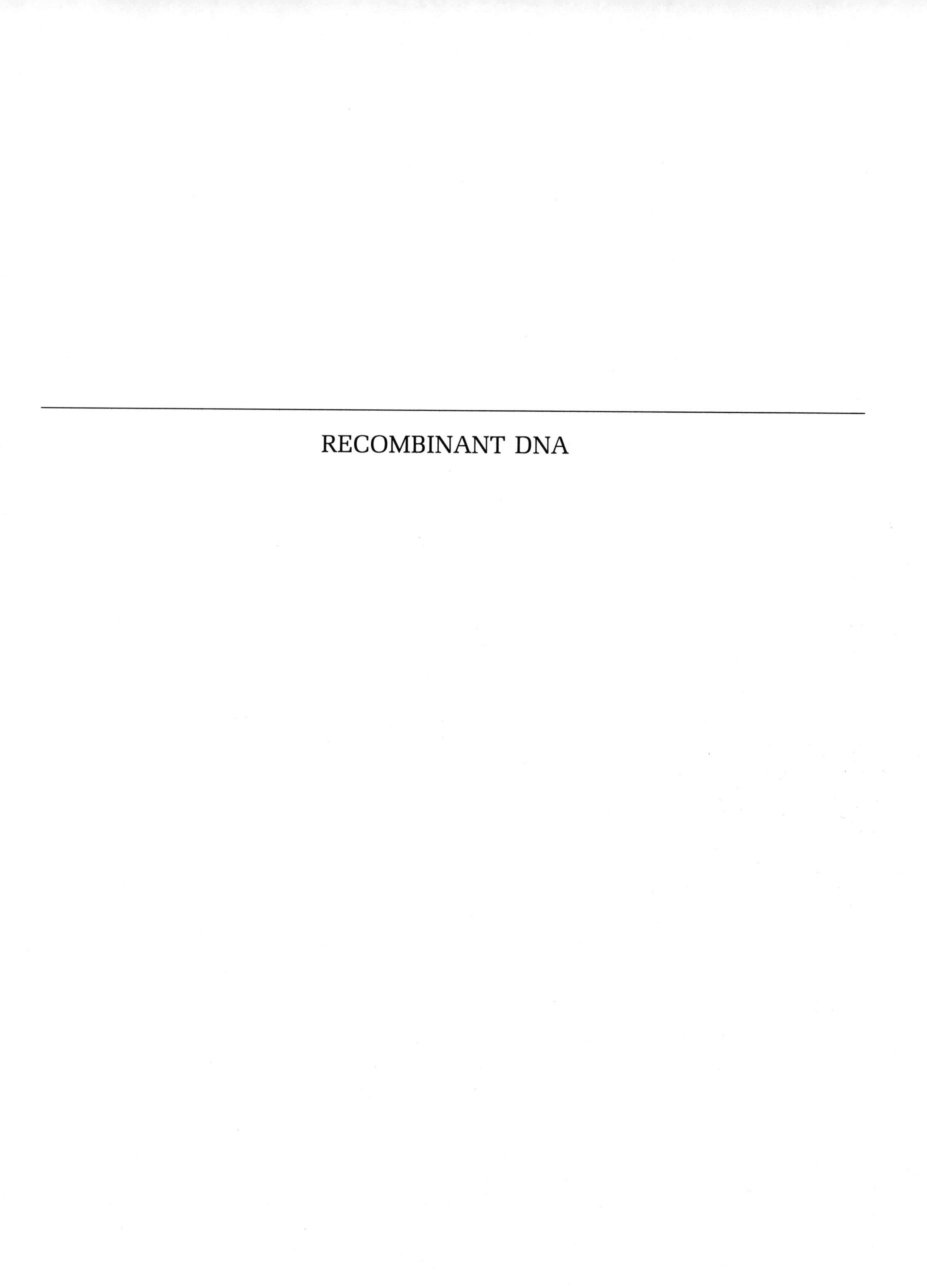

RECOMBINANT DNA

Readings from
SCIENTIFIC AMERICAN

RECOMBINANT DNA

With Introductions by
David Freifelder
Brandeis University

W. H. Freeman and Company
San Francisco

Most of the *Scientific American* articles in RECOMBINANT DNA are available as separate Offprints. For a complete list of more than 1200 articles now available as Offprints, write to W. H. Freeman and Company, 660 Market Street, San Francisco, California 94104.

Library of Congress Cataloging in Publication Data

Main entry under title:

Recombinant DNA.

Bibliography: p. 145
Includes index.
1. Recombinant DNA—Addresses, essays, lectures.
I. Freifelder, David Michael, 1935- II. Scientific American.
QH442.R4 575.2 77-29159
ISBN 0-7167-0093-X
ISBN 0-7167-0092-1 pbk.

Printed in the United States of America

9 8 7 6 5 4 3 2 1

PREFACE

Biological research has many practical objectives: the elimination of disease, for example, increased productivity of the soil, and the general improvement of the quality of life. However an equally motivating one is the satisfaction of man's curiosity about the nature of living things. Until about a century ago the study of biology consisted mainly of simple observations: good eyes, patience, intelligence, and interest were all that were necessary to be a good biologist. Today the research is highly sophisticated and very expensive so that much of it requires government financing; consequently it is the practical objectives that are usually cited as justification for this expenditure of tax money. Although this support, amounting to hundreds of millions of dollars annually, *should* create public interest in biological research, it usually does not, because most of the work is either in basic research, which is not easily understood by the layman, or on medical problems that take years to solve. Consequently the public is unaware of the substantial advances made in our understanding of living systems. Rather the types of items that tend to be reported by the news media are the occasional development of new drugs and clinical techniques and the flashy experiments aimed at creating life in a test tube.

Recently attention has focused on recombinant DNA research, in which genes are manipulated in an elegant way to create new types of microorganisms. The interest has arisen primarily because scientists have recognized the possible existence of potential hazards in the work. These hazards were first discussed in closed meetings and then reported in journals that are widely read by the scientific community. Genuine concern for public safety, coupled with memories of the extreme criticism directed against atomic scientists after the end of World War II, rapidly caused recombinant DNA research to become a subject of controversy, not only within the scientific community but throughout the nation. The issue has been debated in public panel discussions, in the newspapers, on television, and finally in Congressional hearings. Its importance makes it essential that both the scientific community and the lay public understand what recombinant DNA is, and the pros and cons; this is the purpose of this book.

To begin with, let us trace the way in which this technique of manipulating genetic material developed.

In the early period of what we like to call "modern biological research," a new technology developed—the ability to change microorganisms, seemingly at will. This was first done by *mutagenesis,* that is, the introduction of chemical modifications of DNA, so that a *mutation,* or heritable change in the structure of the genetic material, has been induced. The organism with the new character was called a *mutant.* The technique of *genetic recombination* ("crossing over") was then used to create an organism carrying two mutations from two cells each containing a single but different mutation. This technique worked only if the two cells were variants of the same species. It was an extremely important one, and led to the development of a new philosophy in basic biological research—that is, if a biological system is too complicated to study as it exists in nature, it should be genetically altered by mutation and recombination so that it can yield information in simpler experiments. This concept proved especially productive for François Jacob and Jacques Monod who were studying the regulation of the genetic system in *E. coli* responsible for metabolism of the sugar lactose. They created thousands of *E. coli* mutants and recombinants that were altered in lactose metabolism; by studying the properties of the variants, they uncovered many of the general rules that govern gene expression, and for their beautiful work received the Nobel Prize. Their techniques for construction of bacterial strains were rapidly adopted by hundreds of laboratories throughout the world, and it is now possible to construct an *E. coli* strain having any combination of characteristics *already present* in *E. coli.*

In the early 1960's Julius Marmur applied similar methods of strain construction, and as a result isolated recombinants that contained the DNA of two distinct species of bacteria, *E. coli* and *Serratia marcescens.* This work was extended to other bacteria, and new hybrid species were created. Apart from the fact that these hybrid bacteria facilitated the isolation of certain classes of DNA molecules, the work had little impact.

In 1973 the ability to induce genetic exchanges was significantly advanced. Ronald Davis and Janet Mertz at Stanford University were studying a class of *E. coli* enzymes called *restriction endonucleases.* They discovered that these enzymes could be used to break isolated DNA molecules into fragments that could then be reassembled in any desired sequence. This method was applicable to *any* DNA molecule and allowed the splicing together of large blocks of genes from the same or diverse organisms. Shortly afterwards Stanley Cohen, also at Stanford, used this method to transfer genes from one organism to another (e.g., from frog to *E. coli*). His work differed from Marmur's in that it was not only DNA from a *bacterium* that could be introduced into *E. coli,* but the DNA molecule from *any* organism—bacterium, plant, or human.

This technique of producing recombinant DNA is called *genetic engineering,* or *cloning.* It has generated a controversy—possibly of unsurpassed magnitude—between scientists, educators, politicians, and the public. At the one extreme are those who foresee great benefits to mankind ensuing from the ability to create new microorganisms; at the other, those who envision ecological danger, if not calamity, in the production of organisms that have not been screened by a half billion years of Darwinian selection. (In principle this is the same worry that NASA has about introducing Moon bacteria, if they exist, into an Earth environment). Another aspect of the entire issue is the important question of the extent to which the public and the government should be allowed to regulate the activity of scientists.

Now that one city government (Cambridge, Massachusetts) has chosen to regulate recombinant DNA research, and the federal government is considering a similar measure, the issue has become clearly political, and the public, which may have to vote on it some day, ought to have a sound understanding of the various arguments. It is my hope that this collection of articles, originally prepared on a smaller scale for my students at Brandeis University, will provide the necessary background.

This book is divided into three sections. In Section I the fundamental biological concepts are presented, and we see that the information, which determines the properties of all living cells and most viruses, resides in DNA. Thus, if the DNA molecules of two different organisms could be cut with microscissors and the pieces taped together in a new combination, the result would be yet a third organism, with new properties. Section II describes the source of the scissors, how the taping is done, and how genes are isolated. In Section III the value and hazards of genetic engineering are explored, and the controversy analyzed. The purpose of this final section is to present both the facts and the arguments without bias.

April, 1978 *David Freifelder*

CONTENTS

Note on cross-references to SCIENTIFIC AMERICAN *articles:* Articles included in this book are referred to by title and page number; articles not included in this book but available as Offprints are referred to by title and offprint number; articles not included in this book and not available as Offprints are referred to by title and date of publication.

I

BASIC MOLECULAR BIOLOGY

I BASIC MOLECULAR BIOLOGY

INTRODUCTION

The creation of new organisms requires no more than manipulations of DNA molecules, because it is the chemical structure alone of the DNA of a particular cell that determines all the properties and potentials of that cell. This molecule has a remarkable double-stranded helical structure, described by Francis H. Crick in the first article. Each strand consists of a chain of sugar (deoxyribose) molecules, each deoxyribose being separated by a phosphate group. Attached to every sugar is a side group called a *base*. There are four bases that occur in DNA—adenine, cytosine, guanine, and thymine. The order, or sequence, in which these bases occur specifies all of the biological properties of a cell containing that strand of DNA. Biologists prefer to restate this by saying that the total information possessed by a cell or an organism resides in its DNA.

As a first step in obtaining the information from the DNA, the sequence of bases on one of the DNA strands is "read" or *transcribed* by an enzyme, *RNA polymerase,* which synthesizes a *messenger RNA* molecule whose base sequence is complementary to that of the transcribed DNA strand. For example, if the sequence in the DNA strand is adenine, thymine, guanine, cytosine, the corresponding sequence in the RNA will be uracil, adenine, cytosine, guanine. The synthesis of messenger RNA is described in the article by Vincent Allfrey and Alfred Mirsky.

The base sequence of the messenger RNA is *translated* into the amino acid sequence of a protein by means of the *genetic code,* as explained in the second article by Crick. That is, the four bases can be arranged in 64 distinct groups of three bases called *triplets* or *codons.* Sixty-one of these are matched to one of the 20 amino acids (there are more than 20 amino acid codons because many amino acids are matched to codons) and three are *stop codons* signaling the end of a protein molecule. This translation process requires many enzymes and a set of molecules called *transfer RNA* (tRNA) molecules, which actually align the amino acids according to the codon sequence; translation occurs on structures called *ribosomes,* whose properties are described in the article by Masayasu Nomura. Owing to the fidelity of each step in the processes of transcription and translation, the sequence of bases in DNA uniquely specifies the sequence of amino acids of each protein in the cell. In turn, the amino acid sequence of a protein determines all of the chemical and physical properties of the protein, as explained in the article by William H. Stein and Stanford Moore.

Finally the combined properties of all of the proteins determine what the cell is and what it can do. Thus, *one need only introduce a new DNA molecule into a cell to alter the capabilities of that cell.* The single complication, which is only poorly understood, is that, although all organisms—bacteria and mammals—use the same genetic code, the presence of animal DNA in a bacterium does not always lead to protein production. Presumably some of the signals for initiation of transcription and translation are not universal; however, this difficulty has been overcome in some experiments by a kind of genetic trickery, in which recombinant DNA techniques have been used to remove an ineffective initiation signal (the start point for synthesis of messenger RNA) and replace it with an effective one.

SUGGESTED ADDITIONAL READING FOR PART I

BIOCHEMISTRY. L. Stryer. W. H. Freeman and Company, 1975. This book contains beautiful multicolor drawings showing the processes described in Section I.

1 Nucleic Acids

by F. H. C. Crick
September 1957

These polymers appear to carry the pattern of living matter from one generation to the next. Their basic chain consists of sugars joined by phosphates. Attached to the sugars, in turn, are bases

If proteins are the principal stuff of life, the nucleic acids are its blueprints—the molecules on which the Secret of Life, if we may speak of such a thing, is written. The nucleic acids occur in every living cell. It seems, according to our best present information, that they direct the manufacture of proteins and hold the key to the hereditary constitution of all living things. Like the proteins, the nucleic acids are high polymers, but they are polymers with a difference. If we ever achieve a complete understanding of their construction and behavior, we shall probably have the answer to how nature goes about forming each living organism.

In this article I shall give my own view of the meaning of the facts learned about the nucleic acids so far. A great deal of work has been done on these substances, and there is considerable room for disagreement in interpreting the findings. But it is possible to form a general theory which seems to fit most of the known facts and serves as an attractive working hypothesis.

There are two kinds of nucleic acid: DNA, short for deoxyribonucleic acid, and RNA, ribonucleic acid. DNA is always found in the nucleus of the cell, RNA mainly in the cytoplasm outside the nucleus. Chemically they are very similar. Each consists of a long chain of phosphate and sugar molecules with small side groups (called bases) attached to the sugars [*see diagrams on pages 6 and 7*]. In DNA the sugar is deoxyribose; in RNA it is a very slightly different molecule called ribose. The two also differ in one of the four bases attached as side groups. Both contain adenine, guanine and cytosine, but in the case of DNA the fourth base is thymine, whereas in RNA it is uracil.

The bases along the backbone of the nucleic acid do not follow a regular order (such as ABCDABCD and so on). We have some reason to believe that the sequence in each case has a particular meaning and determines the function of the molecule, just as the sequence of letters in this sentence conveys my meaning to you as you read. But more about this later.

Electron-microscope pictures show that pieces of DNA are long and rather stiff, like a piece of cord. And by X-ray analysis it has been found that DNA is actually a double molecule, with one chain twined around the other in helical fashion. The bases of one chain fit neatly onto the bases of the other. But in order to fit, a given base on one chain must be opposite a particular one on the other: guanine pairs only with cytosine and adenine only with thymine. Thus the sequence of bases on one chain determines the sequence on the other.

Much less is known about the structure of RNA, but some progress has been made in creating simpler synthetic analogues. One of these polymers has only the base adenine attached to the backbone, and is known as polyadenylic acid—or Poly A to its friends. Under certain circumstances Poly A probably takes the form of two chains wound around each other, with the adenine of one chain fitting onto the adenine of the other. Another synthetic analogue to RNA has been made with uracil as the base—it is known as Poly U. When Poly U and Poly A are put together in a solution, a remarkable combination takes place: the two chains join in an intertwined helical structure like that of DNA. It seems very likely that the structure is held together mainly by hydrogen bonds between the adenines on one chain and the uracils on the other. Alexander Rich and his colleagues at the National Institutes of Health in Bethesda, Md., who deciphered this structure, have found that in the presence of magnesium the mixture of Poly A and Poly U seems to form a three-chain structure, one Poly A chain joining up with two Poly U chains. The details of this new structure are eagerly awaited. Other interesting questions arise. Does Poly G (made with guanine) pair up with Poly C (containing cytosine), as the DNA pairing would lead us to expect? Unfortunately it seems very difficult to produce a good Poly G, so for the present we cannot answer this question.

There are indications that natural RNA consists of two chains, but it gives rather poor X-ray pictures, suggesting an irregular structure. Whether the disorder is inherent in the molecule or is produced during the extraction of RNA from the cell we cannot say, because we have little information to tell us what RNA is like inside living cells. Perhaps in the cell it assumes an orderly and significant configuration only when it is combined with DNA or protein.

Some progress has recently been made toward synthesizing the nucleic acids with the help of enzymes extracted from living cells. Marianne Grunberg-Manago and Severo Ochoa at New York University discovered an enzyme system in certain bacteria with which they were able to make RNA-like molecules, as well as Poly A and Poly U. Arthur Kornberg and his colleagues at Washington University in St. Louis found a different system, also in bacteria, which produces DNA-like material. The RNA-type polymers were made from diphosphates of the nucleotides corresponding to the four natural bases (adenine, uracil, guanine and cytosine). When all four diphosphates were provided at the same

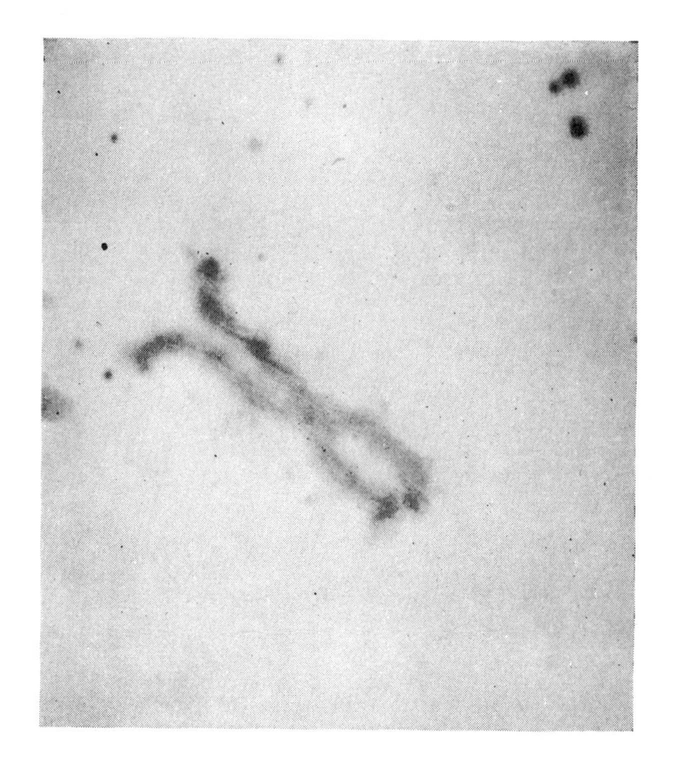

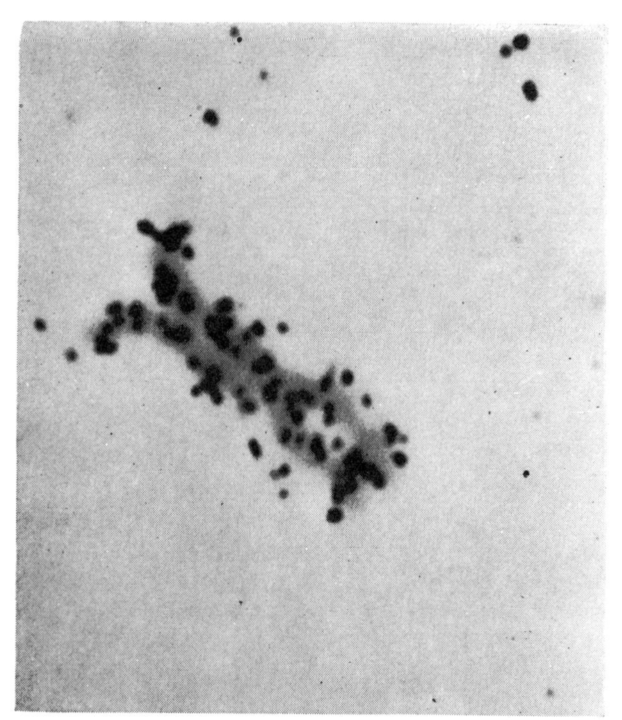

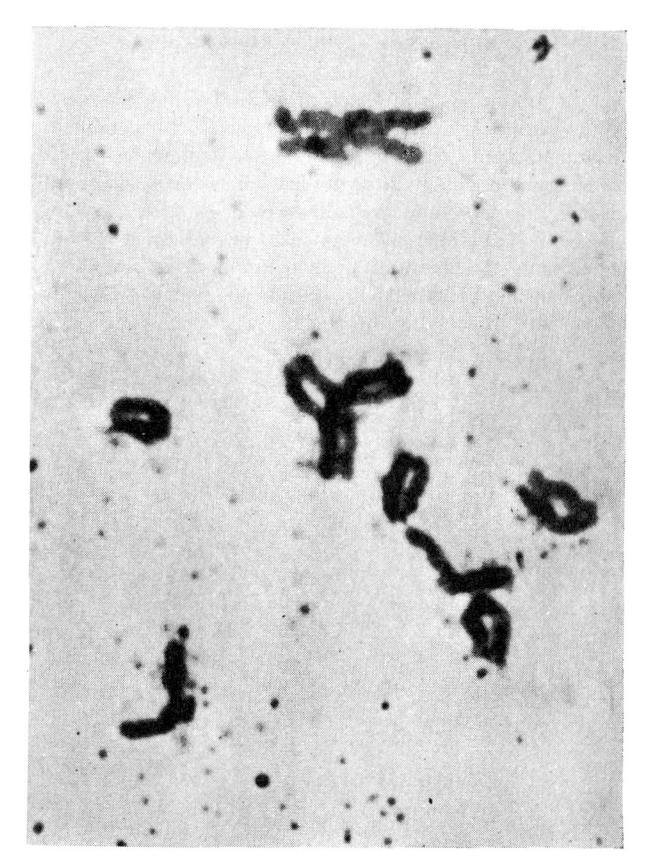

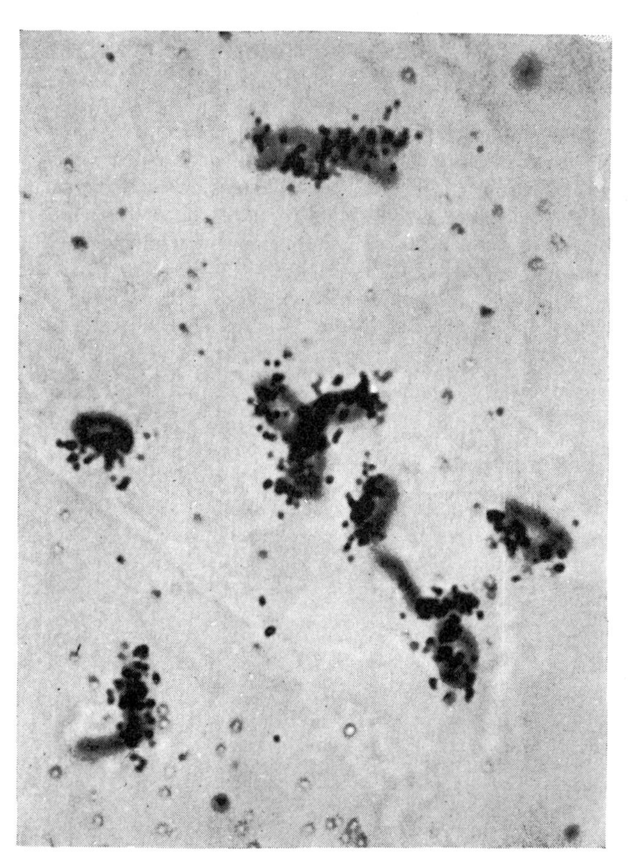

CHROMOSOMES in these photomicrographs are in cells of broad bean seedlings grown in an experiment by J. Herbert Taylor of Columbia University and Philip S. Woods and Walter L. Hughes of Brookhaven National Laboratory. The seedlings were grown in a medium containing thymidine labeled with radioactive hydrogen (tritium). When the cells synthesized new chromosome material, the DNA contained labeled thymine. The photomicrograph at upper left shows a single bean chromosome. The photomicrograph at upper right shows a layer of photographic emulsion above the same chromosome. Each dark spot in the emulsion was made by a particle emitted in the decay of a tritium atom. After the cells had produced one generation of labeled DNA, they were transferred to a medium containing no radioactive thymidine. The photomicrograph at lower left shows several paired chromosomes in a cell two divisions after labeling. The photomicrograph at lower right shows a layer of photographic emulsion above the same paired chromosomes. The spots in this photomicrograph roughly indicate that only one chromosome of each pair is radioactive.

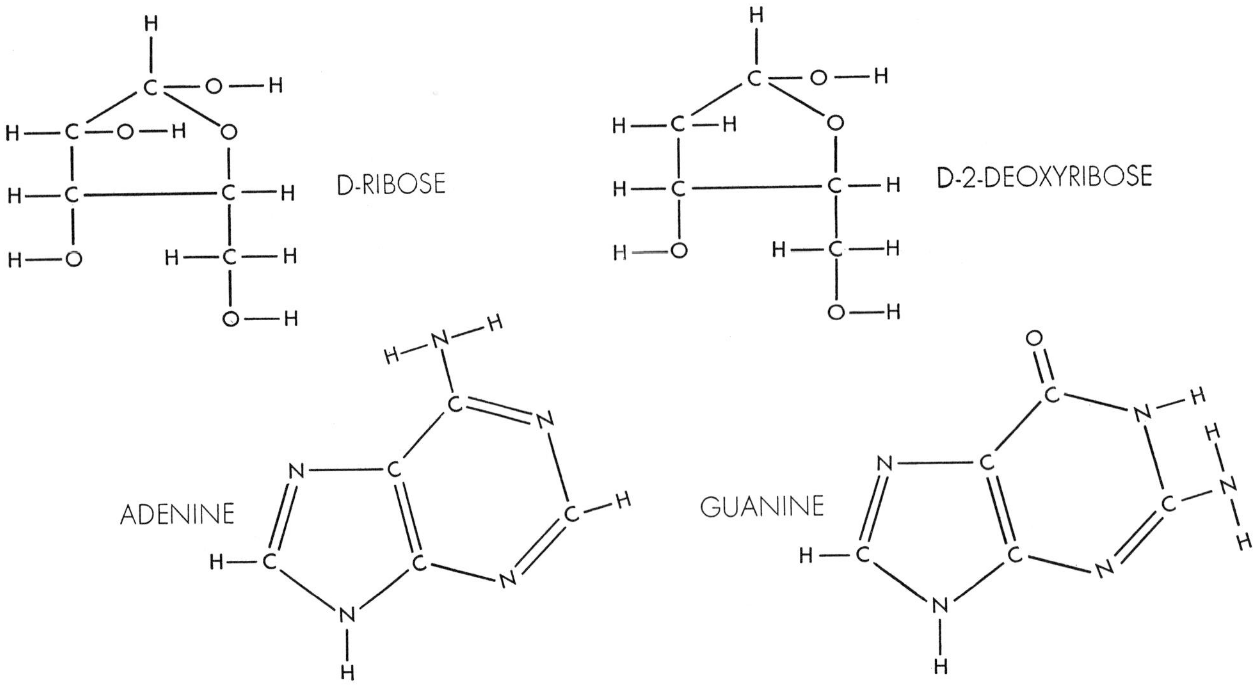

D-RIBOSE

D-2-DEOXYRIBOSE

ADENINE

GUANINE

BASES AND SUGARS forming the basic units of nucleic acids are diagrammed. The D-ribose sugar is the basis of ribonucleic acid (RNA) and gives it its name. Deoxyribonucleic acid contains the deoxyribose sugar. Adenine and guanine, which are found in

time, a product called Poly AUGC was formed. This is very similar to natural RNA, but so far has not shown any biological activity. If the sequence of the bases in natural RNA accounts for its biological behavior, this synthetic material may turn out to be "nonsense RNA." But then perhaps some *natural* RNA is nonsense RNA. Who knows? In any event the enzyme system that makes this substance seems to be widely distributed among bacteria, and is presumably there for some purpose.

The system that synthesized Kornberg's DNA-type material requires the triphosphate instead of the diphosphate of the nucleotides, and of course deoxyribose rather than ribose. The exact requirements of the system were at first obscure: the brilliant work of Kornberg and his colleagues is a fine example of biochemical order being extracted from confusion. Two conditions appear to be necessary. First, all four nucleotides

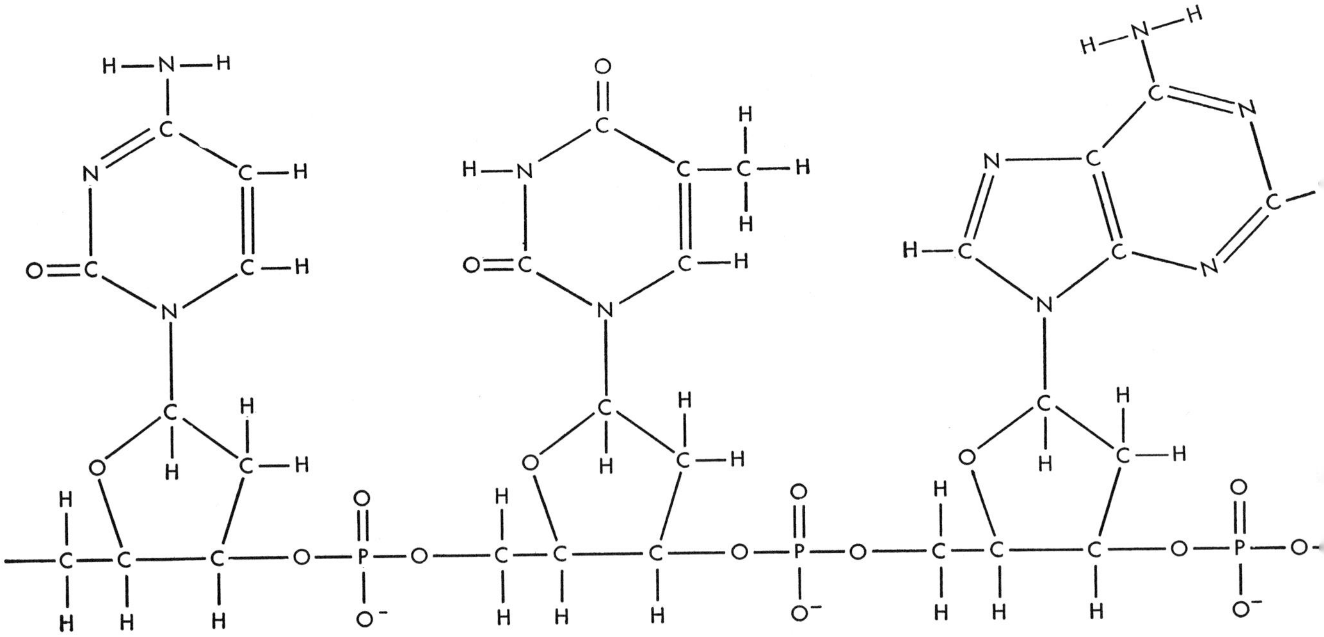

NUCLEIC ACID CHAIN is made up of a backbone of sugar molecules linked by phosphate groups. Each sugar link carries a side chain of a base like those at the top of the page. The chain above has side chains of cytosine, thymine, adenine, thymine, cytosine and

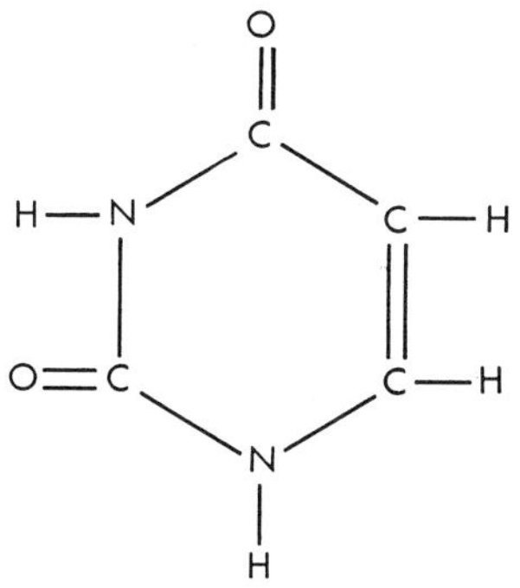

URACIL

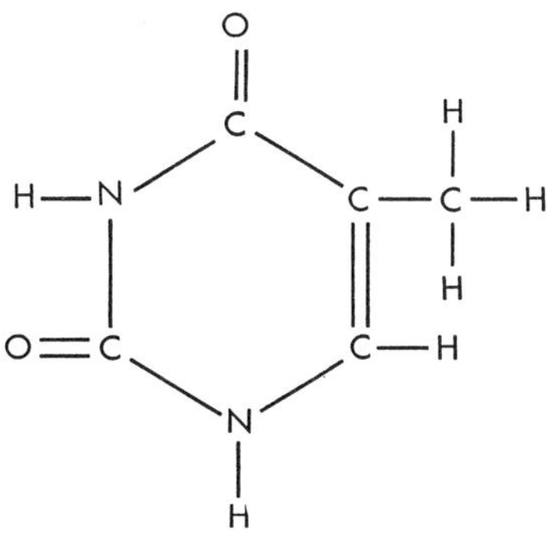

THYMINE

CYTOSINE

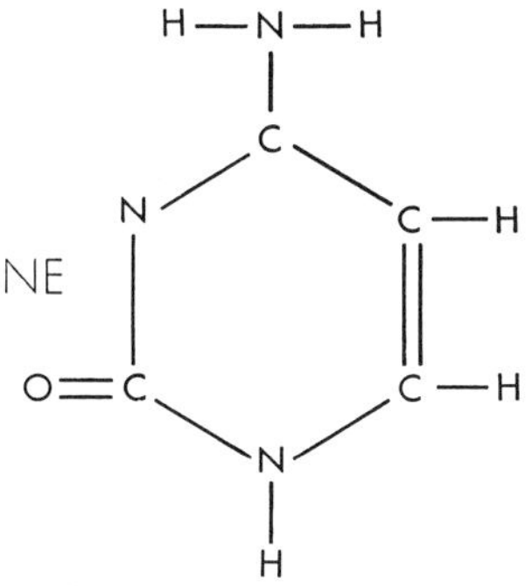

5-METHYL CYTOSINE

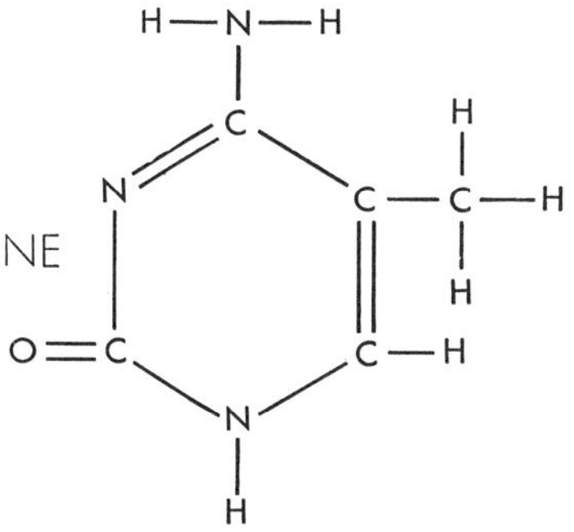

both DNA and RNA, are bases of the purine type. The other bases shown are all pyrimidines. Cytosine is common to both DNA and RNA, but uracil is found only in RNA. In DNA thymine occurs instead of uracil. The 5-methyl cytosine, found in DNA, is less common.

must be present simultaneously: if any one is omitted, little or no synthesis takes place. Secondly, the polymerization will not proceed unless some natural DNA is present as a primer. The experiment has provoked great interest and excitement, because these two conditions suggest the possibility that the synthetic "DNA" is produced by replication of the priming DNA, rather than assembled from the raw materials at random. It has not yet been possible to show that the synthetic DNA is biologically active, but every biochemist arriving in England from the States is eagerly asked if he knows whether Kornberg has done this yet. Perhaps by the time this article is published . . .

There are many reasons to suspect that DNA is either the genetic material of life (what used to be called the genes) or an important part of it. DNA is always associated with chromosomes, and not

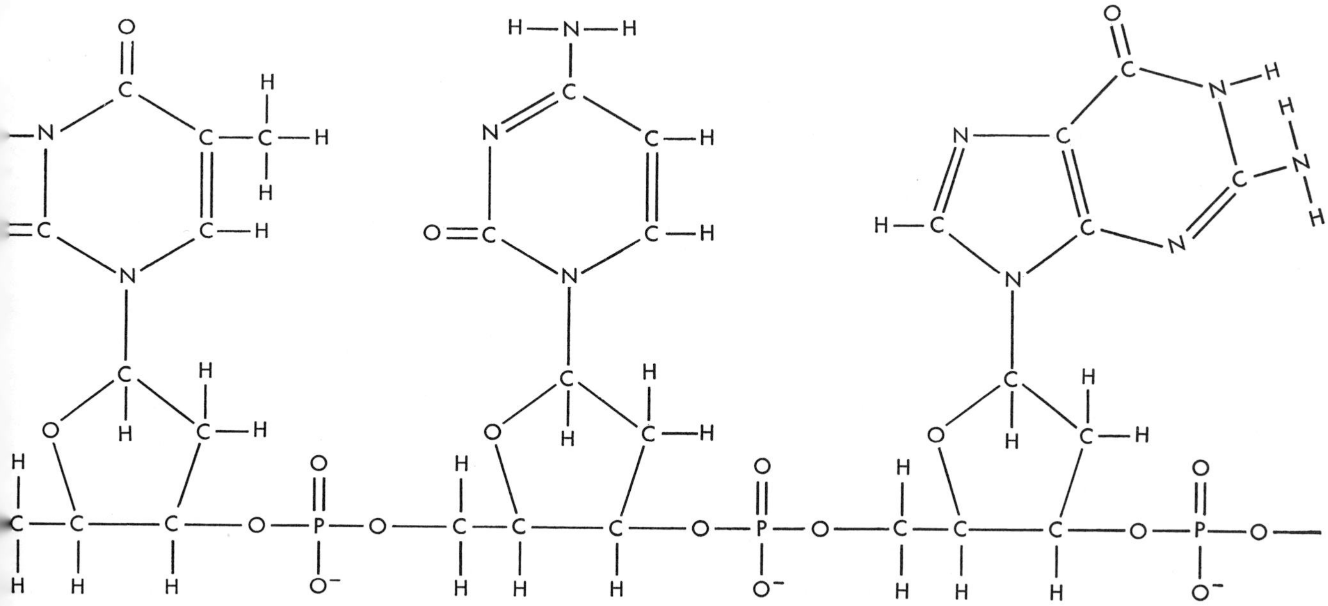

guanine. It represents a fragment of a DNA molecule which consists of thousands of sugar links with side chains in seemingly random order. RNA is similar except that it has ribose as the sugar instead of deoxyribose, and uracil replaces thymine in its side chains.

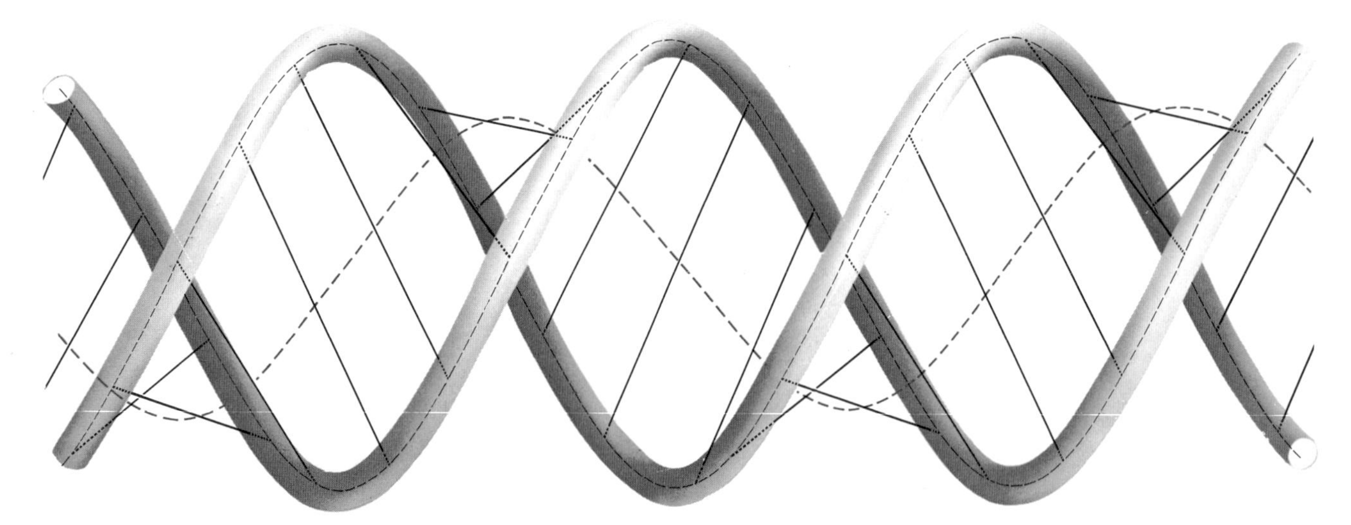

MODEL OF DNA STRUCTURE is a double helix with slanted cross-links connecting the two chains. The nucleic acid chains which form the helixes are identical but head in opposite directions. The cross-links (*black lines*) consist of complementary pairs of side chains connected by hydrogen bonds. The broken colored line traces the centers of the cross-links and also forms a helix.

with any other part of the cell. Each set of chromosomes appears to have a fixed amount of DNA. There are at least two phenomena which give direct evidence that DNA plays a genetic role. Firstly, it has been found that pure DNA extracted from certain bacteria is capable of transferring some of the properties of this strain to a related strain, and the transformed bacteria pass these properties on to their descendants. Secondly, when a bacterial virus infects a bacterium, it is the virus's DNA (not its protein) that enters the bacterial cell, and much of this DNA turns up in the progeny virus produced in the cell.

Taking all the evidence together, it is difficult to resist the conclusion that DNA is genetic material. If that is the case, our problem is to learn how DNA reproduces itself. The double-helical structure of DNA suggests a possible answer, which I have discussed in a previous article [see "The Structure of the Hereditary Material," by F. H. C. Crick; SCIENTIFIC AMERICAN Offprint 5]. The basic idea is that the two chains of

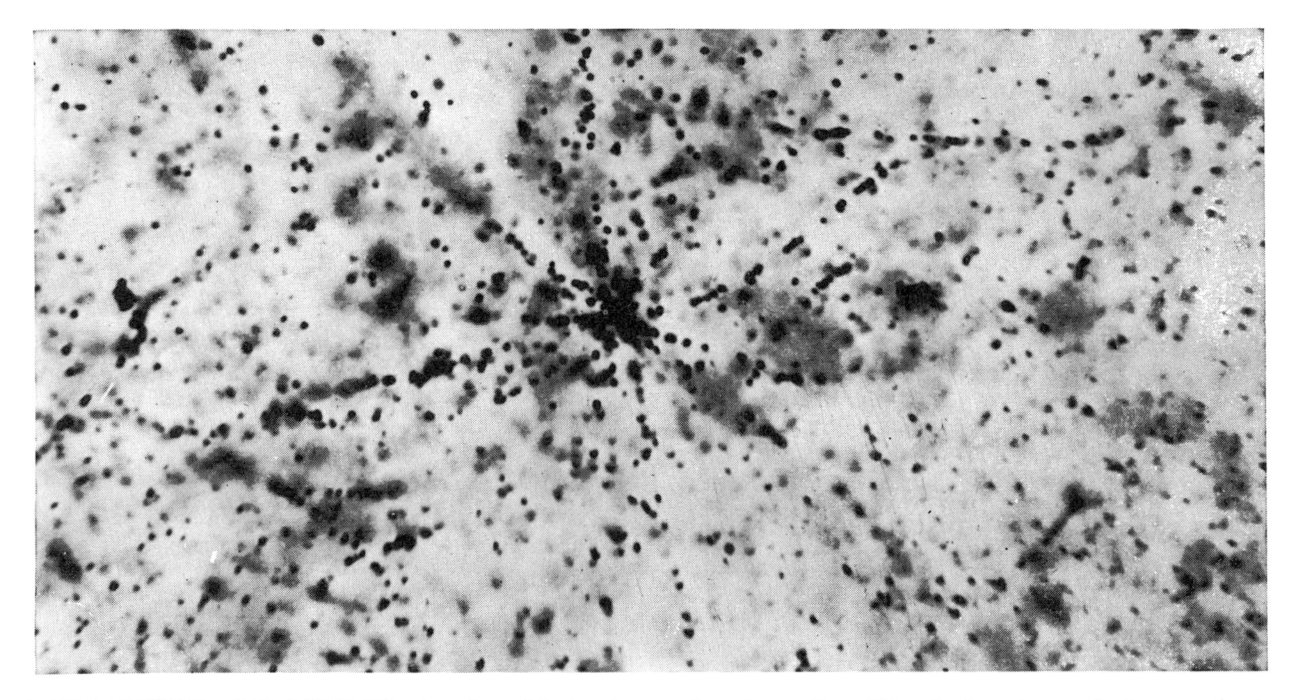

RADIOACTIVITY OF LABELED DNA in a bacterial virus is indicated by this "star" of tracks in a nuclear-emulsion photograph made by Cyrus Levinthal of the Massachusetts Institute of Technology. When bacteria infected with the virus were grown in a medium containing radioactive phosphorus, the viruses used the radioactive material in making new DNA. Then viruses containing radioactive DNA were transferred to a culture of bacteria growing in a nonradioactive medium. When the bacteria were broken open by the virus infection, their contents were mixed with the photographic emulsion. Each track in the developed emulsion is made by a particle emitted in the decay of a phosphorus atom; the number of tracks is proportional to the amount of radioactive phosphorus in the virus. Thus it is possible to trace how the phosphorus in the DNA of first generation of viruses is distributed in later generations.

the DNA, which fit together as a hand fits into a glove, are separated in some way and the hand then acts as a mold for formation of a new glove while the glove acts as a mold for a new hand. Thus we finish up with two gloved hands where we had only one before. In chemical terms we imagine that monomers supplied by the cell align themselves along the mold chain with complementary bases pairing up.

There is experimental evidence which gives some support to this idea. Cyrus Levinthal at the University of Michigan studied the DNA of a bacterial virus (called T2) by a special radioactive tracer method. He discovered that the DNA of the virus is not all in one piece—*i.e.*, it is an aggregation of DNA molecules of different sizes. In each virus there is one "big piece" with a molecular weight of about 42 million, and the rest are considerably smaller. Levinthal concentrated on this big piece and followed it through the reproduction of the virus. He found that after it had been labeled with radioactivity (by growing the virus in a medium containing radioactive phosphorus) its radioactivity dropped to one half in the virus progeny produced later in a nonradioactive medium. Presumably this means that the big piece now was composed of two parts—half radioactive DNA from the parent, and half newly synthesized, nonradioactive DNA. When these progeny reproduced themselves in a nonradioactive medium, some of them again had DNA of the same composition—apparently half original and half new.

It is very tempting to construe this as evidence for the hand-glove hypothesis of the replication of DNA: that is to say, to suppose that the radioactivity represents a chain (hand or glove) which acted as the mold and the other half represents the new chain molded on this. We can suppose further that the big piece is the part of the DNA which carries the virus's genes. But these are as yet only assumptions. All we can say is that the experiments do seem to bear out the idea that DNA carries out an internal duplication. However, I should add that certain experimental findings by Gunther S. Stent of the University of California do not easily fit into this picture.

Recently J. Herbert Taylor of Columbia University and his colleagues made a similar experiment with bean seedlings. They grew the seedlings in a medium containing thymidine labeled with radioactive hydrogen (tritium), and thus tagged the thymine of the new DNA synthesized in the bean cells. Then the seedlings were transferred to a non-

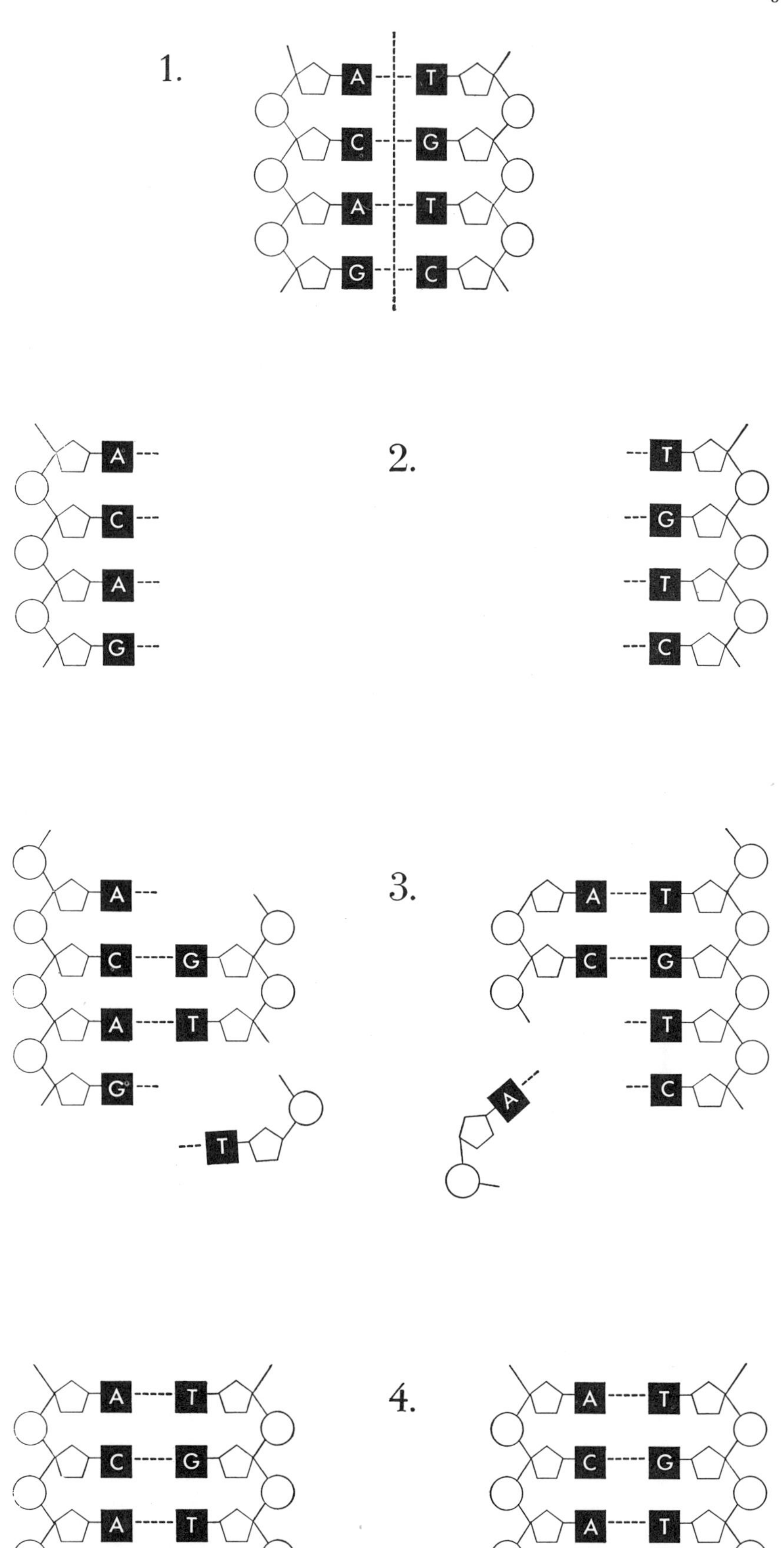

POSSIBLE REPLICATION MECHANISM of DNA is outlined. Two linked chains (1) come apart (2). Monomers then assemble along each chain (3). The result is two pairs of linked chains whose bases (*black squares*) have the same sequence as those of the original pair. The bases are labeled: A, adenine; C, cytosine; T, thymine; and G, guanine.

radioactive medium for further growth. The beauty of bean cells for such an experiment is that their chromosomes are very large, so that under a microscope it is possible to see their various parts and to catch them in the act of dividing [*see photomicrographs on page 5*]. Taylor found that after the second division of the cells one daughter chromosome of a pair tended to be radioactive while the other was not (apart from secondary complications).

This again suggests that the DNA of chromosomes is a two-part structure. However, neither the virus nor the bean-cell experiment really tells us whether the duplicating process operates at the level of the helical chains. Further, we still have the difficult problem of trying to imagine what sort of mechanism the chromosomes can employ to unwind the two DNA chains to free them for replication.

Assuming that DNA does represent the genes, how does it act to determine the make-up of the total organism? Here

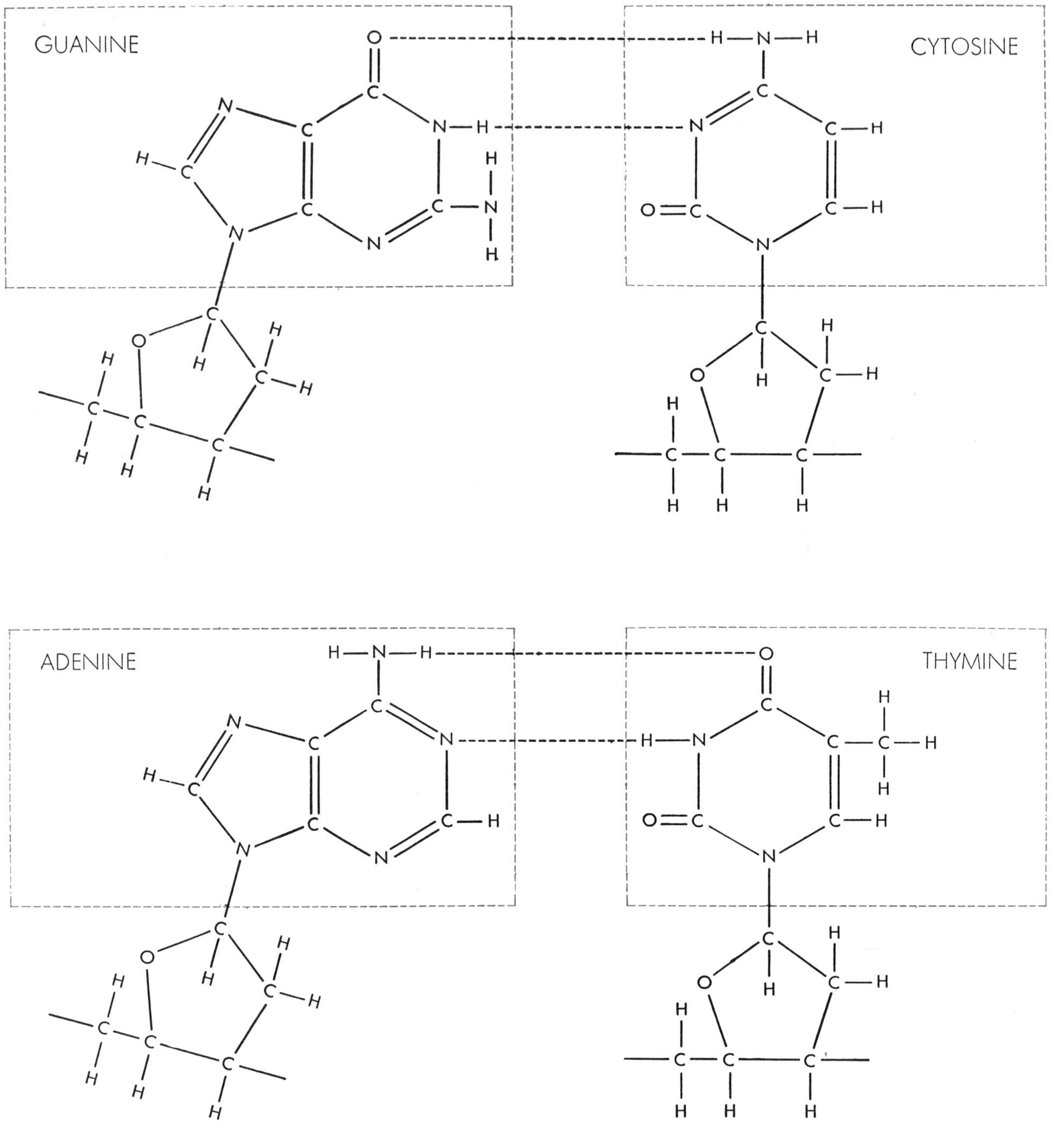

BASES OF ADJACENT DNA CHAINS can only be linked when they are paired as indicated in this diagram. The base guanine pairs with cytosine; the base adenine, with thymine. Each pair of bases is linked in two places by hydrogen bonds (*broken lines*).

again we have a suggestive general theory but very few facts, although the ones we have are interesting enough. The theory says very simply that the nucleic acids control the making of each organism's characteristic living substances—its proteins. DNA bears the master plans, carrying them on from generation to generation; RNA constitutes the working copies used in the actual synthesis of the proteins. This production is controlled in a rather straightforward way by the sequence of bases in each nucleic acid, which determines in what particular order amino acids will be assembled in the polypeptide chains that make up a protein. The order is all-important, for it governs the character and functions of the protein.

I think it very likely that this hypothesis is correct, at least in broad outline. It makes sense of an enormous body of biochemical facts, and it has the beauty which we associate with simplicity and generality. Moreover it has been a powerful guide in attacking the bewildering complexity of biological systems.

Let us consider first the function of RNA. It is found mainly in combination with protein in particles in the cytoplasm of the cell. These particles are believed to contain the templates on which specific proteins are modeled.

The most convincing evidence that RNA is responsible for the specific construction of proteins has come from recent work on the tobacco mosaic virus, done mainly by Heinz Fraenkel-Conrat and his colleagues at the University of California and by Gerhard Schramm and co-workers at the University of Tübingen in Germany. They have separated the RNA from the proteins of the virus and used the RNA, separately and in combination with different proteins, to generate virus progeny. RNA alone, inoculated into a tobacco plant, has proved capable of reproducing the virus. In this case the progeny multiplying in the infected plant had the protein corresponding to the virus strain from which the RNA was taken, although the plant had never seen this protein before. Equally remarkable is the fact that a "mixed virus," made by combining the RNA of one strain with the protein of another, yields progeny which contain not the new protein but the protein with which the RNA was originally associated. In other words, the infected plant manufactures a protein dictated by the RNA, not a copy of the protein actually given to it. This result illustrates very well a hypothesis which my colleagues and I call the Central Dogma: namely, that once information (meaning here the determination of a sequence of units) has been passed into a protein molecule it cannot get out again, either to form a copy of the molecule or to affect the blueprint of a nucleic acid. The idea is not universally accepted, however. In fact, Sir Macfarlane Burnet, the eminent Australian virologist, persuasively argued another point of view in a very interesting little book which he published recently.

The hemoglobin of a horse's blood is not exactly the same as that of human blood; the insulin molecule, too, differs slightly in horses, pigs, sheep, whales and so on. These well-known facts suggest that genes control the amino acid make-up of specific proteins—but they do not necessarily prove it. The proof has now been provided by my colleague Vernon Ingram, working in the Cavendish Laboratory of the University of Cambridge. He discovered that sickle cell hemoglobin (a defect responsible for a fatal disease) is exactly like normal hemoglobin, as far as he can tell, except for a single alteration: in one place where the normal molecule has a glutamic acid, the diseased hemoglobin has a valine unit. Genetic studies have established that this difference is due to a single gene. It is interesting that a change in one amino acid out of nearly 300 results in the fatal sickle cell condition, but it is even more interesting to see that a gene can control such a small change. Genes not only work powerfully but they can also work delicately.

How do they manage to operate so selectively upon the fine-structure of a protein? The most spectacular information on this point has recently been announced by Seymour Benzer of Purdue University. By very elegant genetic techniques he mapped a single "gene" of a bacterial virus, and he was able to distinguish more than 100 different functional sites arranged in a linear order along the length of the "gene." Assuming that genes are made of DNA, we can perhaps trace a correspondence between his map and the DNA molecule. Benzer calculated that the smallest distance between sites on his genetic map would take in just a few base-pairs along the DNA double chain.

From every point of view biology is getting nearer and nearer to the molecular level. Here in the realm of heredity we now find ourselves dealing with polymers, and reducing the decisive controls of life to a matter of the precise order in which monomers are arranged in a giant molecule.

2 How Cells Make Molecules

by Vincent G. Allfrey and Alfred E. Mirsky
September 1961

The inherited master plan of the cell is contained in the molecules of deoxyribonucleic acid. They direct the manufacture of protein enzymes, which in turn engineer all the chemical reactions of life

In our laboratory we have the portraits of Gregor Mendel and Friedrich Miescher side by side. Mendel in 1866 set forth evidence, from his observation of inheritance in the pea plant, for the idea that genetic information is carried in discrete units from one generation to the next. Miescher in 1869 isolated from the nucleus of cells a substance that he called nuclein and that is known today as deoxyribonucleic acid (DNA). He knew that he had in his hands a novel substance containing nitrogen and phosphorus, and he was well aware of its location within the nucleus. But he could have had no idea of what DNA does in the nucleus, because the role of the nucleus in heredity was at that time unknown, even to Mendel. It was well over half a century, long after the death of both men, before their work could be fused. The fusion required much more nucleic-acid and protein chemistry than there was in Miescher's time and a vast amount of new biology, including the unearthing of Mendel's work in 1900.

In the last three decades of the 19th century, work in biology, led by August Weismann, demonstrated the continuity of the germ plasm and showed that the nucleus of the cell plays a central role in heredity. Attention soon focused on the chromosomes. Since sperm and egg nuclei provide equal complements of chromosomes (except for the sex chromosomes) and since there is precisely equal cleavage and distribution of chromosomes at cell division, it seemed clear that chromosomes are concerned with the continuity that is essential in heredity. With the rediscovery of Mendel and the growth of genetics, biologists talked less about the germ plasm and more about the genes as discrete units of the germinal material. It became increasingly clear that each gene is derived from a pre-existing gene. The genes were located in the chromosomes, which now far more conclusively than before were shown to contain the materials determining heredity.

Meanwhile the chemistry of nucleic acids was making progress. But there was little contact between the two movements. The now standard color test for DNA was first demonstrated in 1914 by the German chemist Robert Feulgen in a test tube. Not until 10 years later did Feulgen use the test to stain cells and show that the chromosomes are the locus of DNA concentration in the nucleus. Yet it cannot be said that this experience led to the idea that DNA is the essential gene material.

The evidence that genetic information is carried by DNA came in the late 1940's from a number of sources. André Boivin and Roger and Colette Vendrely of the University of Strasbourg and Alfred E. Mirsky and Hans Ris of the Rockefeller Institute measured the DNA content of nuclei in germ cells and various somatic cells and found that in a given organism the DNA content is constant per set of chromosomes. This constancy pointed to DNA as the essential material of the genes. In experiments on pneumococci Oswald T. Avery, Colin M. MacLeod and Maclyn McCarty of the Rockefeller Institute showed that hereditary traits can be transmitted from one strain of bacteria to another by transferring to cells of the latter DNA extracted from the former. Their experiments conclusively established DNA as the carrier of genetic information.

This development made a great impression on geneticists, for it went a long way toward answering one of their outstanding questions: What is the nature of the gene? The impact on biochemists was far greater; it revealed to them what the problems of biochemistry could be. The principle of genetic continuity, the rule that each gene comes from a pre-existing gene, was now transformed to the biochemical question: How does the molecule of DNA replicate itself? There was also the problem of the passage of genetic information from DNA in the chromosome to the fabric of the cell. As the carrier of genetic information, it was plain, DNA does much more than replicate itself. It plays an active role, directing the life of the cell. Biochemists could now study the molecular basis of gene action, tracing the effect of DNA in discrete observable events in the synthesis of the molecules of which the cell is made.

The fusion of biochemistry and cell biology has brought an accelerating growth of understanding over the past 20 years. It is now possible to answer at the molecular level some of the fundamental questions of genetics and cell biology that go back to the era of Weismann. Some of the most significant knowledge of DNA activity has come in recent months. It is difficult to believe that the discovery of this substance goes back to 1868!

When the chromosomes are coiled up tightly in stumpy rods, the DNA they contain is the repository of genetic information, but it is inert. When DNA is actively communicating its information to the cell, the chromosomes have an entirely different appearance. Then, in their "lampbrush" configuration, the chromosomes uncoil into delicate filaments forming a lacelike structure in which DNA and other components are readily accessible for interaction with the surrounding medium [*see illustration on page 20*].

Chemists have shown that the molecule of DNA consists of a long, un-

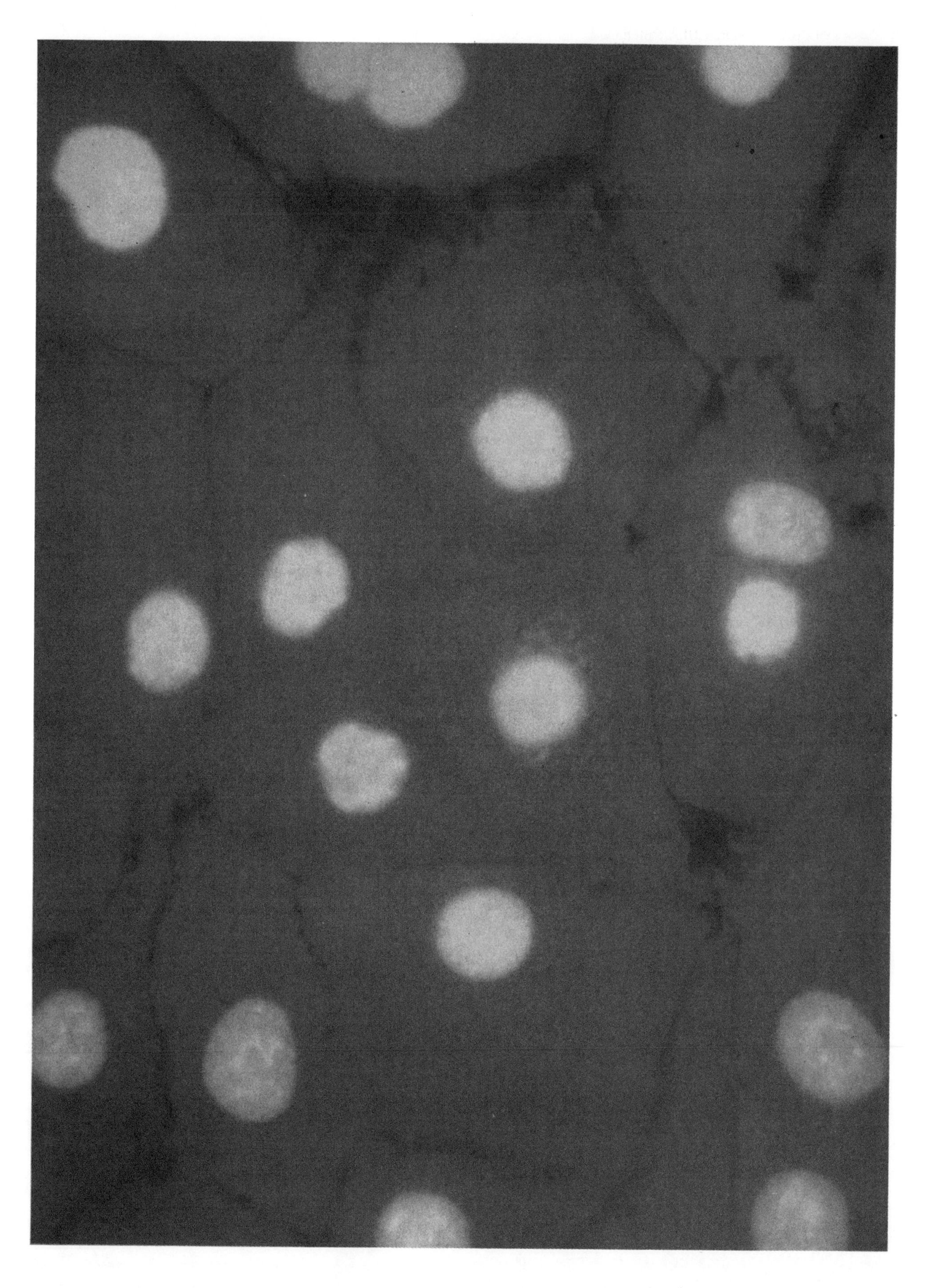

NUCLEIC ACIDS, the "blueprints" and "templates" of protein synthesis, are visualized in this photomicrograph of human amnion cells made by Suydam Osterhout at the Rockefeller Institute. Stained with acridine orange, the deoxyribonucleic acid (DNA) fluoresces yellow-green and is seen to be localized in the nuclei. Ribonucleic acid (RNA) fluoresces orange-red and is seen throughout the cytoplasm. There is RNA in the nucleus also, but it is obscured by the DNA. The magnification is 3,500 diameters.

branched chain, the backbone of which is made up of alternate five-carbon sugar (deoxyribose) and phosphate groups. To each sugar is attached a nitrogenous base; in most DNA's there are four such bases: adenine, guanine, thymine and cytosine. The unit in the chain, consisting of phosphate-sugar-base, is called a nucleotide. From X-ray crystallography, particularly the work of F. H. C. Crick and J. D. Watson, came the understanding that a DNA molecule consists not of a single polynucleotide chain but of two, twined around each other in a double helix and held together by hydrogen bonds between the bases. The companion bases are never identical but are always specifically complementary, with adenine joined to thymine and guanine to cytosine. This is demonstrated by experiments with simple, synthetic polynucleotides. Thus a synthetic DNA, or polynucleotide, made up exclusively of thymine bases (polythymidylic acid) binds a complementary chain made up exclusively of adenine bases (polyadenylic acid). In chains made up of all four bases, as in the natural molecule, the sequence of bases in one chain governs the sequence in the other; if the actual order of bases in one chain of a DNA were known, one could write down the order of its complementary chain. To complete the picture, Crick and Watson proposed that the genetic information carried by the molecule is encoded in the sequence of its bases.

The usefulness of this work to the classical concerns of genetics began to be demonstrated in 1957 when Arthur L. Kornberg, then at Washington University in St. Louis, brought about the synthesis of DNA in a cell-free system containing the four nucleotides, the enzyme polymerase and DNA. In the presence of polymerase the nucleotides linked together to form long chains of DNA. More important, Kornberg found that the polymerization of the four nucleotides will proceed only if a small amount of DNA is present to "prime" the reaction. He soon found that the over-all proportions of the bases in the product DNA paralleled the base composition of the primer used. Early this year Kornberg made the point more strongly. As primers he used a number of different DNA's prepared from virus, bacterial and animal sources, each DNA having its own characteristic sequence of nucleotides. Using a statistical method (known as nearest-neighbor nucleotide frequencies) to analyze the products of these reactions, he found that each primer DNA directs the polymerization so that the sequence of the nucleotides in the enzymatically synthesized DNA is the same as its own.

According to the Watson-Crick model, when DNA primes the making of more DNA, the double helix uncoils. Then along each chain a complementary chain is formed. In Kornberg's experiments the pattern of nearest-neighbor frequencies in every case showed the pairing of adenine to thymine and guanine to cytosine, just as in the model. A geneticist can now see how it is that each gene is derived from a pre-existing gene.

It was of course the more general activity of the genes—by which they bring about the expression of hereditary traits in the organism—that led to their discovery. The way to an understanding

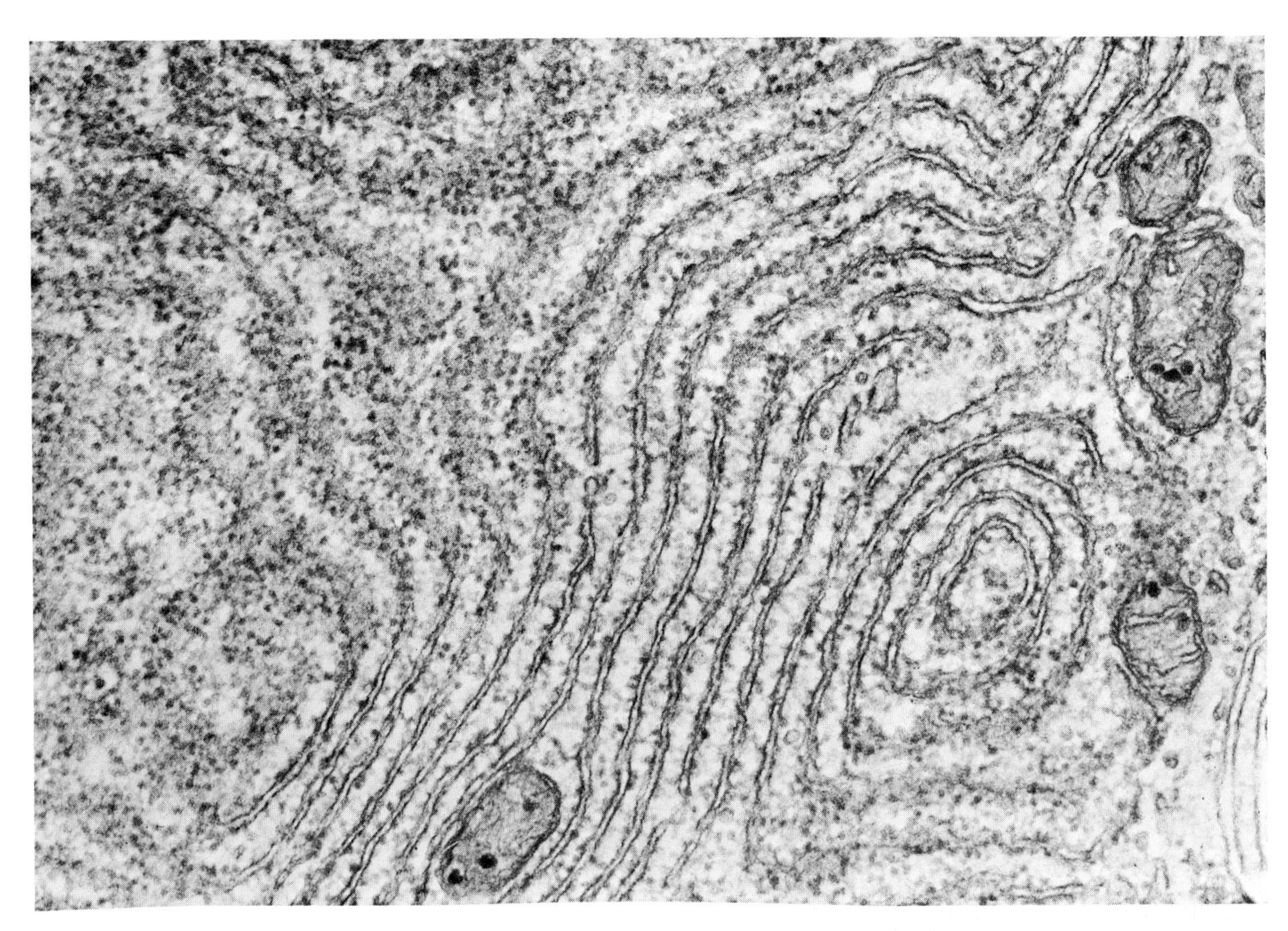

RIBOSOMES, the sites of protein synthesis, are shown in this electron micrograph made by Bernard Tandler of the Sloan-Kettering Institute. The cytoplasmic membrane system of a human submaxillary gland cell is enlarged 70,000 times. The ribosomes are the small, dark particles, each about .000015 millimeter in diameter, lining the membranes. The large oblong bodies are mitochondria.

of this activity at the molecular level—how DNA governs the biosynthetic processes of the cell—has become clear only in the last 10 years.

The heritable changes that geneticists first studied were necessarily those most readily observed. Many of these were changes in color. One of the early Mendelians was the English physician Archibald Garrod, who made a penetrating study of a rare hereditary disease in man characterized by the appearance of a black pigment, alcapton, in the urine. The pigment is formed because there is a derangement in the metabolism of the amino acid tyrosine; one particular reaction that normally occurs fails to occur. This failure, Garrod perceived as far back as 1909, is due to the absence of an enzyme that is normally present. The role of the normal gene, therefore, is to determine the production of a particular enzyme, and this is what the abnormal gene fails to do.

The idea that the action of a gene is concerned with the formation of a particular enzyme was ignored by most geneticists for some 30 years. It was revived by George W. Beadle and Edward L. Tatum when they showed the principle at work in heritable metabolic derangements of the red bread mold *Neurospora*. This time the idea made a deep impression on geneticists and biochemists, in part because the chemical nature of enzymes had meanwhile been revealed. Between 1926 and 1930 James B. Sumner of Cornell University and John H. Northrop of the Rockefeller Institute had shown that enzymes are proteins. The biochemical function of the gene was now to be looked for in the more general function of protein synthesis. It was firmly demonstrated in the early 1950's, when other examples of the determination of protein structure by genes were found in many animals, fungi and bacteria.

But the decisive new experiments were again, as in Garrod's time, on man—on human hemoglobin. Beginning with the hemoglobin of sickle-cell anemia, studied by Linus Pauling and his colleagues at the California Institute of Technology, quite a number of hereditary hemoglobin anomalies were discovered in an enterprising world-wide search. In several cases it was found that a gene mutation produces a single amino acid substitution in one location of the peptide chains, containing about 150 amino acids, that make up the hemoglobin molecule. So much precise information has by now been acquired linking genes and the amino acid composition in hemoglobins that there is no longer any doubt that genes determine protein synthesis.

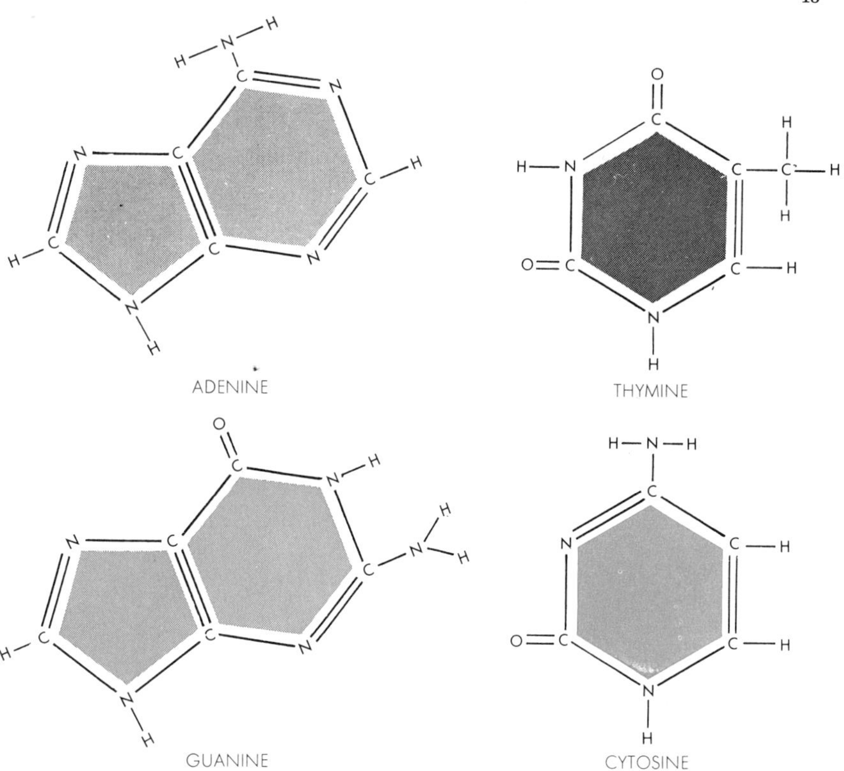

DNA, the carrier of genetic information that serves as the "blueprint" for protein synthesis, has as its key components four nitrogenous bases. Their formulas are diagramed above and their arrangement in the complex DNA molecule is illustrated on the following two pages.

One of the best examples of the control of protein synthesis by the gene concerns the synthesis of DNA, the substance of the gene itself. When a bacterial virus that contains DNA enters a bacterial cell, a large quantity of virus DNA is soon synthesized. At the same time there is a five-to-ten-fold increase in activity of the enzyme polymerase, the enzyme that polymerizes mononucleotides and so produces DNA. But the polymerase in this case is the virus-induced polymerase, and it is distinctly different from the polymerase present in the bacterium before entry of the virus. This is to be expected since the genetic information for the new enzyme comes from virus DNA and not from the bacterial genes.

The understanding of how DNA directs protein synthesis in the cell involves among other things a vast lore of knowledge about protein synthesis and structure gained without reference to the cell. But the primary action of DNA also raises a logical problem that should be dealt with at the outset. Only four different nucleotides make up the long chains of DNA. Some 20 amino acids make up the long polypeptide chains of proteins. The genetic information in DNA is therefore spelled out in a four-letter alphabet. But the information in this molecular script is communicated to another in which the message must be translated into a 20-letter alphabet, the letters of which are entirely different from those in the four-letter alphabet. How is the information encoded and conveyed? The most plausible answer is that groups of three or four nucleotides are arranged in different sequences, each corresponding to a particular amino acid. By such an arrangement the four-letter DNA alphabet is able to determine the spelling out of protein structure in the 20-letter alphabet.

All cells synthesize protein, some continuously, others for only a part of their life cycle. The proteins they make are enormously varied, differing in size, shape, over-all chemical composition and physical properties. But whatever their function, and regardless of their size, shape, solubility or enzyme activity, all proteins have an underlying similarity in constitution: they are all made up of the relatively simple molecular units of amino acid. The synthesis of a protein from these smaller units is conceptually a simple process, involving the joining of the individual amino acids to form long chains. The length of the chain and the sequence of the amino acids vary, of course, from one protein to another. But the essential unit of structure, the link that prolongs the chain, is ubiquitous; it is the peptide bond, the chemical union between the carboxyl group (COOH) of one amino acid and the amino group (NH_2) of the next amino acid in the chain.

These bonds, connecting amino acids in various sequences, are the key linkages to be created in carrying out the synthesis of proteins or smaller polypeptides, either in the cell or in the laboratory. Since peptide bonds do not form spontaneously when amino acids are mixed, other chemical means have to be used to drive the formation of the bonds. The synthetic system of the cell begins with a reaction that "activates" the carboxyl group of the amino acids.

This reaction, originally discovered by Mahlon B. Hoagland of the Harvard Medical School in 1955, derives the energy necessary for the activation from adenosine triphosphate (ATP), the main energy currency of the cell [see "How Cells Transform Energy," Offprint 91]. The ATP is cleaved to release two of its three phosphate groups; the remaining fragment, adenosine monophosphate (AMP), joins up with the acid group of the amino acid. In this way the amino acid is potentiated for the formation of the peptide bond. The enzymes that carry out this activation have great specificity: in general they react with only one type of amino acid. It is probable that most cells have activating enzymes for at least 20 amino acids. In animal cells these enzymes have been found in the nucleus as well as in the cytoplasm, and there is good evidence that they play a role in the synthesis of the nuclear proteins, including the proteins of the chromosomes.

The activating enzymes mediate only the first step in a very complex and precisely ordered chain of reactions. The order is supplied by the information encoded in the DNA molecule. To convey the information from the DNA in the nucleus to the site of most protein synthesis in the cytoplasm, the cell employs another polynucleotide–ribonucleic acid (RNA)–as an intermediary. In RNA the sugar is ribose instead of deoxyribose; the main RNA bases are adenine, guanine, cytosine and, instead of thymine, uracil. As in DNA, the different nucleotides are linked together through their phosphate groups to form long chains.

The need for ribonucleic acid in protein synthesis was suggested 20 years ago, when Torbjörn O. Caspersson in Stockholm and Jean Brachet in Brussels showed that tissues that synthesize large amounts of protein, whether for growth or multiplication, are always rich in RNA. One of the highest RNA concentrations is found in the cells of the spinning gland of silkworms, which produce the proteins fibroin and sericin of the silk thread. In mammals, high RNA concentrations occur in such specialized cells as those of the pancreas and liver, in which many proteins are synthesized for transport to other parts of the organism.

When cells synthesize protein on this scale, it is usually observed that the synthetic machinery is highly organized into a lamellar network of membranes and RNA-rich particles. The biochemical knowledge of these structures comes largely from studies of cell homogenates, from which different intracellular structures can be isolated by centrifuging at different speeds; that knowledge is now significantly supplemented by electron micrographs of the cytoplasmic membrane system [*see illustration on page 14*]. Particular interest attaches to the particles, which, in a liver cell, are

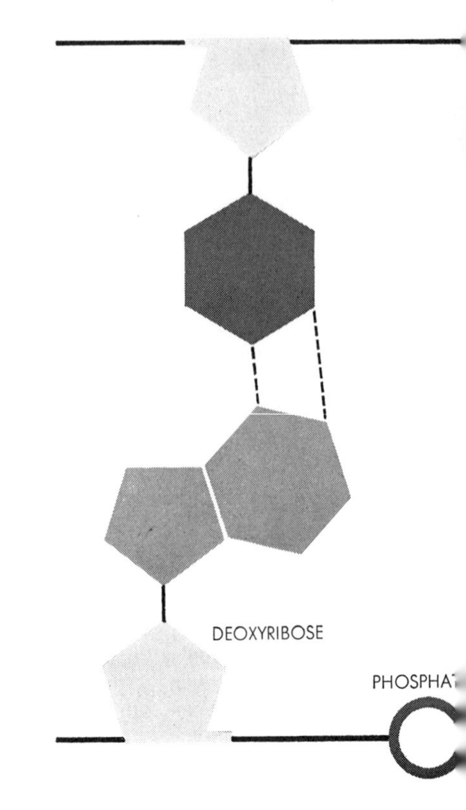

DNA MOLECULE is in the form of a long double chain of nucleotides—phosphate-linked deoxyribose sugar groups to each of

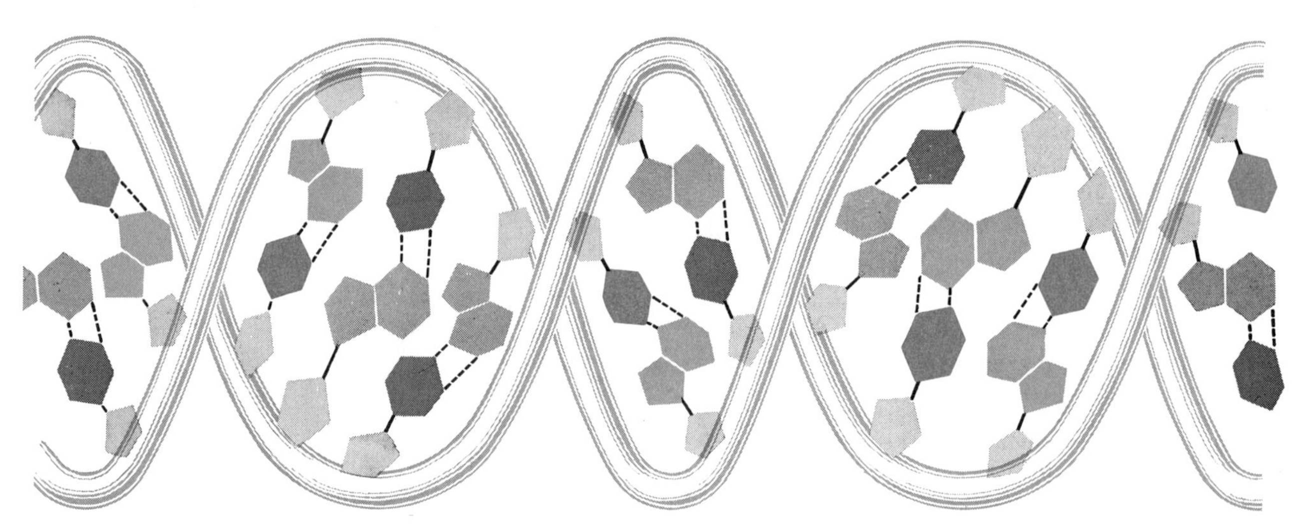

DNA MOLECULE, shown at the top of these pages as a straight ladder, is actually twisted into a double helix, according to the generally accepted Watson-Crick model. In this drawing the phosphate groups are not shown. The sugar and base molecules are shown diagrammatically; in the actual three-dimensional model the base pairs that make up the crosslinks all lie in parallel planes.

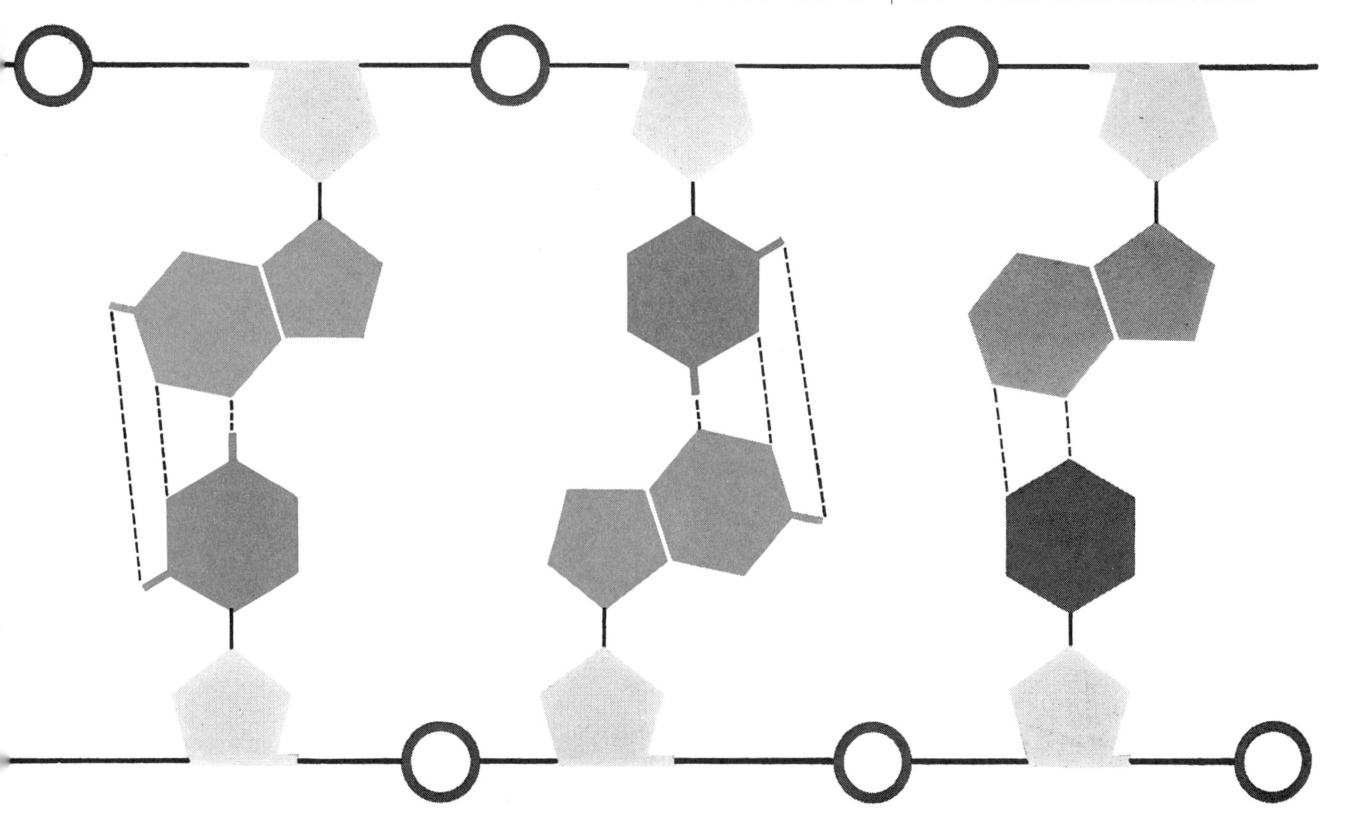

which one of the four bases is attached as a side group. Hydrogen bonds (*broken lines*) link pairs of bases to form the double chain. The bases are always paired as shown: adenine with thymine and guanine with cytosine. The sequence of pairs, however, can be varied infinitely. The sequence encodes the information that determines what kinds of protein shall be synthesized by the cell.

roughly spherical and about .000015 millimeter in diameter. In fully differentiated cells of the pancreas or liver, most of these particles are attached to membrane surfaces, but in embryonic tissues the particles appear free in the cytoplasm. George E. Palade of the Rockefeller Institute has suggested that the particles, not the membranes, are the primary sites of protein synthesis in these cytoplasmic systems. This view has been proved correct by biochemical studies of the particles themselves, isolated not only from pancreas and liver but also from tumor cells, plants, yeasts and bacteria. Most workers in the field now refer to the particles as ribosomes.

Whatever cell they come from, ribosomes are extraordinarily rich in RNA. Bacterial ribosomes, for example, have a molecular weight of nearly three million, of which at least 60 per cent is RNA. Ribosomes prepared from liver or yeast have a molecular weight of about four million, of which more than 40 per cent is RNA. Few purified ribosome preparations contain less RNA than this. Protein, often rich in basic amino acids, constitutes most of the remaining mass of the isolated ribosomes.

The first chemical indication that ribonucleoproteins play such a direct role in the synthesis of other proteins came from "tracer" experiments in which living animals were given isotopically labeled amino acids (that is, amino acids containing atoms of nitrogen 15 or carbon 14 instead of the nitrogen 14 or carbon 12 atoms usually present). Henry Borsook of the California Institute of Technology and Tore J. M. Hultin of the Wenner-Gren Institute in Stockholm in 1950 independently showed that when the cells of a labeled tissue were broken and fractionated, the highest concentration of labeled amino acids showed up in the microsome fraction. This fraction was subsequently shown by Philip Siekevitz and Palade to contain the ribosomal particles attached to membrane fragments. In 1953 Marie Maynard Daly and the authors made careful kinetic studies of the rate of nitrogen-15 amino acid uptake into different protein fractions of the pancreas cell. The results made it very likely that some of the protein attached to ribonucleic acid was a direct precursor of the enzyme proteins that are found free in the cell. Moreover, we found that an attack on the RNA by a specific enzyme, ribonuclease, stopped protein synthesis in isolated subcellular fractions. The experiments made on whole animals and in isolated cells were soon supplemented by studies of cell-free systems containing RNA, which would incorporate amino acids into protein in the test tube. The development of these systems was largely due to the experiments of Paul C. Zamecnik and his colleagues at the Massachusetts General Hospital.

Although many experiments suggested a direct role for RNA in protein synthesis, it has only recently been shown that the function of RNA is to supply the information necessary to organize the sequence of the amino acids in peptide chains. The argument is clinched by experiments in which a modification of the RNA has brought a change in the protein product. Working with a plant virus, Gerhard Schramm and his co-workers at the University of Tübingen succeeded in substituting a hydroxyl for an amino group in the viral RNA; the substitution led to the formation of a different viral protein. By growing bacteria in the presence of 5-fluoro-uracil, François Gros of the Pasteur Institute in Paris has caused this substance to replace the normal base uracil in the bacterial RNA; the bacteria thereupon synthesized an abnormal protein instead of the enzyme *beta*-galactosidase.

From all that is known, it is now supposed that RNA's in the ribosomes act as templates that determine the se-

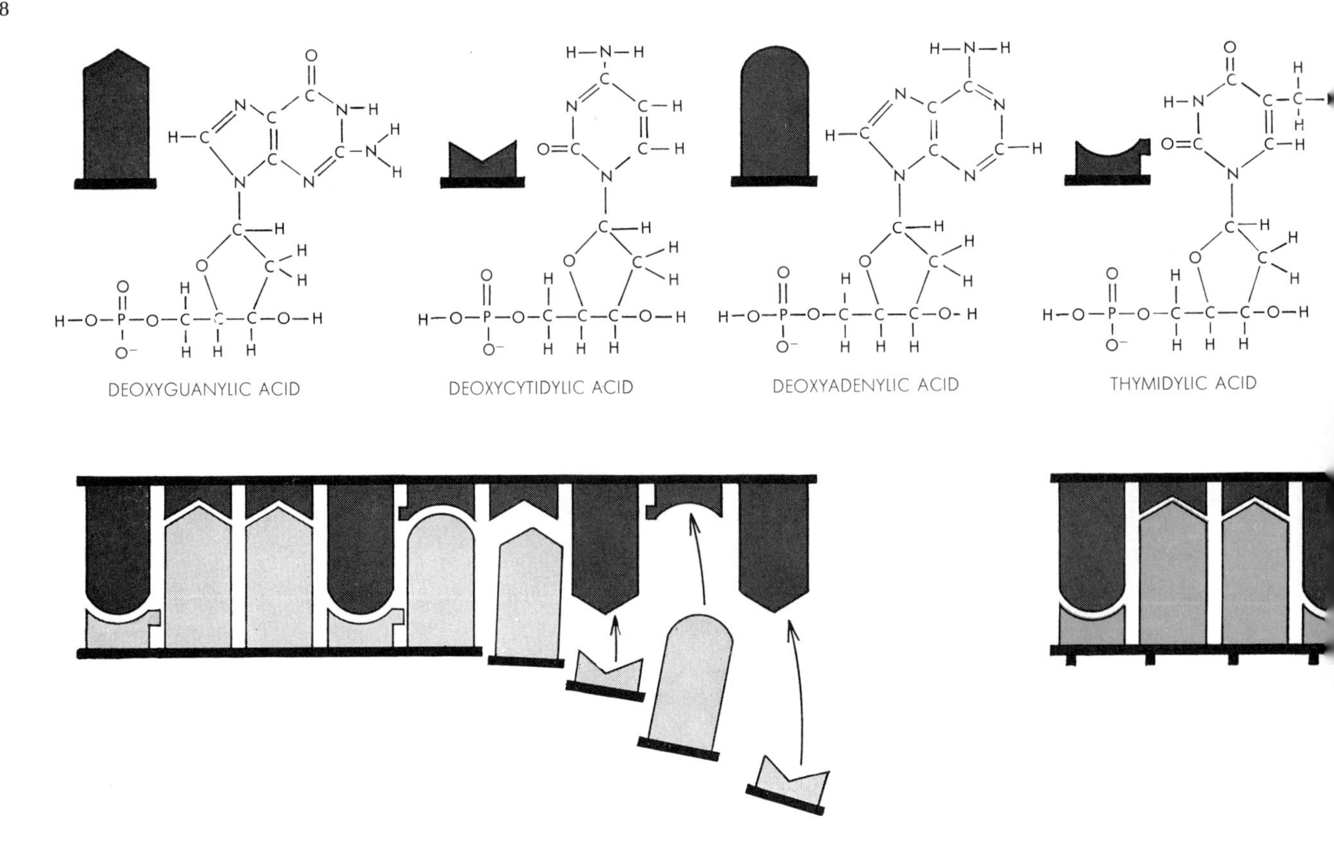

"BASE PAIRING" is the process by which DNA is replicated in cell division and by which it makes RNA, the closely related nucleic acid that in turn synthesizes proteins. The four nucleotides of DNA are symbolized (*upper left*) by four building blocks shaped as complementary pairs. RNA (*upper right*) differs only slightly: its sugar has an extra oxygen atom, and uridylic acid replaces DNA's thymidylic acid. The lower diagrams show how DNA can either replicate or form RNA carrying the same genetic information.

quence in which amino acids are linked together in protein chains. They reproduce in their nucleotide sequences the information encoded in the master templates of the DNA molecules in the cell nucleus.

But how do amino acids get to the RNA templates from their activating enzymes? Here too ribonucleic acids play a role. In 1957 several laboratories announced the discovery of low-molecular-weight RNA's that transfer activated amino acids to ribosomes. The function of these "transfer" RNA's is to get amino acids properly lined up on the RNA template of the ribosome. This view, put forward by Crick and others, assumes that there are 20 or so transfer RNA's, each specific for a particular amino acid and also capable of recognizing certain specific sites on the template. It is, of course, the nucleotide sequence in the shorter chains of these low-molecular-weight RNA's that capacitates them for

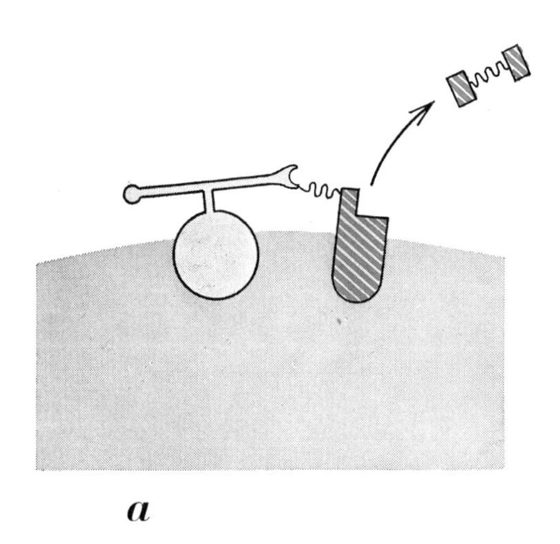

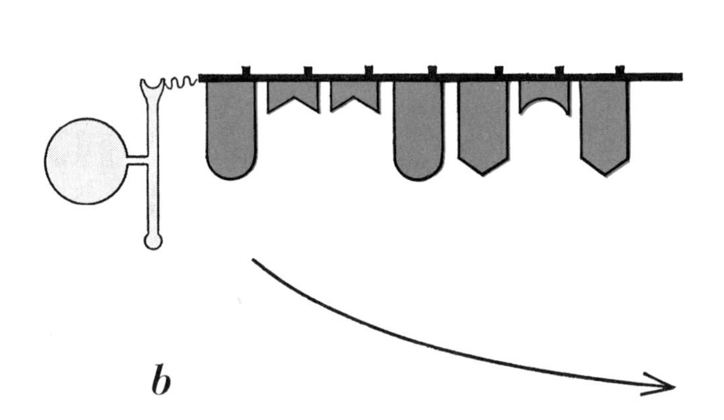

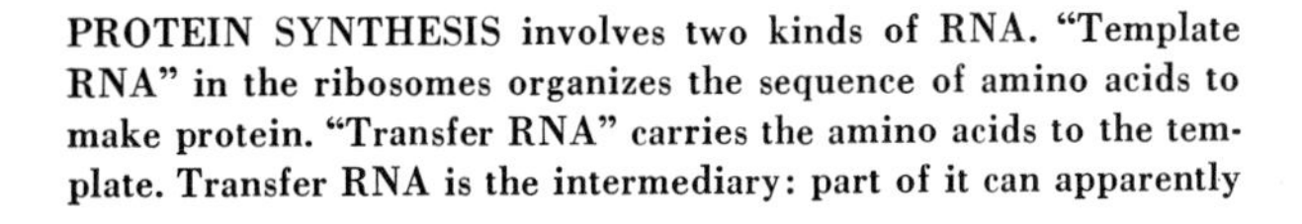

PROTEIN SYNTHESIS involves two kinds of RNA. "Template RNA" in the ribosomes organizes the sequence of amino acids to make protein. "Transfer RNA" carries the amino acids to the template. Transfer RNA is the intermediary: part of it can apparently recognize a specific amino acid and part is coded to seek the proper site on the template. The process begins (*a*) with activation of an amino acid by adenosine triphosphate (ATP), the cellular energy carrier (*hatched*). Two phosphate groups drop from ATP and, with

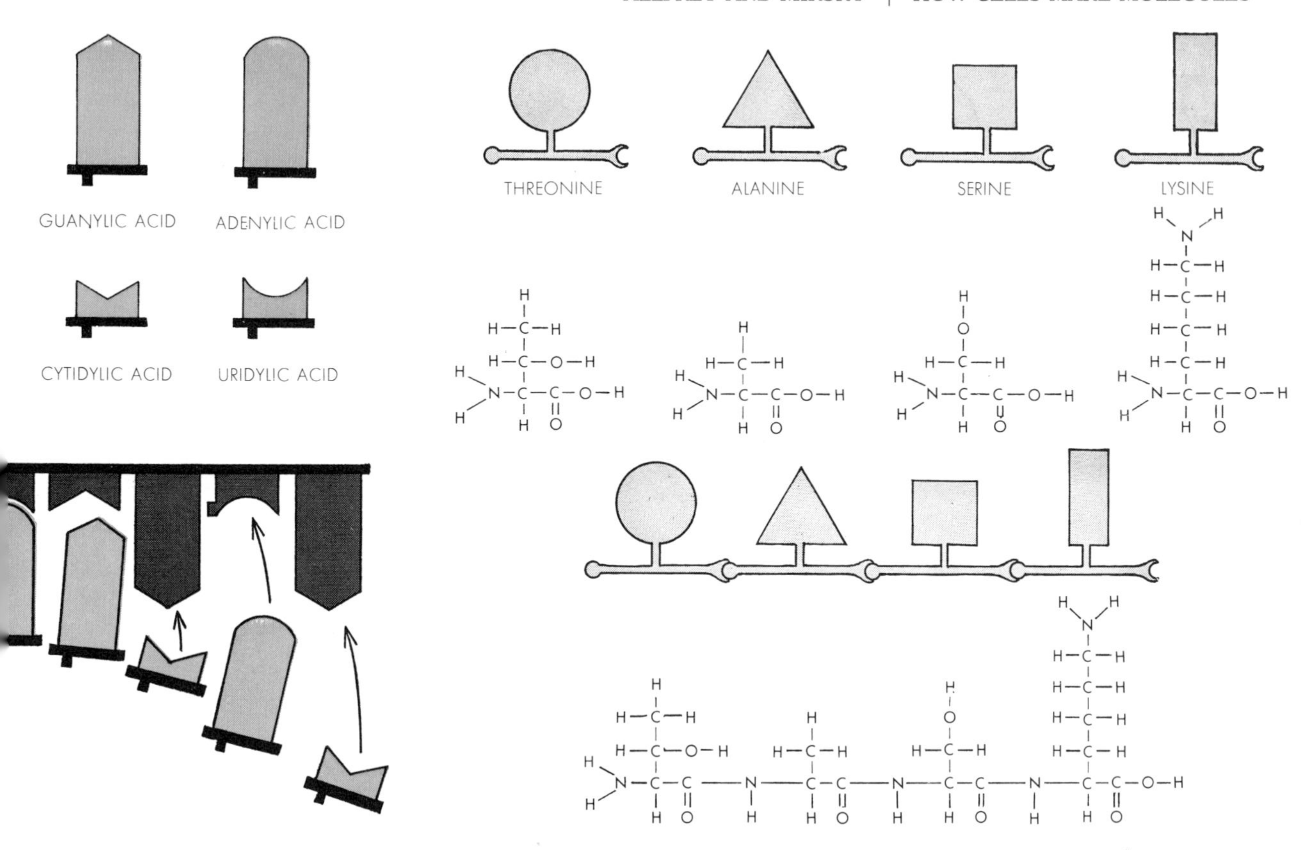

A single strand of DNA (*dark gray*) assembles new DNA nucleotides (*light gray*) or RNA nucleotides (*color*) that match its complementary strand.

AMINO ACIDS are the constituents of proteins. Here four amino acids (of 20-odd that are known) are represented by building blocks. In the top row the symbols and formulas of the four are shown separately; in the bottom row they have been linked by peptide bonds, in which H_2O is dropped from adjacent COOH and NH_2 groups to form a fragment of protein.

both of these highly specific reactions. According to this scheme the amino acids are transferred upon activation to the appropriate transfer-RNA chain. The transfer-RNA molecule then combines with its complementary sequence of nucleotides in the template RNA. This transfer reaction is known to be mediated by an enzyme requiring guanosine triphosphate. Other transfer RNA's, carrying other amino acids, take their appropriate places on the template, and the amino acids are now aligned and held in proper sequence to form the specific polypeptide chain of the protein.

Though this picture of transfer RNA as an adapter molecule is still tentative, something is known about the coupling of amino acids in the ribosome. Recent work on hemoglobin synthesis, by Richard Schweet of the University of Kentucky Medical School and by Howard M. Dintzis at the Massachusetts Institute of Technology, indicates that

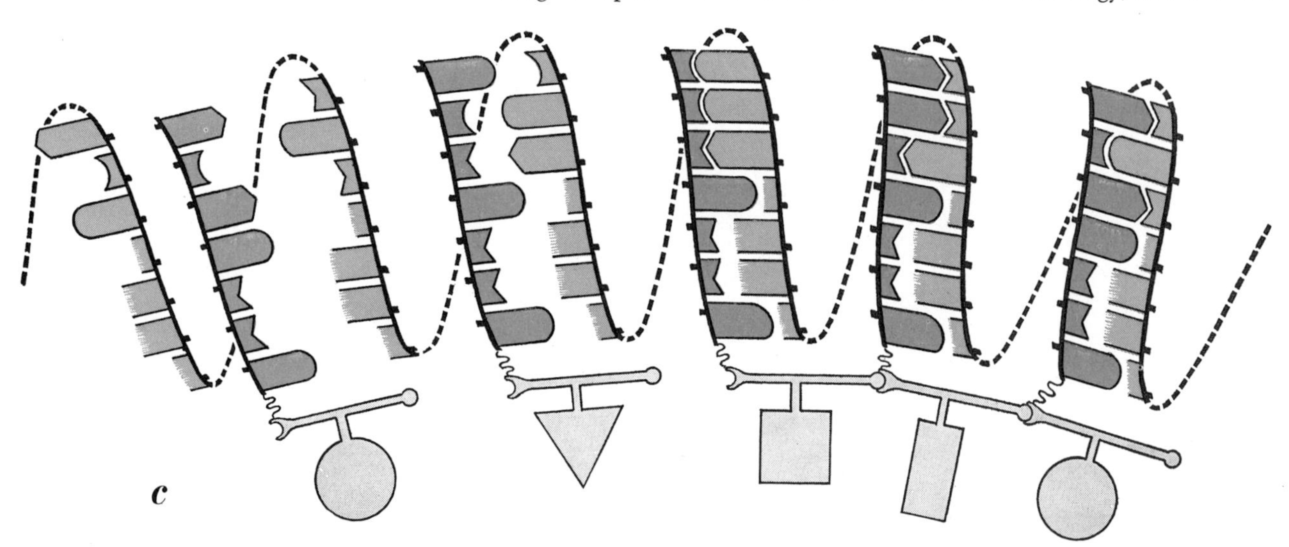

the help of an enzyme (*gray*), the amino acid is attached to the remaining adenylic acid by a high-energy bond (*wavy line*). Then a transfer RNA molecule moves in (*dark-colored symbol at "b"*). Each transfer RNA has an adenylic acid at one end; that end takes over the bond to the amino acid. Finally, the transfer RNA carries the selected amino acid to a ribosome (*c*). There the coded section of the transfer RNA finds its place on the template (*light color*) and positions the amino acid to join with others in a protein chain.

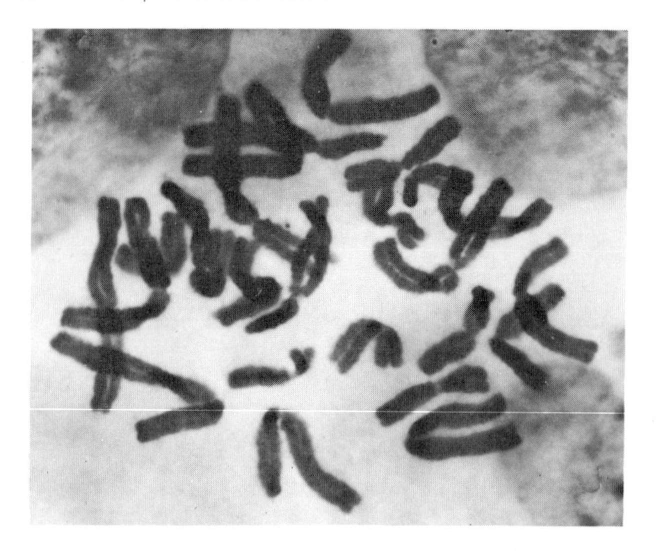

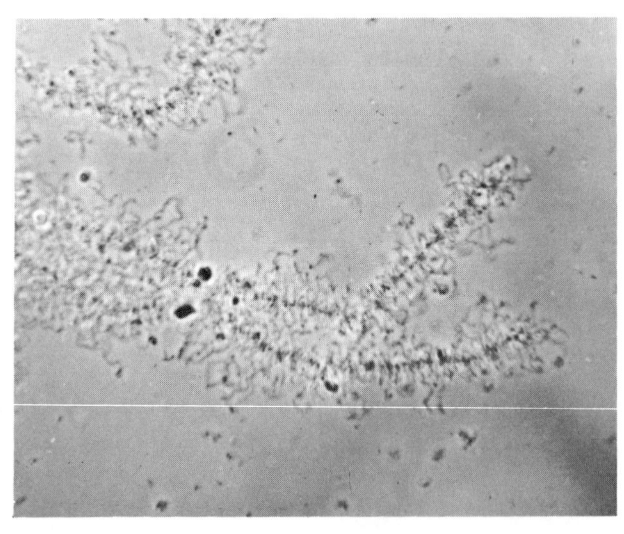

DNA IN CHROMOSOMES is an inert repository of genetic information when the chromosomes are tightly coiled rods during mitosis (*left*). Between cell divisions, when DNA is doing its genetic work, the chromosomes are greatly extended, as in the "lampbrush" configuration (*right*). These photomicrographs of newt chromosomes were taken by H. G. Callan of St. Andrews University.

peptide-bond formation proceeds much like a zipper, starting with the amino acid valine at one end of the chain and closing bond after bond until the protein molecule is finished. New amino acids are added to the growing chain at the rate of about two per second, finishing the protein molecule (of 150 amino acids) in 1.5 minutes. This impressive feat, the synthesis of a finished protein molecule in less than two minutes, is testimony to the efficiency of the protein-synthetic mechanism of the cell.

One question now remains: How is the information encoded in the master template, DNA, transmitted to ribonucleic acids? The most suggestive clues come from experiments on enzyme systems that synthesize RNA. Samuel Weiss at the University of Chicago and Jerard Hurwitz at New York University have described enzyme systems isolated from animal cells and bacteria that utilize all four nucleotides (as triphosphates) for RNA synthesis. Although the nucleic acid being synthesized is RNA, DNA must be present for synthesis to occur. What is more remarkable, the nucleotide composition of the DNA determines what kind of RNA will be formed. The base-pairing rules of the Watson-Crick DNA model appear to apply with equal rigor to the formation of these DNA-RNA hybrids. DNA templates induce the synthesis of complementary RNA molecules only.

This control over RNA synthesis by DNA is also seen in living cells. Elliot Volkin of the Oak Ridge National Laboratory found in 1958 that when bacterial viruses infect bacteria, an RNA is formed that resembles the virus DNA and not that of the host in its base composition. Benjamin D. Hall and Sol Spiegelman of the University of Illinois then showed that the sequence of the nucleotides in the new RNA molecule is complementary to that of the RNA of the virus.

So the story comes full circle. Specific genetic information resides in the nucleotide sequences in DNA. By means of base-pairing mechanisms these sequences are copied to produce either new DNA molecules for new cells or the RNA templates needed for protein synthesis. Specific nucleotide sequences in the ribosome templates encode the amino acid sequence for particular proteins. The transfer RNA's recognize these sequences and bring amino acids into the proper alignment. Peptide bonds then form with great specificity and rapidity, putting together the protein molecules characteristic of the species. These proteins, many of them enzymes, are the tools with which the cell synthesizes the host of other molecules (purines, pyrimidines, amino acids, carbohydrates, fats, sterols, pigments and so on) necessary to its structure and function.

What has been said so far makes it clear that there is a transmission of genetic information from the DNA in the chromosomes to the sites of protein synthesis. Textbooks often show this flow as an arrow leading from the nucleus to the cytoplasm. There is no corresponding arrow from the cytoplasm back to the nucleus. Absence of the return arrow might suggest to a biologist that there is something essential missing in the scheme, for in all biological systems that have been carefully studied there is a feedback. In the cell there is indeed evidence for a feedback control directed from the cytoplasm to the chromosomes. In some cases the feedback comes quickly and lasts for only a short time. In the pancreas, for example, when the cells are stimulated so that their cytoplasm synthesizes digestive enzymes, tracer experiments show that within a few minutes there is a rise in the uptake of amino acids into proteins of the chromosomes. There are also less immediate and more enduring cytoplasmic influences on chromosomes. Among these are the profound changes associated with cell differentiation [see "How Cells Specialize," Offprint #94].

But although protein synthesis in chromosomes has been shown to be subject to feedback control, there is at present no evidence that the sequence of bases in DNA can be altered by feedback. The chromosomes of germ cells have changed in the course of evolution so that they carry genetic information that is effective in adapting an organism to its environment. The DNA of a germ cell has been shaped by evolution so that it can determine the synthesis of enzymes and other proteins that make for a viable organism. The important point here is that the changes that have taken place during the course of evolution in the DNA molecules of the germ cells of an organism are not the direct result of a feedback from the cytoplasm to the chromosomes in the nucleus. Changes in DNA itself, according to the generally held views of biologists today, are due to mutation and selection.

The Genetic Code: III

3

by F. H. C. Crick
October 1966

The central theme of molecular biology is confirmed by detailed knowledge of how the four-letter language embodied in molecules of nucleic acid controls the 20-letter language of the proteins

The hypothesis that the genes of the living cell contain all the information needed for the cell to reproduce itself is now more than 50 years old. Implicit in the hypothesis is the idea that the genes bear in coded form the detailed specifications for the thousands of kinds of protein molecules the cell requires for its moment-to-moment existence: for extracting energy from molecules assimilated as food and for repairing itself as well as for replication. It is only within the past 15 years, however, that insight has been gained into the chemical nature of the genetic material and how its molecular structure can embody coded instructions that can be "read" by the machinery in the cell responsible for synthesizing protein molecules. As the result of intensive work by many investigators the story

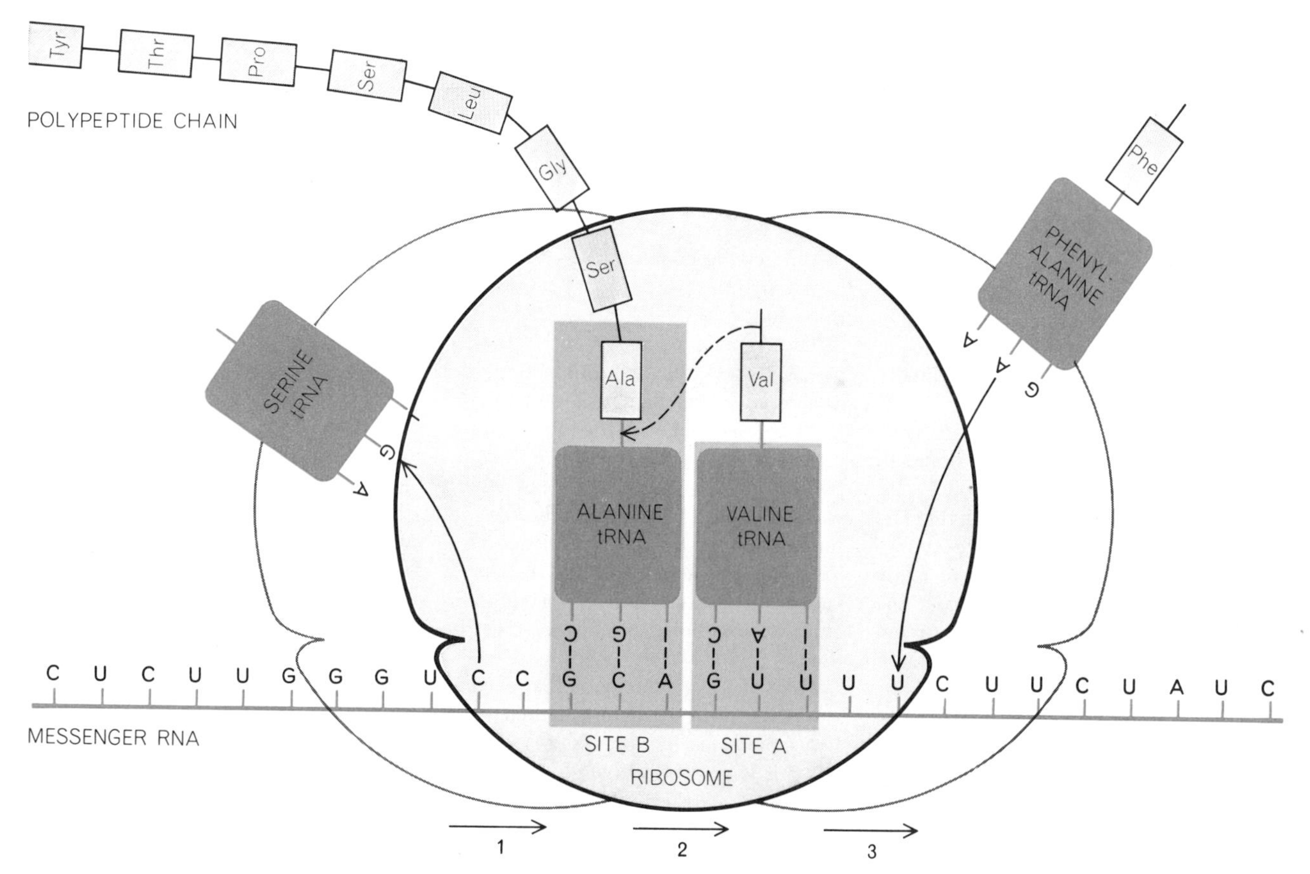

SYNTHESIS OF PROTEIN MOLECULES is accomplished by the intracellular particles called ribosomes. The coded instructions for making the protein molecule are carried to the ribosome by a form of ribonucleic acid (RNA) known as "messenger" RNA. The RNA code "letters" are four bases: uracil (U), cytosine (C), adenine (A) and guanine (G). A sequence of three bases, called a codon, is required to specify each of the 20 kinds of amino acid, identified here by their abbreviations. (A list of the 20 amino acids and their abbreviations appears on the next page.) When linked end to end, these amino acids form the polypeptide chains of which proteins are composed. Each type of amino acid is transported to the ribosome by a particular form of "transfer" RNA (tRNA), which carries an anticodon that can form a temporary bond with one of the codons in messenger RNA. Here the ribosome is shown moving along the chain of messenger RNA, "reading off" the codons in sequence. It appears that the ribosome has two binding sites for molecules of tRNA: one site (*A*) for positioning a newly arrived tRNA molecule and another (*B*) for holding the growing polypeptide chain.

AMINO ACID	ABBREVIATION
ALANINE	Ala
ARGININE	Arg
ASPARAGINE	AspN
ASPARTIC ACID	Asp
CYSTEINE	Cys
GLUTAMIC ACID	Glu
GLUTAMINE	GluN
GLYCINE	Gly
HISTIDINE	His
ISOLEUCINE	Ileu
LEUCINE	Leu
LYSINE	Lys
METHIONINE	Met
PHENYLALANINE	Phe
PROLINE	Pro
SERINE	Ser
THREONINE	Thr
TRYPTOPHAN	Tryp
TYROSINE	Tyr
VALINE	Val

TWENTY AMINO ACIDS constitute the standard set found in all proteins. A few other amino acids occur infrequently in proteins but it is suspected in each case that they originate as one of the standard set and become chemically modified after they have been incorporated into a polypeptide chain.

of the genetic code is now essentially complete. One can trace the transmission of the coded message from its original site in the genetic material to the finished protein molecule.

The genetic material of the living cell is the chainlike molecule of deoxyribonucleic acid (DNA). The cells of many bacteria have only a single chain; the cells of mammals have dozens clustered together in chromosomes. The DNA molecules have a very long backbone made up of repeating groups of phosphate and a five-carbon sugar. To this backbone the side groups called bases are attached at regular intervals. There are four standard bases: adenine (A), guanine (G), thymine (T) and cytosine (C). They are the four "letters" used to spell out the genetic message. The exact sequence of bases along a length of the DNA molecule determines the structure of a particular protein molecule.

Proteins are synthesized from a standard set of 20 amino acids, uniform throughout nature, that are joined end to end to form the long polypeptide chains of protein molecules [*see illustration at left*]. Each protein has its own characteristic sequence of amino acids. The number of amino acids in a polypeptide chain ranges typically from 100 to 300 or more.

The genetic code is not the message itself but the "dictionary" used by the cell to translate from the four-letter language of nucleic acid to the 20-letter language of protein. The machinery of the cell can translate in one direction only: from nucleic acid to protein but not from protein to nucleic acid. In making this translation the cell employs a variety of accessory molecules and mechanisms. The message contained in DNA is first transcribed into the similar molecule called "messenger" ribonucleic acid—messenger RNA. (In many viruses—the tobacco mosaic virus, for example—the genetic material is simply RNA.) RNA too has four kinds of bases as side groups; three are identical with those found in DNA (adenine, guanine and cytosine) but the fourth is uracil (U) instead of thymine. In this first transcription of the genetic message the code letters A, G, T and C in DNA give rise respectively to U, C, A and G. In other words, wherever A appears in DNA, U appears in the RNA transcription; wherever G appears in DNA, C appears in the transcription, and so on. As it is usually presented the dictionary of the genetic code employs the letters found in RNA (U, C, A, G) rather than those found in DNA (A, G, T, C).

The genetic code could be broken easily if one could determine both the amino acid sequence of a protein and the base sequence of the piece of nucleic acid that codes it. A simple comparison of the two sequences would yield the code. Unfortunately the determination of the base sequence of a long nucleic acid molecule is, for a variety of reasons, still extremely difficult. More indirect approaches must be used.

Most of the genetic code first became known early in 1965. Since then additional evidence has proved that almost all of it is correct, although a few features remain uncertain. This article describes how the code was discovered and some of the work that supports it.

Scientific American has already presented a number of articles on the genetic code. In one of them [" The Genetic Code," Offprint 123] I explained that the experimental evidence (mainly indirect) suggested that the code was a triplet code: that the bases on the messenger RNA were read three at a time and that each group corresponded to a particular amino acid. Such a group is called a codon. Using four symbols in groups of three, one can form 64 distinct triplets. The evidence indicated that most of these stood for one amino acid or another, implying that an amino acid was usually represented by several codons. Adjacent amino acids were coded by adjacent codons, which did not overlap.

In a sequel to that article ["The Genetic Code: II," Offprint 153]. Marshall W. Nirenberg of the National Institutes of Health explained how the composition of many of the 64 triplets had been determined by actual experiment. The technique was to synthesize polypeptide chains in a cell-free system, which was made by breaking open cells of the colon bacillus (*Escherichia coli*) and extracting from them the machinery for protein synthesis. Then the system was provided with an energy supply, 20 amino acids and one or another of several types of synthetic RNA. Although the exact sequence of bases in each type was random, the proportion of bases was known. It was found that each type of synthetic messenger RNA directed the incorporation of certain amino acids only.

By means of this method, used in a quantitative way, the *composition* of many of the codons was obtained, but the *order* of bases in any triplet could not be determined. Codons rich in G were difficult to study, and in addition a few mistakes crept in. Of the 40 codon compositions listed by Nirenberg in his article we now know that 35 were correct.

The Triplet Code

The main outlines of the genetic code were elucidated by another technique invented by Nirenberg and Philip Leder. In this method no protein synthesis occurs. Instead one triplet at a time is used to bind together parts of the machinery of protein synthesis.

Protein synthesis takes place on the comparatively large intracellular structures known as ribosomes. These bodies travel along the chain of messenger RNA, reading off its triplets one after another and synthesizing the polypeptide chain of the protein, starting at the amino end (NH_2). The amino acids do not diffuse to the ribosomes by themselves. Each amino acid is joined chemically by a special enzyme to one of the codon-recognizing molecules known both as soluble RNA (sRNA) and transfer RNA (tRNA). (I prefer the latter designation.) Each tRNA molecule has its own triplet of bases, called an anticodon, that recognizes the relevant codon on the messenger RNA by pairing bases with it [*see illustration on page 21*].

Leder and Nirenberg studied which amino acid, joined to its tRNA molecules, was bound to the ribosomes in the presence of a particular triplet, that is, by a "message" with just three letters. They did so by the neat trick of passing the mixture over a nitrocellulose filter that retained the ribosomes. All the tRNA molecules passed through the filter except the ones specifically bound to the ribosomes by the triplet. Which they were could easily be decided by using mixtures of amino acids in which one kind of amino acid had been made artificially radioactive, and determining the amount of radioactivity absorbed by the filter.

For example, the triplet GUU retained the tRNA for the amino acid valine, whereas the triplets UGU and UUG did not. (Here GUU actually stands for the trinucleoside diphosphate GpUpU.) Further experiments showed that UGU coded for cysteine and UUG for leucine.

Nirenberg and his colleagues synthesized all 64 triplets and tested them for their coding properties. Similar results have been obtained by H. Gobind Khorana and his co-workers at the University of Wisconsin. Various other groups have checked a smaller number of codon assignments.

Close to 50 of the 64 triplets give a clearly unambiguous answer in the binding test. Of the remainder some evince only weak binding and some bind more than one kind of amino acid. Other results I shall describe later suggest that the multiple binding is often an artifact of the binding method. In short, the binding test gives the meaning of the majority of the triplets but it does not firmly establish all of them.

The genetic code obtained in this way, with a few additions secured by other methods, is shown in the table below. The 64 possible triplets are set out in a regular array, following a plan

FIRST LETTER	SECOND LETTER: U		C		A		G		THIRD LETTER
U	UUU	Phe	UCU	Ser	UAU	Tyr	UGU	Cys	U
	UUC		UCC		UAC		UGC		C
	UUA	Leu	UCA		UAA	OCHRE	UGA	?	A
	UUG		UCG		UAG	AMBER	UGG	Tryp	G
C	CUU	Leu	CCU	Pro	CAU	His	CGU	Arg	U
	CUC		CCC		CAC		CGC		C
	CUA		CCA		CAA	GluN	CGA		A
	CUG		CCG		CAG		CGG		G
A	AUU	Ileu	ACU	Thr	AAU	AspN	AGU	Ser	U
	AUC		ACC		AAC		AGC		C
	AUA		ACA		AAA	Lys	AGA	Arg	A
	AUG	Met	ACG		AAG		AGG		G
G	GUU	Val	GCU	Ala	GAU	Asp	GGU	Gly	U
	GUC		GCC		GAC		GGC		C
	GUA		GCA		GAA	Glu	GGA		A
	GUG		GCG		GAG		GGG		G

GENETIC CODE, consisting of 64 triplet combinations and their corresponding amino acids, is shown in its most likely version. The importance of the first two letters in each triplet is readily apparent. Some of the allocations are still not completely certain, particularly for organisms other than the colon bacillus (*Escherichia coli*). "Amber" and "ochre" are terms that referred originally to certain mutant strains of bacteria. They designate two triplets, UAA and UAG, that may act as signals for terminating polypeptide chains.

that clarifies the relations between them.

Inspection of the table will show that the triplets coding for the same amino acid are often rather similar. For example, all four of the triplets starting with the doublet AC code for threonine. This pattern also holds for seven of the other amino acids. In every case the triplets *XYU* and *XYC* code for the same amino acid, and in many cases *XYA* and *XYG* are the same (methionine and tryptophan may be exceptions). Thus an amino acid is largely selected by the first two bases of the triplet. Given that a triplet codes for, say, valine, we know that the first two bases are GU, whatever the third may be. This pattern is true for all but three of the amino acids. Leucine can start with UU or CU, serine with UC or AG and arginine with CG or AG. In all other cases the amino acid is uniquely related to the first two bases of the triplet. Of course, the converse is often not true. Given that a triplet starts with, say, CA, it may code for either histidine or glutamine.

Synthetic Messenger RNA's

Probably the most direct way to confirm the genetic code is to synthesize a messenger RNA molecule with a strictly defined base sequence and then find the amino acid sequence of the polypeptide produced under its influence. The most extensive work of this nature has been done by Khorana and his colleagues. By a brilliant combination of ordinary chemical synthesis and synthesis catalyzed by enzymes, they have made long RNA molecules with various repeating sequences of bases. As an example, one RNA molecule they have synthesized has the sequence UGUGUGUGUGUG.... When the biochemical machinery reads this as triplets the message is UGU–GUG–UGU–GUG.... Thus we expect that a polypeptide will be produced with an alternating sequence of two amino acids. In fact, it was found that the product is Cys–Val–Cys–Val.... This evidence alone would not tell us which triplet goes with which amino acid, but given the results of the binding test one has no hesitation in concluding that UGU codes for cysteine and GUG for valine.

In the same way Khorana has made chains with repeating sequences of the type *XYZ*... and also *XXYZ*.... The type *XYZ*... would be expected to give a "homopolypeptide" containing one amino acid corresponding to the triplet *XYZ*. Because the starting point is not clearly defined, however, the homopolypeptides corresponding to *YZX*... and *ZXY*... will also be produced. Thus poly-AUC makes polyisoleucine, polyserine and polyhistidine. This confirms that AUC codes for isoleucine, UCA for serine and CAU for histidine. A repeating sequence of four bases will yield a single type of polypeptide with a repeating sequence of four amino acids. The general patterns to be expected in each case are set forth in the table on this page. The results to date have amply demonstrated by a direct biochemical method that the code is indeed a triplet code.

Khorana and his colleagues have so far confirmed about 25 triplets by this method, including several that were quite doubtful on the basis of the binding test. They plan to synthesize other sequences, so that eventually most of the triplets will be checked in this way.

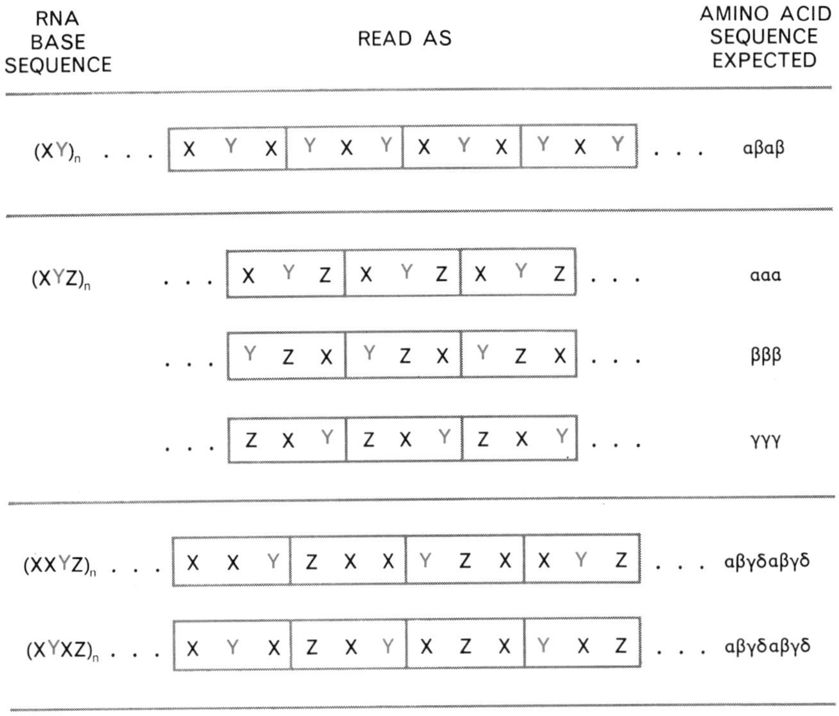

RNA BASE SEQUENCE	READ AS	AMINO ACID SEQUENCE EXPECTED
$(XY)_n$ . . .	X Y X | Y X Y | X Y X | Y X Y . . .	αβαβ
$(XYZ)_n$	. . . X Y Z | X Y Z | X Y Z . . .	ααα
	. . . Y Z X | Y Z X | Y Z X . . .	βββ
	. . . Z X Y | Z X Y | Z X Y . . .	γγγ
$(XXYZ)_n$ . . .	X X Y | Z X X | Y Z X | X Y Z . . .	αβγδαβγδ
$(XYXZ)_n$ . . .	X Y X | Z X Y | X Z X | Y X Z . . .	αβγδαβγδ

VARIETY OF SYNTHETIC RNA's with repeating sequences of bases have been produced by H. Gobind Khorana and his colleagues at the University of Wisconsin. They contain two or three different bases (*X*, *Y*, *Z*) in groups of two, three or four. When introduced into cell-free systems containing the machinery for protein synthesis, the base sequences are read off as triplets (*middle*) and yield the amino acid sequences indicated at the right.

The Use of Mutations

The two methods described so far are open to the objection that since they do not involve intact cells there may be some danger of false results. This objection can be met by two other methods of checking the code in which the act of protein synthesis takes place inside the cell. Both involve the effects of genetic mutations on the amino acid sequence of a protein.

It is now known that small mutations are normally of two types: "base substitution" mutants and "phase shift" mutants. In the first type one base is changed into another base but the total number of bases remains the same. In the second, one or a small number of bases are added to the message or subtracted from it.

There are now extensive data on base-substitution mutants, mainly from studies of three rather convenient proteins: human hemoglobin, the protein of tobacco mosaic virus and the *A* protein of the enzyme tryptophan synthetase obtained from the colon bacillus. At least 36 abnormal types of human hemoglobin have now been investigated by many different workers. More than 40 mutant forms of the protein of the tobacco mosaic virus have been examined by Hans Wittmann of the Max Planck Institute for Molecular Genetics in Tübingen and by Akita Tsugita and Heinz Fraenkel-Conrat of the University of California at Berkeley [see "The Genetic Code of a Virus," by Heinz Fraenkel-Conrat; SCIENTIFIC AMERICAN Offprint 193]. Charles Yanofsky and his group at Stanford University have characterized about 25 different mutations of the *A* protein of tryptophan synthetase.

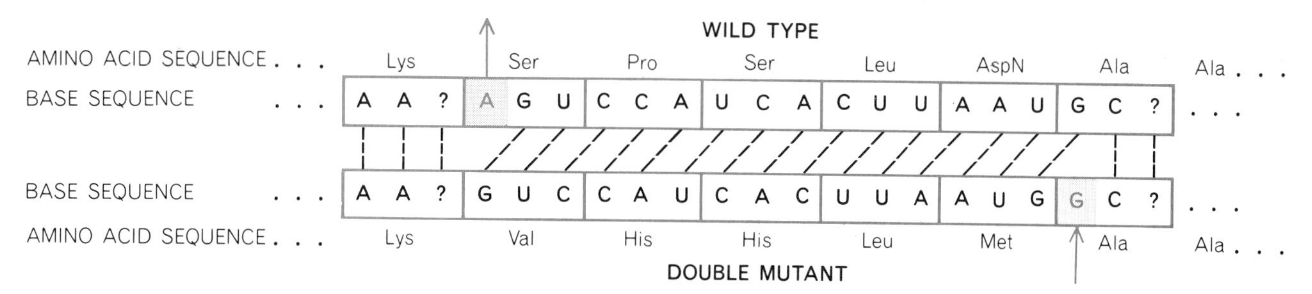

"PHASE SHIFT" MUTATIONS help to establish the actual codons used by organisms in the synthesis of protein. The two partial amino acid sequences shown here were determined by George Streisinger and his colleagues at the University of Oregon. The sequences are from a protein, a type of lysozyme, produced by the bacterial virus T4. A pair of phase-shift mutations evidently removed one base, A, and inserted another, G, about 15 bases farther on. The base sequence was deduced theoretically from the genetic code.

The remarkable fact has emerged that in every case but one the genetic code shows that the change of an amino acid in a polypeptide chain could have been caused by the alteration of a single base in the relevant nucleic acid. For example, the first observed change of an amino acid by mutation (in the hemoglobin of a person suffering from sickle-cell anemia) was from glutamic acid to valine. From the genetic code dictionary on page 23 we see that this could have resulted from a mutation that changed either GAA to GUA or GAG to GUG. In either case the change involved a single base in the several hundred needed to code for one of the two kinds of chain in hemoglobin.

The one exception so far to the rule that all amino acid changes could be caused by single base changes has been found by Yanofsky. In this one case glutamic acid was replaced by methionine. It can be seen from the genetic code dictionary that this can be accomplished only by a change of *two* bases, since glutamic acid is encoded by either GAA or GAG and methionine is encoded only by AUG. This mutation has occurred only once, however, and of all the mutations studied by Yanofsky it is the only one not to back-mutate, or revert to "wild type." It is thus almost certainly the rare case of a double change. All the other cases fit the hypothesis that base-substitution mutations are normally caused by a single base change. Examination of the code shows that only about 40 percent of all the possible amino acid interchanges can be brought about by single base substitutions, and it is only these changes that are found in experiments. Therefore the study of actual mutations has provided strong confirmation of many features of the genetic code.

Because in general several codons stand for one amino acid it is not possible, knowing the amino acid sequence, to write down the exact RNA base sequence that encoded it. This is unfortunate. If we know which amino acid is changed into another by mutation, however, we can often, given the code, work out what that base change must have been. As an example, glutamic acid can be encoded by GAA or GAG and valine by GUU, GUC, GUA or GUG. If a mutation substitutes valine for glutamic acid, one can assume that only a single base change was involved. The only such change that could lead to the desired result would be a change from A to U in the middle position, and this would be true whether GAA became GUA or GAG became GUG.

It is thus possible in many cases (not in all) to compare the nature of the base change with the chemical mutagen used to produce the change. If RNA is treated with nitrous acid, C is changed to U and A is effectively changed to G. On the other hand, if double-strand DNA is treated under the right conditions with hydroxylamine, the mutagen acts only on C. As a result some C's are changed to T's (the DNA equivalent of U's), and thus G's, which are normally paired with C's in double-strand DNA, are replaced by A's.

If 2-aminopurine, a "base analogue" mutagen, is added when double-strand DNA is undergoing replication, it produces only "transitions." These are the same changes as those produced by hydroxylamine—plus the reverse changes. In almost all these different cases (the exceptions are unimportant) the changes observed are those expected from our knowledge of the genetic code.

Note the remarkable fact that, although the code was deduced mainly from studies of the colon bacillus, it appears to apply equally to human beings and tobacco plants. This, together with more fragmentary evidence, suggests that the genetic code is either the same or very similar in most organisms.

The second method of checking the code using intact cells depends on phase-shift mutations such as the addition of a single base to the message. Phase-shift mutations probably result from errors produced during genetic recombination or when the DNA molecule is being duplicated. Such errors have the effect of putting out of phase the reading of the message from that point on. This hypothesis leads to the prediction that the phase can be corrected if at some subsequent point a nucleotide is deleted. The pair of alterations would be expected not only to change two amino acids but also to alter all those encoded by bases lying between the two affected sites. The reason is that the intervening bases would be read out of phase and therefore grouped into triplets different from those contained in the normal message.

This expectation has recently been confirmed by George Streisinger and his colleagues at the University of Oregon. They have studied mutations in the protein lysozyme that were produced by the T4 virus, which infects the colon bacillus. One phase-shift mutation involved the amino acid sequence ...Lys—Ser—Pro—Ser—Leu—AspN—Ala—Ala—Lys.... They were then able to construct by genetic methods a double phase-shift mutant in which the corresponding sequence was ...Lys—Val—His—His—Leu—Met—Ala—Ala—Lys....

Given these two sequences, the reader should be able, using the genetic code dictionary on page 23, to decipher uniquely a short length of the nucleic acid message for both the original protein and the double mutant and thus deduce the changes produced by each of the phase-shift mutations. The correct result is presented in the illustration above. The result not only confirms several rather doubtful codons, such as UUA for leucine and AGU for serine, but also shows which codons are actually involved in a genetic message. Since the technique is difficult, however, it may not find wide application.

Streisinger's work also demonstrates what has so far been only tacitly as-

ANTICODON	CODON
U	A G
C	G
A	U
G	U C
I	U C A

"WOBBLE" HYPOTHESIS has been proposed by the author to provide rules for the pairing of codon and anticodon at the *third* position of the codon. There is evidence, for example, that the anticodon base I, which stands for inosine, may pair with as many as three different bases: U, C and A. Inosine closely resembles the base guanine (G) and so would ordinarily be expected to pair with cytosine (C). Structural diagrams for standard base pairings and wobble base pairings are illustrated at the bottom of this page.

sumed: that the two languages, both of which are written down in a certain direction according to convention, are in fact translated by the cell in the same direction and not in opposite directions. This fact had previously been established, with more direct chemical methods, by Severo Ochoa and his colleagues at the New York University School of Medicine. In the convention, which was adopted by chance, proteins are written with the amino (NH_2) end on the left. Nucleic acids are written with the end of the molecule containing a "5 prime" carbon atom at the left. (The "5 prime" refers to a particular carbon atom in the 5-carbon ring of ribose sugar or deoxyribose sugar.)

Finding the Anticodons

Still another method of checking the genetic code is to discover the three bases making up the anticodon in some particular variety of transfer RNA. The first tRNA to have its entire sequence worked out was alanine tRNA, a job done by Robert W. Holley and his collaborators at Cornell University [see "The Nucleotide Sequence of a Nucleic Acid," by Robert W. Holley; SCIENTIFIC AMERICAN Offprint 1033]. Alanine tRNA, obtained from yeast, contains 77 bases. A possible anticodon found near the middle of the molecule has the sequence IGC, where I stands for inosine, a base closely resembling guanine. Since then Hans Zachau and his colleagues at the University of Cologne have established the sequences of two closely related serine tRNA's from yeast, and James Madison and his group at the U.S. Plant, Soil and Nutrition Laboratory at Ithaca, N.Y., have worked out the sequence of a tyrosine tRNA, also from yeast.

A detailed comparison of these three sequences makes it almost certain that the anticodons are alanine–IGC, serine–IGA and tyrosine–GΨA. (Ψ stands for pseudo-uridylic acid, which can form the same base pairs as the base uracil.) In addition there is preliminary evidence from other workers that an anticodon for valine is IAC and an anticodon for phenylalanine is GAA.

All these results would fit the rule that the codon and anticodon pair in an antiparallel manner, and that the pairing in the first two positions of the codon is of the standard type, that is, A pairs with U and G pairs with C. The pairing in the third position of the codon is more complicated. There is now good experimental evidence from both Nirenberg and Khorana and their co-workers that one tRNA can recognize several codons, provided that they differ only in the last place in the codon. Thus Holley's alanine tRNA appears to recognize GCU, GCC and GCA. If it recognizes GCG, it does so only very weakly.

The "Wobble" Hypothesis

I have suggested that this is because of a "wobble" in the pairing in the third place and have shown that a reasonable theoretical model will explain many of the observed results. The suggested rules for the pairing in the third position of the anticodon are presented in the table at the top of this page, but this theory is still speculative. The rules for the first two places of the codon seem reasonably secure, however, and can be used as partial confirmation of the genetic code. The likely codon-anticodon pairings for valine, serine, tyrosine, alanine and phenylalanine satisfy the standard base pairings in the first two places and the wobble hypothesis in the third place [*see illustration on page 27*].

Several points about the genetic code remain to be cleared up. For example, the triplet UGA has still to be allocated.

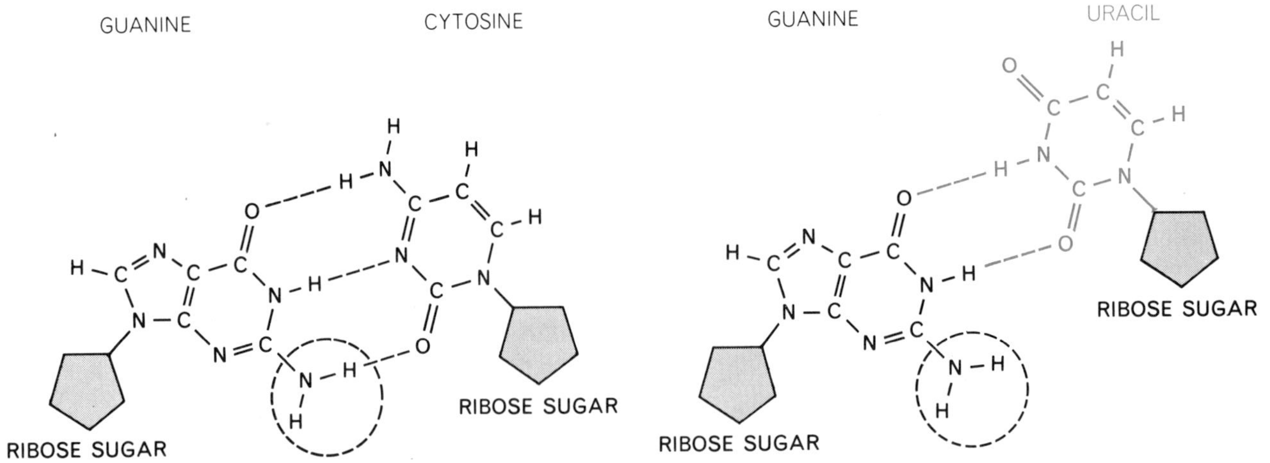

GUANINE CYTOSINE GUANINE URACIL

RIBOSE SUGAR RIBOSE SUGAR RIBOSE SUGAR RIBOSE SUGAR

STANDARD AND WOBBLE BASE PAIRINGS both involve the formation of hydrogen bonds when certain bases are brought into close proximity. In the standard guanine-cytosine pairing (*left*) it is believed three hydrogen bonds are formed. The bases are shown as they exist in the RNA molecule, where they are attached to 5-carbon rings of ribose sugar. In the proposed wobble pairing (*right*) guanine is linked to uracil by only two hydrogen bonds. The base inosine (I) has a single hydrogen atom where guanine has an amino (NH_2) group (*broken circle*). In the author's wobble hypothesis inosine can pair with U as well as with C and A (*not shown*).

The punctuation marks—the signals for "begin chain" and "end chain"—are only partly understood. It seems likely that both the triplet UAA (called "ochre") and UAG (called "amber") can terminate the polypeptide chain, but which triplet is normally found at the end of a gene is still uncertain.

The picturesque terms for these two triplets originated when it was discovered in studies of the colon bacillus some years ago that mutations in other genes (mutations that in fact cause errors in chain termination) could "suppress" the action of certain mutant codons, now identified as either UAA or UAG. The terms "ochre" and "amber" are simply invented designations and have no reference to color.

A mechanism for chain initiation was discovered fairly recently. In the colon bacillus it seems certain that formylmethionine, carried by a special tRNA, can initiate chains, although it is not clear if all chains have to start in this way, or what the mechanism is in mammals and other species. The formyl group (CHO) is not normally found on finished proteins, suggesting that it is probably removed by a special enzyme. It seems likely that sometimes the methionine is removed as well.

It is unfortunately possible that a few codons may be ambiguous, that is, may code for more than one amino acid. This is certainly not true of most codons. The present evidence for a small amount of ambiguity is suggestive but not conclusive. It will make the code more difficult to establish correctly if ambiguity can occur.

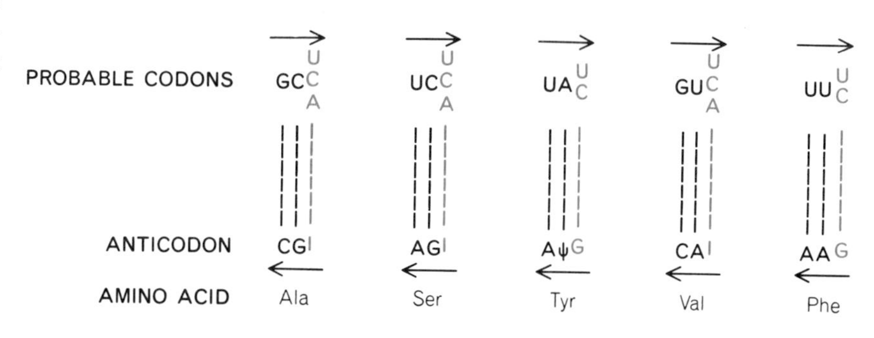

CODON-ANTICODON PAIRINGS take place in an antiparallel direction. Thus the anticodons are shown here written backward, as opposed to the way they appear in the text. The five anticodons are those tentatively identified in the transfer RNA's for alanine, serine, tyrosine, valine and phenylalanine. Color indicates where wobble pairings may occur.

Problems for the Future

From what has been said it is clear that, although the entire genetic code is not known with complete certainty, it is highly likely that most of it is correct. Further work will surely clear up the doubtful codons, clarify the punctuation marks, delimit ambiguity and extend the code to many other species. Although the code lists the codons that *may* be used, we still have to determine if alternative codons are used equally. Some preliminary work suggests they may not be. There is also still much to be discovered about the machinery of protein synthesis. How many types of tRNA are there? What is the structure of the ribosome? How does it work, and why is it in two parts? In addition there are many questions concerning the control of the rate of protein synthesis that we are still a long way from answering.

When such questions have been answered, the major unsolved problem will be the structure of the genetic code. Is the present code merely the result of a series of evolutionary accidents, so that the allocations of triplets to amino acids is to some extent arbitrary? Or are there profound structural reasons why phenylalanine has to be coded by UUU and UUC and by no other triplets? Such questions will be difficult to decide, since the genetic code originated at least three billion years ago, and it may be impossible to reconstruct the sequence of events that took place at such a remote period. The origin of the code is very close to the origin of life. Unless we are lucky it is likely that much of the evidence we should like to have has long since disappeared.

Nevertheless, the genetic code is a major milestone on the long road of molecular biology. In showing in detail how the four-letter language of nucleic acid controls the 20-letter language of protein it confirms the central theme of molecular biology that genetic information can be stored as a one-dimensional message on nucleic acid and be expressed as the one-dimensional amino acid sequence of a protein. Many problems remain, but this knowledge is now secure.

4

The Chemical Structure of Proteins

by William H. Stein and Stanford Moore
February 1961

The atom-by-atom structural formulas of three proteins have now been worked out. This knowledge should help in understanding how the structures of proteins are related to their biological activity

Six years ago, after a decade of pioneering research, Frederick Sanger and his colleagues at the University of Cambridge were able to write the first structural formula of a protein—the hormone insulin. They had discovered the precise order in which the atoms are strung together in the long chains that make up the insulin molecule. With this truly epochal achievement, for which Sanger received a Nobel prize in 1958, there opened a new chapter in protein chemistry.

Some idea of the significance of this chapter for chemistry and biology can be gained from the realization that proteins comprise more than half the solid substance in the tissues of man and other mammals. Proteins are important mechanical elements and perform countless essential catalytic and protective functions. Only when the structures of large numbers of proteins have been worked out will biochemists be in a position to answer many of the fundamental questions they have long been asking. The goal is still far off, but there has been much progress in this challenging field.

Before embarking on a summary of some of the recent developments, it is well to point out that the chemical approach does not provide a complete solution to the problem of protein structure. The order of links in the chain is not the whole story. Each chain is coiled and folded in a three-dimensional pattern, no less important than the atom-by-atom sequence in determining its biological activity. Chemical methods can provide only a partial insight into this three-dimensional, or "tertiary," structure. In the past few years the spatial problem has begun to yield to X-ray analysis. The present article, however, is primarily concerned with the chemical rather than the physical line of attack.

Proteins are very large, they are complex and they are fragile. The molecule of insulin, one of the smallest proteins, contains 777 atoms. Some protein molecules are thought to be 50 times bigger, although even the size of the largest ones is not definitely settled. Fortunately most of them are constructed on the same gen-

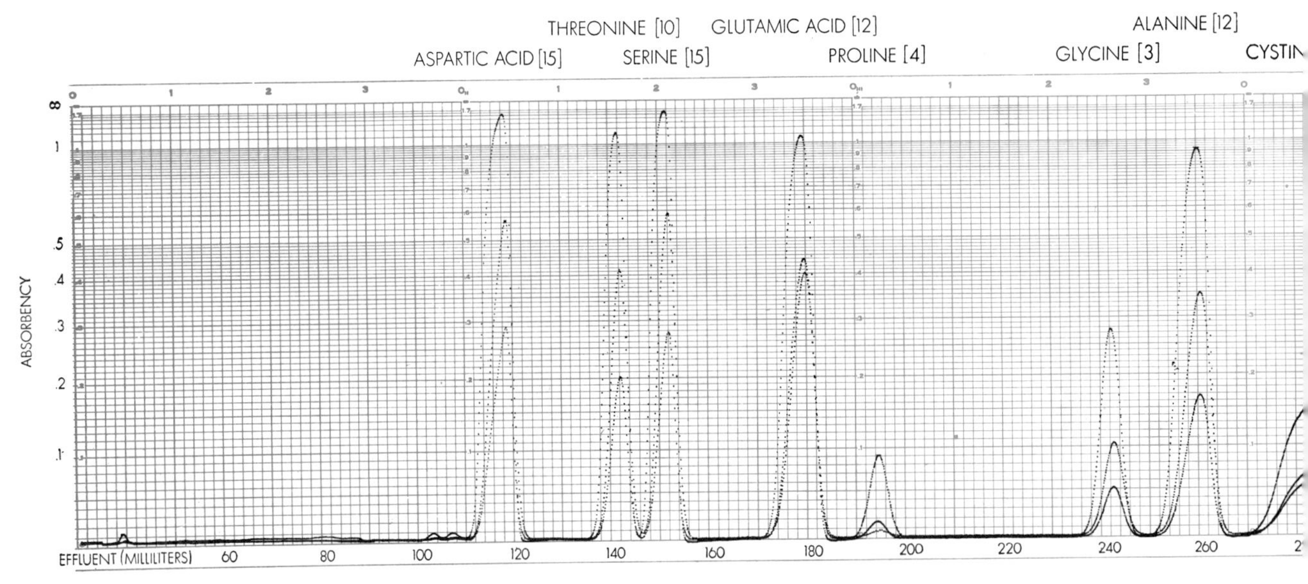

AMINO ACID ANALYSIS is recorded as a series of peaks by an automatic analyzer. The trace reproduced here was obtained in the analysis of a hydrolyzed sample of ribonuclease. The names of the amino acids that were found appear above their corresponding peaks. The number of residues of each in the ribonuclease molecule, determined by the intensity of blue color formed with a special reagent, is shown in brackets. (Proline gives proportionately less color than the other amino acids.) Histidine, lysine, arginine

AMINO ACID ANALYZER is photographed in the authors' laboratory at the Rockefeller Institute. Vertical tubes in center are ion-exchange columns. Photometer unit is enclosed in case at top left; recorder is at bottom left. Next to it on the bench is heating bath.

eral plan. The links, or building blocks, in the molecular chains are amino acids. In their uncombined form each of these substances consists of an amino group (NH_2) and a carboxyl group (COOH), both attached to the same carbon atom. Also attached to this carbon are a hydrogen atom and one of 24 different "side groups." When the amino acids link together in a protein molecule, they join end to end, the carboxyl group of one combining with the amino group of the next to form a "peptide bond" (–CO–NH–) and a molecule of water (HOH). Since each component has lost a molecule of water, it is called an amino acid "residue." Under the action of acids, alkalis or certain enzymes, peptide chains break apart, the –NH– groups regaining a hydrogen atom and the –CO– groups a hydroxyl group (OH).

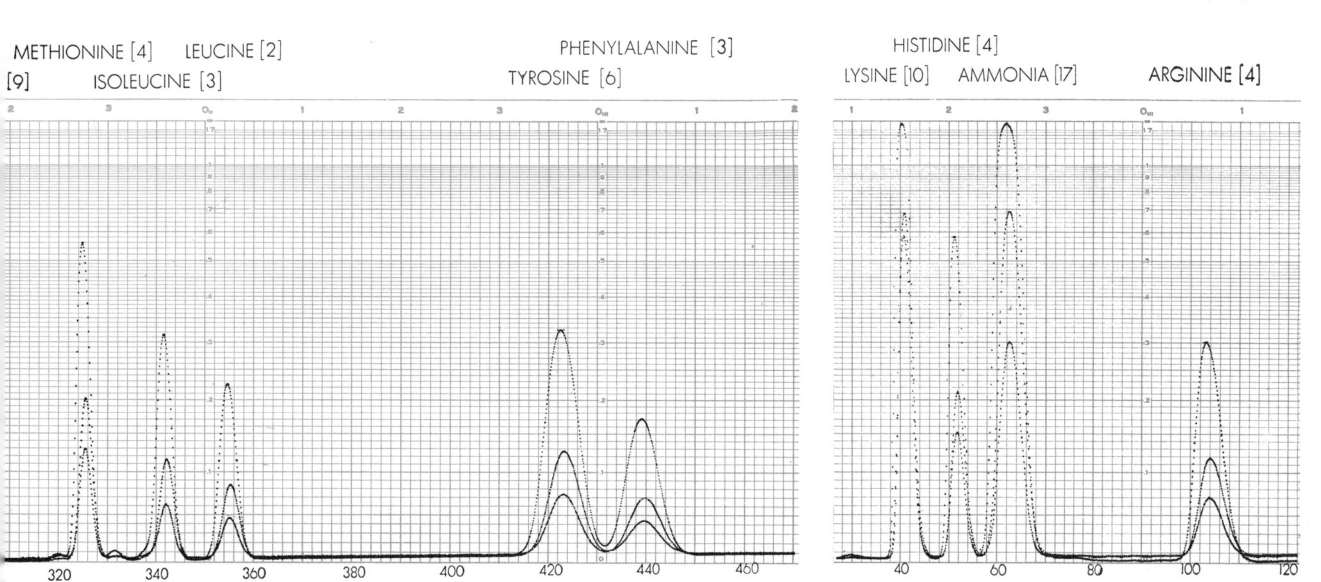

and ammonia are determined with a separate ion-exchange column and recorded separately (*right*). The ammonia is released by the hydrolysis of residues of asparagine and glutamine in the intact ribonuclease molecule. In the process these substances are converted to aspartic acid and glutamic acid respectively. Horizontal scale measures milliliters of solution to have passed through the columns; vertical scale, color intensity. Different curves show absorbencies at different wave lengths and depths of solution.

Thus each residue regains a molecule of water and the peptide bonds are said to have been hydrolyzed.

Among the amino acids the one called cystine is unique. Its molecule is a sort of Siamese twin containing two $-NH_2$ and two $-COOH$ groups, with identical halves of the molecule joined by a disulfide bond (–S–S–). One cystine molecule can therefore enter into two separate peptide chains, cross-linking them by means of the disulfide bond. Cystine can also cause a single chain to fold back on itself.

The Insulin Molecule

Much of the chemistry of proteins had been laboriously uncovered in many

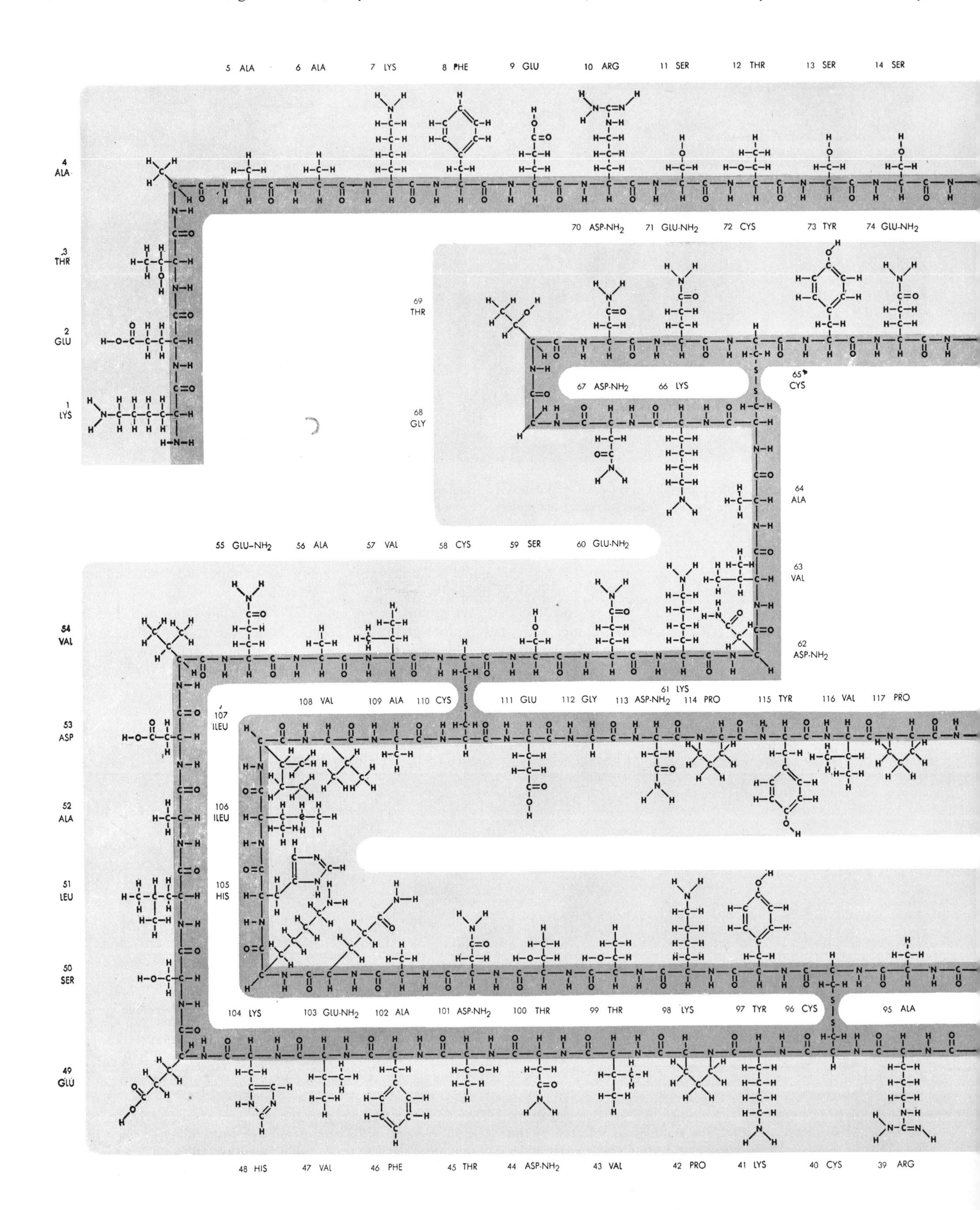

years of research before Sanger took up the study of insulin. His success rested largely on two advances in technique. One was the development of paper chromatography by the British chemists A. J. P. Martin and R. L. M. Synge. By means of this elegant method tiny samples of complex mixtures can be fractionated on a piece of filter paper and their components identified [see "Chromatography," by William H. Stein and Stanford Moore; SCIENTIFIC AMERICAN Offprint 81]. The second key to the problem was Sanger's discovery of a way to label the amino group at the end of a peptide chain. He found that a dinitrophenyl (DNP) group could be attached to free amino groups to form a yellow compound. Even when the peptide is fragmented into separate amino acids, the DNP group remains attached to the residue at the end of the chain, thus making it possible to identify this residue.

The details of Sanger's analysis have already been described in this magazine by one of his co-workers ["The Insulin Molecule," by E. O. P. Thompson; May, 1955]. Here we shall review their work very briefly. With the help of the DNP method they first established that the insulin molecule consists of two chains. These chains are held together by the –S–S– bonds of cystine residues. By treating the hormone with a mild oxidizing agent the experimenters were able to open the disulfide bonds and thus separate the two intact chains. One proved to contain 21 amino acids; the

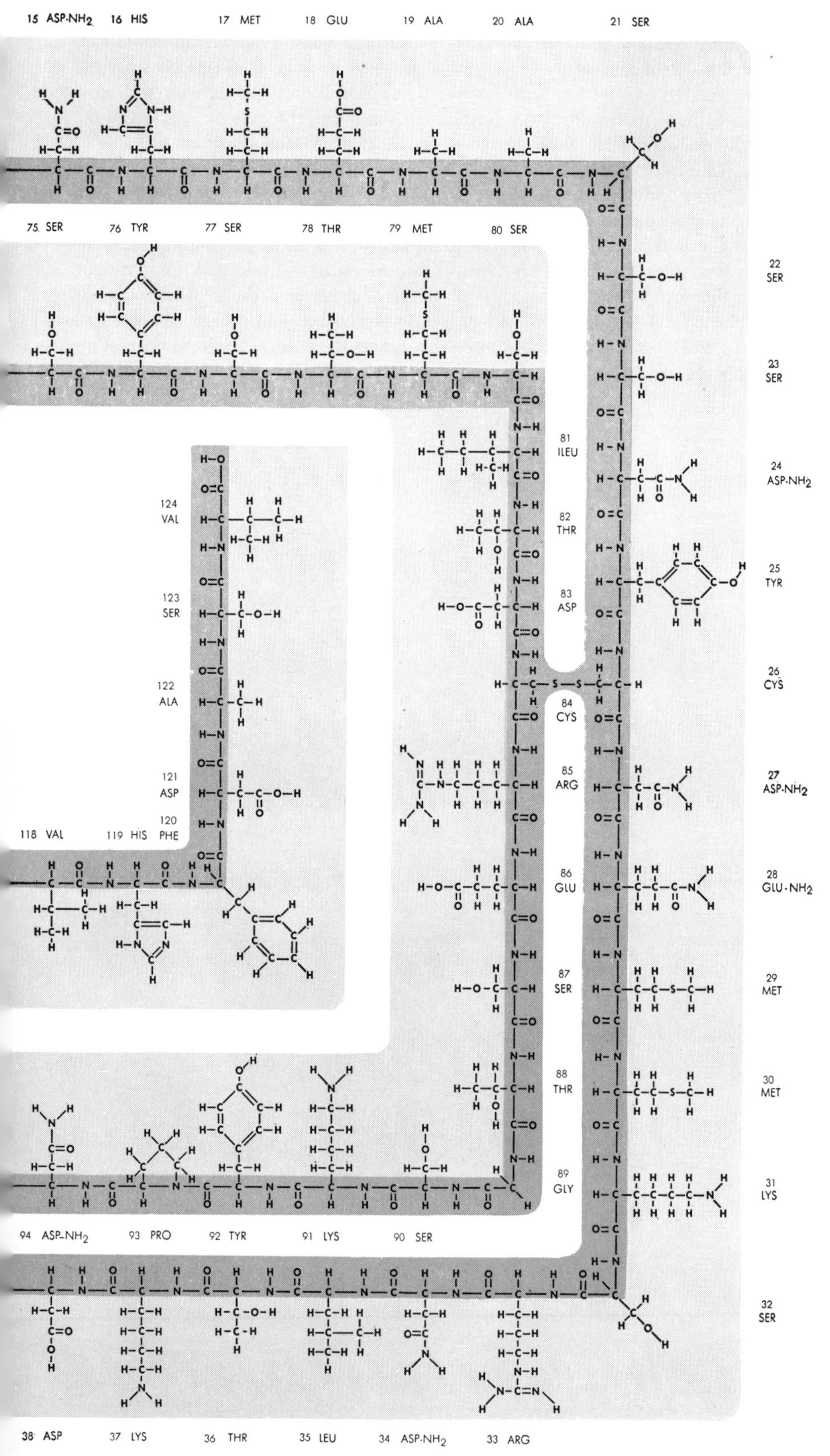

ALA	ALANINE
ARG	ARGININE
ASP	ASPARTIC ACID
ASP-NH2	ASPARAGINE
CYS	CYSTINE
GLU	GLUTAMIC ACID
GLU-NH2	GLUTAMINE
GLY	GLYCINE
HIS	HISTIDINE
ILEU	ISOLEUCINE
LEU	LEUCINE
LYS	LYSINE
MET	METHIONINE
PHE	PHENYLALANINE
PRO	PROLINE
SER	SERINE
THR	THREONINE
TYR	TYROSINE
VAL	VALINE

MOLECULE OF RIBONUCLEASE, an enzyme that digests the cellular substance ribonucleic acid (RNA), is diagramed in two dimensions on these two pages. In this structural formula are 1,876 atoms: 587 of carbon (C), 909 of hydrogen (H), 197 of oxygen (O), 171 of nitrogen (N) and 12 of sulfur (S). The backbone of the chain of amino acid residues is in the darker shaded area; the side chains characteristic of the various amino acids are in the lighter shaded area. The amino acid residues are numbered from 1 to 124, beginning at the amino end of the chain. Abbreviations for amino acids appearing in the diagram are indicated above.

other, 30. They then proceeded to cleave each chain into smaller pieces by treating it with acid, which hydrolyzes peptide bonds more or less at random. The fragments were separated by chromatography and by other means, labeled at their amino ends by the DNP method, broken down further, separated again, relabeled and so on. In this way the order of amino acids in a large number of small pieces was established. By shattering the chain many times and noting overlapping sequences in the various fragments the Cambridge group at last deduced the complete succession of amino acid residues in each part of the molecule.

After the completion of this Herculean task an almost equal effort was required simply to determine the pairing of the half-cystine residues. One chain was found to contain four half-residues of cystine and the other chain two. To find out which ones were paired it was necessary to break the molecule into smaller fragments containing different pairs of half-cystine residues with their disulfide bonds intact. In the process, however, the cystine halves tended to trade partners and produce spurious pairings. Finally a way around the difficulty was worked out, and the disulfide bonds were unequivocally pinned down.

Once Sanger had shown that the problem was solvable a number of workers began to attack the structure of larger proteins. In the six years that have passed two more molecules have been completely delineated and others are on the way. Last year the analysis of the structure of ribonuclease, an enzyme that digests ribonucleic acid, was completed in the authors' laboratory. Ribonuclease contains 124 amino acid residues. As this article was in preparation Gerhard Schramm and his associates in Germany and Heinz L. Fraenkel-Conrat, Wendell M. Stanley and their colleagues at the University of California announced that they had finished working out the structure of the 158-amino-acid-residue protein in tobacco mosaic virus.

At first sight it might seem that any protein should yield to a massive effort along the lines used so successfully with insulin. As Sanger and others realized, however, it is not so simple as that—if simple is the word for 10 years of unremitting work. Problems multiply rapidly with increasing molecular size, and an approach that was difficult and time-consuming in the case of insulin can become fruitless and interminable.

One aid to further progress has been the development of more precise methods for identifying and measuring small quantities of amino acids. Several years ago the authors undertook to apply column chromatography for this purpose. Instead of filter paper we use a five-foot column of an ion-exchange resin. One or two milligrams of an amino acid mix-

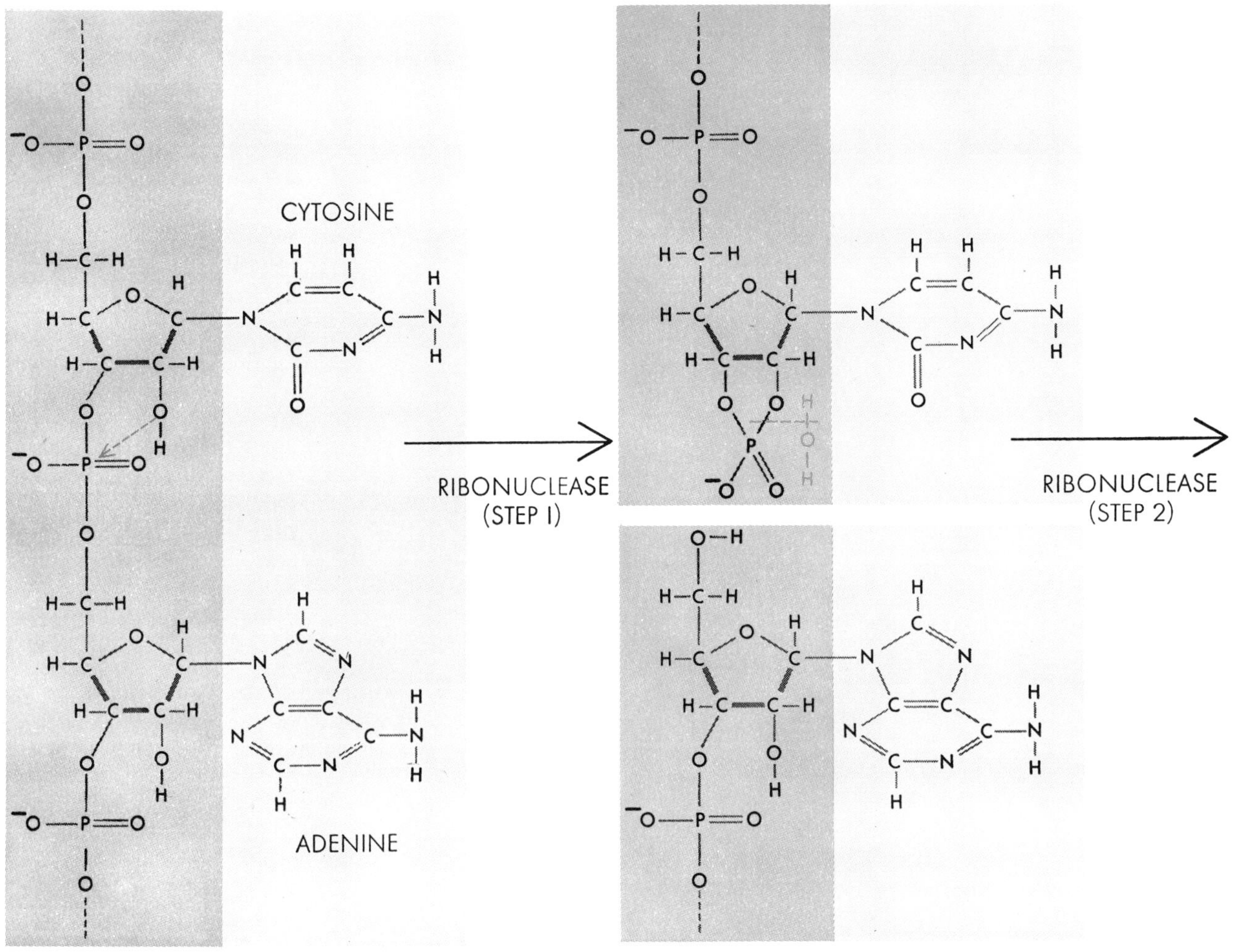

SPLITTING OF RIBONUCLEIC ACID molecule by ribonuclease takes place in two steps. Backbone of the molecule, of which a segment is shown in dark shaded area, is attacked at the phosphorus atom following cytosine (or uracil) but not adenine (or guanine). In the first step the bond between phosphorus and the oxygen atom below it opens, splitting the molecule, a different oxygen uniting

ture placed at the top of the column are washed down through the column by solutions of varying acidity. Depending on their relative affinity for the solutions and for the resin, the individual amino acids move down the column at different rates. By proper choice of salt solutions, acidity and temperature, it is possible to adjust the rates of travel so that the separate amino acids emerge from the bottom of the column at predetermined and well-spaced intervals. To detect the colorless amino acids we heat them with ninhydrin, a reagent that yields a blue color. The intensity of the color is proportional to the amount of amino acid.

In the final version of the device, developed in collaboration with D. H. Spackman, the amino acid analysis is accomplished automatically. The outflow from the column is continuously mixed with ninhydrin, sent through a heating bath and then analyzed by a photometer attached to a recorder. As it flows out of the bath and into the photometer the solution is alternately colorless (when it contains no amino acid) and blue (when an amino acid is present). A continuous plot of the intensity of the blue color shows a series of peaks, each corresponding to a particular amino acid, the area under the peak indicating the amount of that amino acid in the sample. With the automatic amino acid analyzer one operator, working part time, can carry out a complete quantitative analysis of the amino acids from a hydrolyzed protein in 24 hours. The device played an essential role in the work on ribonuclease.

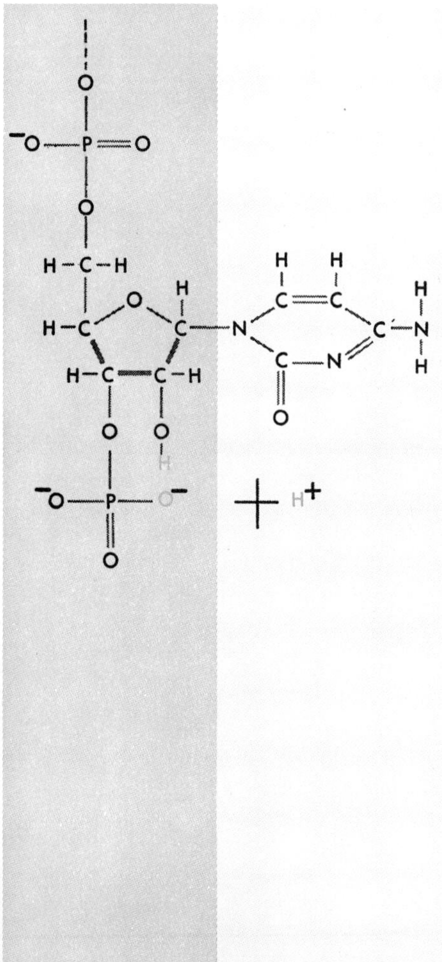

We chose ribonuclease for study for several reasons. The protein, first isolated from beef pancreas in 1920, was available in relatively pure form, having been crystallized by Moses Kunitz of the Rockefeller Institute in 1940. It was known to be a rather small protein, with a molecular weight of about 14,000. (The molecular weight of insulin is 5,733.) Moreover, the manner in which ribonuclease breaks down the ribonucleic acid (RNA) molecule had been worked out in several laboratories. As a result there was a chance of correlating its structure, once that was established, with its biochemical function.

As a first step in the program C. H. W. Hirs, together with the authors, further purified crystalline ribonuclease with the aid of ion-exchange resins. When a hydrolyzed sample was analyzed, it proved to contain a total of 124 amino acid residues of 17 different kinds, plus a quantity of ammonia [*see illustration on pages 28 and 29*]. The presence of ammonia indicated that some of the aspartic acid and glutamic acid that showed up in the analysis came not from residues of these amino acids in the protein molecule but from the related amino acids, asparagine and glutamine. When the last two are hydrolyzed, they are converted to aspartic and glutamic acids and ammonia. Thus the intact ribonuclease molecule contains 19 different kinds of amino acid residue.

There were four residues of cystine and therefore four disulfide bonds. At about that time Christian B. Anfinsen of the National Heart Institute demonstrated, by the DNP method, that ribonuclease consists of a single peptide chain. Therefore the disulfide bonds must cause the folding together of sections of the same chain.

After opening up the disulfide bonds by oxidation, Hirs proceeded to break the 124-link chain into smaller pieces. Here Sanger's method of random hydrolysis with acid did not seem promising. It probably would produce so many small fragments that they would be almost impossible to separate. And even if they were isolated and their structure determined, it seemed unlikely that the sequence of amino acids in the complete chain could be deduced from these small bits and pieces. For a more selective method of dissection we turned to the protein-splitting enzymes. A number of these had been purified and their mode of attack on peptide chains elucidated at the Rockefeller Institute by Max Bergmann and Joseph S. Fruton and their colleagues. The most specific is trypsin, which cleaves only bonds involving the carboxyl groups of the amino acids arginine and lysine. Others, such as chymotrypsin and pepsin, also confine their activity to certain bonds, though not so selectively as does trypsin.

Since there are 10 lysine and four arginine residues in ribonuclease, Hirs first treated the protein with trypsin. The products were separated from one another by the use of columns of ion-exchange resins, and each was then analyzed for amino acids. All told, 13 peptide fragments were isolated, ranging from peptides with only two amino acid residues to some containing more than 20. Among them they accounted for all of the 124 amino acid residues of ribonuclease. To obtain additional fragments Hirs and J. L. Bailey also split the molecule at other points with chymotrypsin and with pepsin.

The amino acid composition of all the fragments was determined, and the residue at the amino end identified in some of them. With this knowledge it was possible, in a sort of crossword-puzzle fashion, to derive a partial structural formula that showed the order in which the various peptides produced by the different enzymatic cleavages must have been arranged in the parent molecule.

The next step was to determine the sequence of amino acid residues within the fragments. One extremely valuable tool, developed by Pehr Edman in Sweden, was a reaction that can clip off one amino acid unit at a time from the amino end of a peptide sequence. There was also a pair of enzymes that can do the same sort of job in some cases, one at the amino and the other at the carboxyl end. Using these and other means, Hirs analyzed 24 peptides completely and examined parts of many more. Each step of each manipulation was monitored on the amino acid analyzer.

After five years of work the complete sequence of the ribonuclease chain was finally established [*see illustration on*

with the phosphorus (*dashed arrow*). In second step another phosphorus-oxygen bond is cleaved with addition of water (H-O-H).

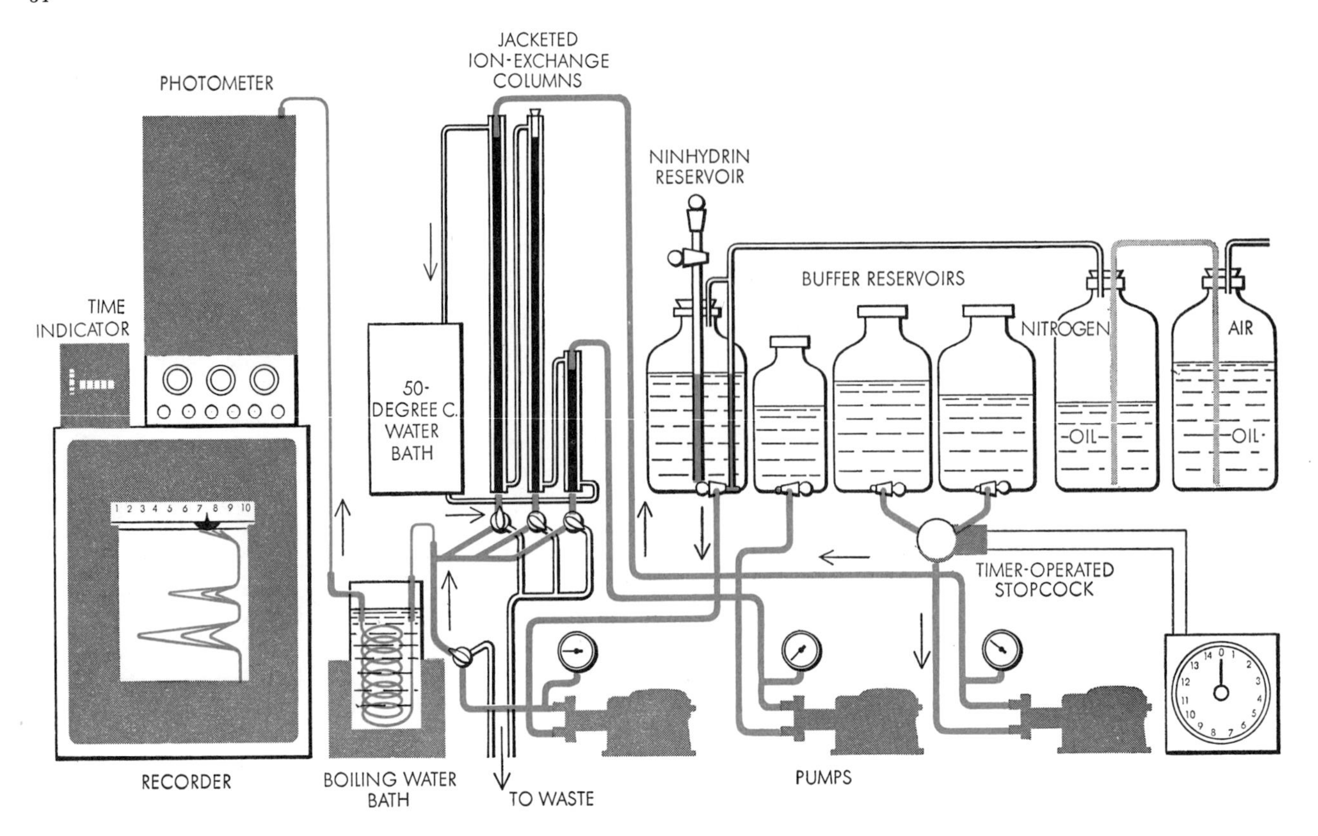

AUTOMATIC ANALYZER for amino acids is diagramed schematically. The two pumps at lower right drive salt solutions through ion-exchange columns. The third pump delivers color reagent (ninhydrin) to a stream emerging from bottom of column. The mixture passes through boiling water bath, where blue color is developed. Intensity of the color is measured by the photometer.

pages 30 and 31]. Spackman then undertook to determine the arrangement of the four disulfide bonds. As in the case of insulin, the problem turned out to be difficult. Two of the bonds were particularly fragile. Only after a long study did Spackman find out how to break ribonuclease into peptides that contained cystine residues with their disulfide bonds intact. Once this had been accomplished the crosslinks could be located and the complete formula written down.

Here we must emphasize that, while the complete formula is certainly correct in most respects, it must still be considered a working hypothesis. Although a great deal of quantitative data supports it, and although Anfinsen's laboratory has derived about a quarter of it independently, we still cannot be completely certain of the results. Degradative experiments, in which molecules are broken down, do not offer final proof for an organic chemical structure; the last word comes when a postulated structure is synthesized and then shown to have all the same properties as the natural product. In spite of the substantial advances of the past few years in the synthesis of complex peptides, it will probably be some time before a molecule of the size of ribonuclease is put together. Until then we must be on the lookout for surprises, because it is entirely possible that ribonuclease contains chemical linkages that are not revealed by the degradative techniques we employed. (For example, unusual linkages have already been found in peptide antibiotics by Lyman C. Craig and his associates at the Rockefeller Institute.)

In any case, our formula has one important deficiency: it is two-dimensional. As we have mentioned, the biological activity of a protein molecule usually depends not only on the sequence of its amino acids but also on how the peptide chain is coiled. Ribonuclease is no exception: it is inactivated by disruption of its three-dimensional structure. Of course, the order of amino acids in any peptide chain must influence its spatial arrangement. Certain sequences are known to preclude certain kinds of folding. To what extent a given sequence may require a given kind of folding is not yet clear.

Although much structural information can be obtained through chemical techniques, they can go only part of the way with a molecule as big as that of a protein. For many years X-ray crystallography has been applied to the problem, recently with striking success. A description of this work must await a separate article. Here we shall merely mention that in a series of brilliant investigations British groups headed by J. C. Kendrew and Max F. Perutz have worked out, respectively, the complete spatial arrangement of the peptide chains in the oxygen-carrying proteins myoglobin and hemoglobin. The sequences of amino acids in these molecules have not yet been determined, although they doubtless will be soon. Indeed, it may prove possible to discover the order by X-ray methods alone, which would provide a valuable check on the methods of organic chemistry. When the sequences are found, it will be possible to place each amino acid residue in its proper position on the models of the coiled chains derived from X-ray studies. At that exciting moment the first true picture of a protein will have been drawn. Unfortunately the methods used by Kendrew and Perutz have not yet been successful with insulin or ribonuclease. Sooner or later, however, the difficulties will be

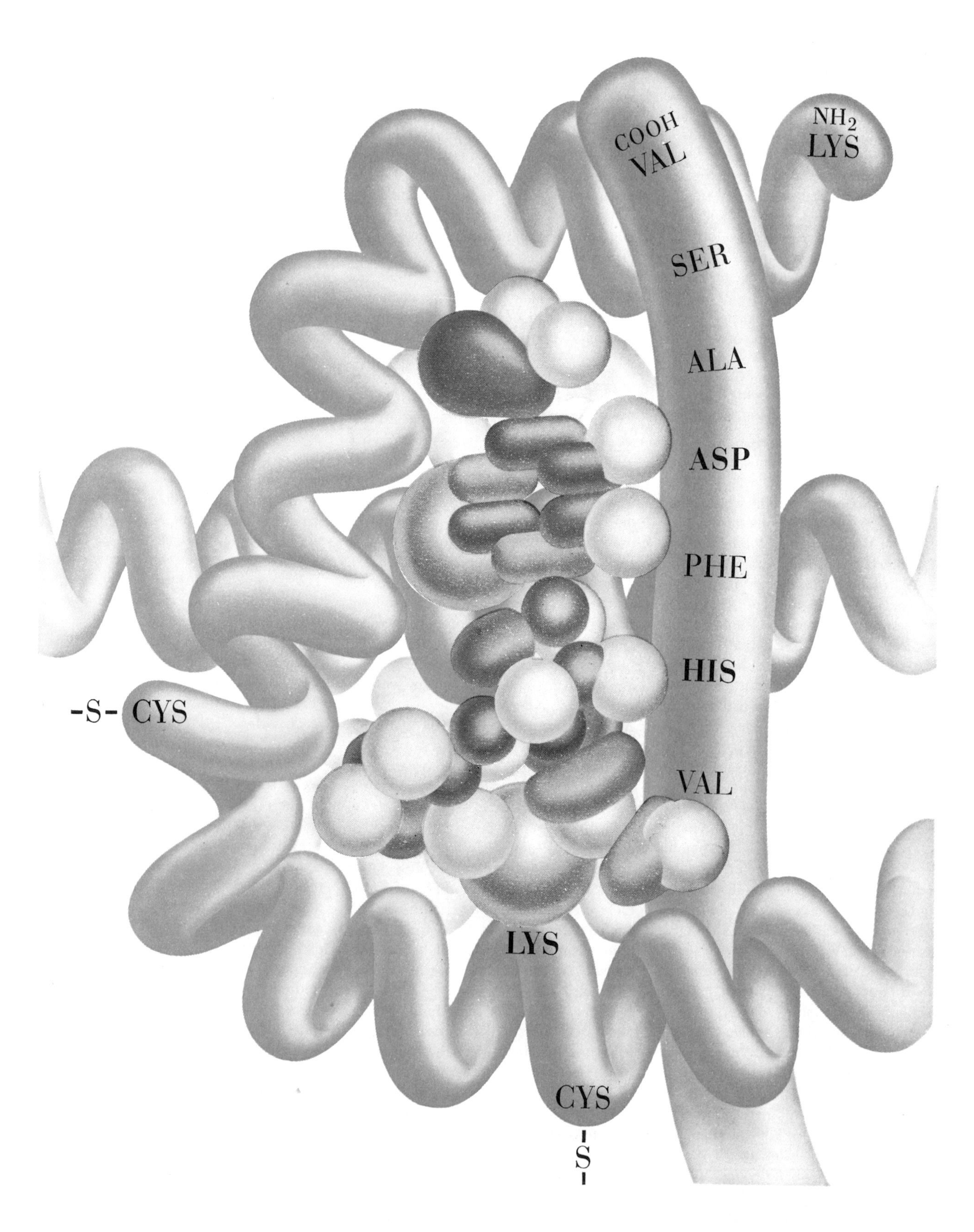

ACTIVE SITE of ribonuclease is represented in this strictly hypothetical conception. A small molecule on which the enzyme acts is shown in gray, nested in a form-fitting cavity on the enzyme surface. Peptide chain, some of which is believed to be helical and some not, is shown in color. Some of the amino acid residues thought to form part of the active site are shown in bold type.

resolved and we shall have the complete portraits of these molecules as well.

Active Sites of Enzymes

In the meantime the techniques of organic chemistry are throwing considerable light on one aspect of the relation between the structure of proteins and their function. That is the question of the "active site" of enzymes. Being proteins, enzymes are all large molecules. Often their substrates—the substances they act on—are very much smaller. For example, ribonuclease splits not only ribonucleic acid but also the comparatively small molecule cyclic cytidylic acid [*see illustration on page 35*]. In such a case only a small portion of the enzyme molecule can be in contact with the substrate when the two are joined together, as they are during the reaction catalyzed by the enzyme. This portion is known as the active site. The concept of an active site does not imply that the rest of the enzyme has no function. Sometimes part of the molecule is dispensable; sometimes it is not. The whole molecule may perhaps be something like a precision lathe, where a ton or more of machinery is required to bring a few ounces of metal in the cutting tool to bear accurately on the work. Similarly, the bulk of the enzyme molecule may be required to bring the active site into proper contact with the substrate. The problem facing the chemist is to discover which amino acid residues make up the active site and just what biological function the rest of the molecule performs.

Now that the structure of an enzyme —ribonuclease—is known, one can hope to find its active site and see how that section is related to the rest of the molecule. Workers in a number of laboratories are engaged in the effort and are making good progress. Before considering their results, it is appropriate to look at some of the earlier work on the active sites of enzymes.

A particularly illuminating series of investigations began in a most unlikely fashion with the study of nerve gases. Developed by the Germans during World War II but happily never used, these substances are phosphate esters of various organic alcohols. (An ester is the compound formed by the reaction of an acid and an alcohol.) They are extremely toxic, rapidly causing death by respiratory paralysis when they are inhaled or even absorbed through the skin. Studies of their physiological action showed that they inactivate choline esterase, an enzyme that breaks down acetylcholine in the body. This last substance plays an essential part in regulating the transmission of nerve impulses. When allowed to accumulate in excessive amounts, it deranges the nervous system, destroying its control over the breathing apparatus.

As soon as the relationship between nerve gas intoxication and choline esterase was established, several laboratories set out to discover the nature of the reaction between the two substances. It was quickly discovered that one molecule of nerve gas combined with one molecule of choline esterase to give an inactive product. The next question was: What part of the large protein molecule is involved? This could not be answered directly because the choline esterase molecule is too large to be studied in detail by the methods available. Besides, it has not been isolated in sufficient amounts. Irwin B. Wilson and David Nachmansohn of the Columbia University College of Physicians and Surgeons approached the problem from the other side; they determined how the reaction between the enzyme and the nerve gas or other inhibitors was influenced by the exact chemical structure of the much smaller organic phosphate molecules. They could then deduce many properties of the enzyme surface. As a result of these investigations a compound was devised that could displace the nerve gas from its combination with choline esterase and so reactivate the enzyme.

Then came an important observation. A. K. Balls and E. F. Jansen of Purdue University found that an organic phosphate called DFP (for diisopropylphosphorofluoridate) also inhibits the enzyme chymotrypsin. This protein was available in large quantities and it had already been extensively studied. On examination of the reaction with DFP, Balls and Jansen found that again one molecule of phosphate combined with one enzyme molecule. Subsequently the point of attachment was identified as a specific serine residue in the enzyme molecule.

There must be something special about this particular serine residue: DFP does not react with the free amino acid serine. No more than one molecule of DFP combines with each molecule of chymotrypsin in spite of the fact that the enzyme has 25 serine residues. And any treatment that destroys the enzymatic activity of chymotrypsin, even temporarily, also destroys its ability to react with DFP. Concluding from this evidence that the reactive serine is part of the active site of the enzyme, several groups of chemists have investigated the sequence of amino acids around it. They labeled the serine with DFP containing radioactive phosphorus and then broke down the inactivated chymotrypsin with other enzymes. From the mixture of fragments a peptide containing the labeled serine was isolated. The serine residue proved to be flanked by an aspartic acid residue on one side and a glycine residue on the other.

Meanwhile several other enzymes had been found to be sensitive to DFP. In all of them the inactivation proved to involve a reaction with a single serine residue. Analysis disclosed that in two of the enzymes the serine had the same neighbors as in chymotrypsin. In two others the sequence proved to be glutamic acid followed by serine and alanine. The two arrangements are in fact much alike because glutamic acid and aspartic acid are closely related chemically, as are glycine and alanine.

Finding the same type of sequence in so many enzymes strikingly confirms a major tenet of biochemistry: a similarity in function must reflect a similarity in structure. But this cannot be the whole story. Although the enzymes are alike in some respects, they are not identical. They do not all catalyze exactly the same reactions and so their active sites must differ in some ways. Of course the sequence aspartic acid, serine and glycine is not by itself biochemically active, nor are the larger peptide fragments that contain it. The serine in them does not react with DFP. Additional residues are required, but which ones, and how they are oriented in space with respect to the active serine, is still largely unknown.

There is persuasive evidence that the active site of these enzymes also includes a histidine residue. Yet in two of the enzyme molecules quite a number of residues on each side of the crucial trio have been identified, and histidine is not among them. If histidine does form part of the active site and is near the active serine, it must be brought there by a folding of the chain. Although still hypothetical, the idea that the amino acid units in an active site are brought into juxtaposition from different sections of a peptide chain by three-dimensional coiling is attractive. Among other things, this hypothesis explains why disrupting only the three-dimensional structure of an enzyme leads to loss of activity. The picture will become much clearer when the complete structures of the DFP-sensitive enzymes are worked out.

Ribonuclease is not sensitive to DFP, but it too can be inactivated by a specific chemical reaction involving a particular amino acid residue. The inactivating agent is either iodoacetic acid or bromoacetic acid, and it combines with a histidine residue. Two investigators in England, E. A. Barnard and W. D. Stein, have obtained evidence that the histidine residue concerned is the one at position 119, six residues from the carboxyl end of the ribonuclease chain. Neither reagent reacts with any of the other three histidines in the molecule. Unfolding the chain or otherwise destroying the activity of ribonuclease prevents the reaction with this residue. Thus the histidine residue at position 119 seems almost surely a part of the active site of ribonuclease. Other studies implicate the aspartic acid at position 121 (removing it together with the last three residues inactivates the molecule, whereas splitting off the last three alone does not) and perhaps also the lysine at position 41.

Some remarkable experiments by F. M. Richards at Yale University have provided much information about the relationship between the activity of the enzyme and its over-all structure. Using a bacterial enzyme called subtilisin, Richards succeeded in splitting the ribonuclease molecule at a single point—the bond between the alanine at position 20 and the serine at position 21. Although the peptide link had been broken, the two fragments did not separate from each other nor did the combination lose enzymatic activity. Treatment with mild acid, however, separated the altered ribonuclease into two parts, one a peptide of 20 residues, the other a large fragment of 104 residues. Neither fragment by itself exhibited activity. But when dissolved together in a neutral solution they recombined instantly, and enzymatic activity was regained. The peptide bond did not re-form under these gentle conditions. Instead the two fragments were held together by so-called secondary forces, probably analogous to those that unite enzymes and their substrates. The simplest explanation, though not the only one, for these results is that activity depends on the juxtaposition of one or more amino acid residues from among the first 20 with one or more in the rest of the molecule.

This brief account of the research on active sites by no means exhausts the work being done on ribonuclease, nor does it exhaust the list of enzymes under investigation. To mention another, the protein-digesting enzyme papain (found in the papaya) has been studied for years at the University of Utah School of Medicine by Emil L. Smith and his colleagues. They have gone far toward determining its structure and have, in addition, made the striking observation that more than 100 amino acid residues of the 185 in the native enzyme can be whittled away without abolishing its activity. The structures of trypsin and chymotrypsin are also being energetically studied, particularly by Hans Neurath and his associates at the University of Washington.

As structural chemistry advances, answers to a number of other fascinating biological problems begin to appear. One can study the differences among analogous protein molecules produced by different species. Sanger is the pioneer in this field, with his studies of the structure of the insulin produced by the cow, the pig and the sheep. Also of great interest are the differences in protein molecules produced by members of the same species under the influence of genetic mutation. In this area a good deal has already been learned about abnormal hemoglobins, and much more will be known when the complete structure of the hemoglobin molecule has been worked out.

Finally, the recent elucidation of the complete amino acid sequence in the protein of tobacco mosaic virus opens the way to a deeper understanding of the natural synthesis of proteins. The virus consists of protein and ribonucleic acid. According to present theory, the protein is synthesized by infected tobacco-plant cells under the direction of the viral nucleic acid. Now it may be possible to follow this process in detail.

The next few years of research on proteins should be exciting ones, both for chemists and biologists.

5 Ribosomes

by Masayasu Nomura
October 1969

They are the organelles that conduct the synthesis of proteins in the living cell. Their structure and functioning are studied by taking them apart and seeing how they reassemble themselves

It is one thing to discover the basic principles of a life process and quite another to know in detail the chemical mechanisms that underlie it. In order to genuinely understand a cellular function one must study the machinery that performs it, and in many cases that means studying a highly organized cellular element, or organelle, that provides the machinery. One must first determine the organelle's structure and learn how it operates and then find out how the organelle itself is generated in the cell. In this article I shall relate how the structural and functional description of the ribosome, the organelle that conducts protein synthesis, has been attempted and is even now being achieved.

The story of the ribosome goes back to the discovery some years ago that the capacity of various types of cell to synthesize proteins was correlated with the cells' content of ribonucleic acid (RNA), and that most of the cellular RNA was in the form of small particles (then known as microsomes) in the cytoplasm of the cell. This suggested that the particles must play some role in protein synthesis, but the real importance of ribosomes emerged only after intensive biochemical investigation.

The pioneer work was done by Paul C. Zamecnik and his collaborators at the Massachusetts General Hospital in the 1950's [see "The Microsome," by Paul C. Zamecnik; SCIENTIFIC AMERICAN Offprint 52]. They homogenized rat-liver cells, added amino acids labeled with atoms of the radioactive isotope carbon 14, fortified the homogenate with adenosine triphosphate (ATP) to provide chemical energy and were able to detect the formation of small amounts of protein. By a process of elimination they established that several cellular organelles, including the nucleus and the mitochondrion, were not necessary for protein synthesis but that the microsomes were essential. They were able to identify other cellular components required for protein synthesis, including the small RNA molecules called transfer RNA and enzymes that attach amino acids to transfer-RNA molecules. These early test-tube assembly systems, however, made only very small amounts of protein. Then in 1961 Marshall W. Nirenberg and J. Heinrich Matthaei of the National Institutes of Health found that in order to obtain intensive synthesis of protein in cell-free extracts of the bacterium *Escherichia coli* it was necessary to include a third type of RNA, called messenger RNA, that had been postulated by François Jacob and Jacques Monod of the Pasteur Institute in Paris.

Once a complete cell-free protein-synthesizing system could be assembled it was possible to study the functioning of its several components. One of the most interesting of these was the ribosome. It was now clear that this particle coordinates the translation of the genetic information in the sequence of nucleotide bases in the messenger RNA (transcribed from the DNA molecule, the gene) to the sequence of amino acids in each protein manufactured by the cell [*see illustration on page 40*]. The first systematic studies of ribosomes were initiated about 1957 by several groups, notably one at Harvard University led by Alfred Tissières and James D. Watson and one in the Carnegie Institution of Washington that included Ellis T. Bolton, Roy Britten and Richard B. Roberts. (I should add that my association with Watson's group at that time, although it was brief, had a great influence on my later research on the ribosome.) The initial studies were done mainly on *E. coli* ribosomes, which consist of two subunits of unequal size that are designated 30S and 50S. The size is determined by the rate, measured in Svedberg units (S), at which a particle sediments when it is spun at high speed in an ultracentrifuge. Together these particles constitute the functional unit in protein synthesis: the 70S ribosome. (The reason the two S values are not additive is that the shape of a particle influences its rate of sedimentation.) In each of these subunits proteins represent about a third of the total mass; the rest is RNA. The 50S ribosome subunit contains a 23S RNA molecule and a 5S RNA molecule. The 30S ribosome subunit incorporates one 16S RNA molecule. In 1961 J. P. Waller and J. I. Harris of Harvard observed that the ribosome contains different kinds of protein molecules, indicating that its structure must be quite complex. Subsequent experiments conducted by many workers, including those in my own group at the University of Wisconsin, show that the 30S ribosomal subunit includes either 19 or 20 different protein molecules and that the 50S subunit apparently has more than 30 protein molecules.

As the early work on the structure of the ribosome was proceeding, a general picture of its functional properties had begun to emerge. The existence of specific ribosomal binding sites for transfer and messenger RNA was demonstrated, forcing the conclusion that the ribosome plays an active role in protein synthesis and is not merely an inert workbench on whose surface amino acids are assembled. The observed physical complexity of the ribosome must therefore reflect the complexity of its function. What we needed to establish was the relation between structure and function. Yet as far as the actual roles of the RNA and proteins and their critical interrelations were concerned, the ribosome was still a mysterious "black box." How was one to understand this complicated piece of machinery? One could take it apart and try to reassemble it, but what tools were delicate enough to avoid destroying the machine in the process?

In 1961 Jacob, Sidney Brenner and Matthew S. Meselson, in a paper describing the classic experiments that proved the messenger-RNA theory, noted the presence of two kinds of ribonucleoprotein particle in mixtures that were centrifuged in a solution of the salt cesium chloride, which forms a density gradient in the centrifuge cell. (Density-gradient centrifugation, originally developed by Meselson, Franklin W. Stahl and Jerome R. Vinograd at the California Institute of Technology, separates large molecules on the basis of their different buoyancy in such a solution.) When they centrifuged bacterial extracts containing ribosomes, they observed two bands containing ribosomal particles. The lighter band (the *B* band), corresponding to a density of 1.61, contained messenger RNA as well as proteins being synthesized; the heavier band (the *A* band), corresponding to the density of 1.65, did not.

The presence of the *A* band was not relevant to the main theme of the paper, and no reason was given for its presence. At the time the paper was published I was working at the University of Osaka on both ribosomes and messenger RNA, and I was quite curious about this phenomenon. We knew then that the 30*S* and 50*S* subunits and their aggregate, the 70*S* ribosome, all have the same chemical composition: 65 percent RNA and 35 percent protein. Since the buoyancy of complex molecules in solution usually reflects their chemical composition, why should density-gradient centrifugation reveal two kinds of ribonucleoprotein particle?

In the summer of 1962 I had an opportunity to visit Meselson's laboratory at Harvard to look into the question. When we recovered the particles from the two bands, we found that the *B* band contained undegraded 50*S* and 30*S* ribosomal subunits. The denser *A* band, however, consisted of a mixture of smaller 40*S* and 23*S* "core" particles that had been created from the usual ribosomal subunits by the splitting off of about 40 percent of the protein during the density-gradient centrifugation; the split proteins could be found in a protein fraction at the top of the gradient. The explanation for the *B* band was apparently that in crude bacterial extracts some of the ribosomes are resistant to this splitting, perhaps because they are stabilized by messenger RNA and growing protein chains.

On returning to Osaka I continued experiments with Robert K. Fujimura to characterize these core particles. Initially we prepared 40*S* and 23*S* particles (the latter are not to be confused with 23*S* RNA molecules) by respectively centrifuging purified 50*S* and 30*S* ribo-

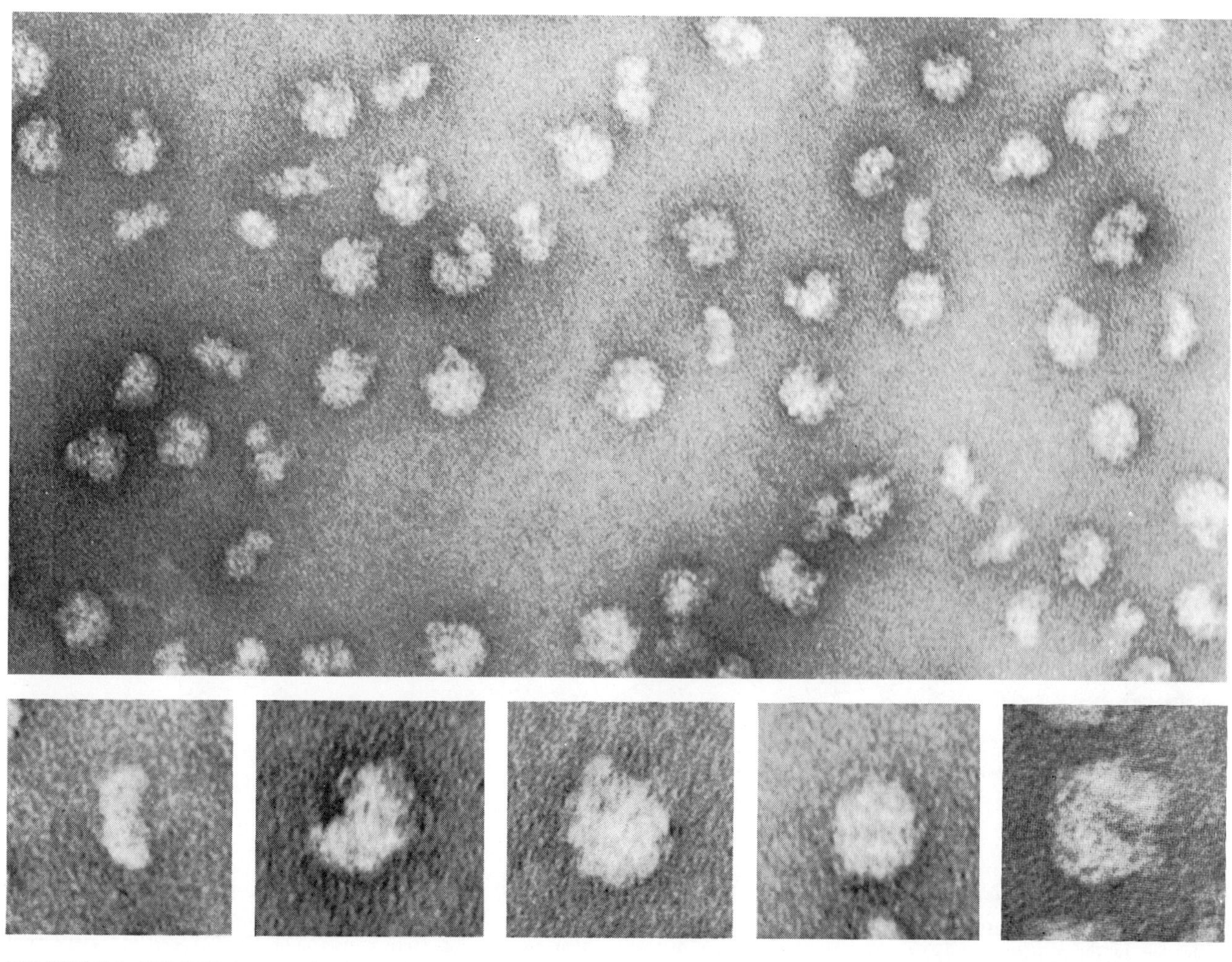

RIBOSOMAL SUBUNITS are enlarged 450,000 diameters in an electron micrograph (*top*) made by Martin Lubin of the Dartmouth Medical School. To prepare them, intact (70*S*) ribosomes were dissociated in solutions with a low magnesium concentration and the subunits were negatively stained with uranyl formate. Individual particles are enlarged 800,000 diameters (*bottom*). The smaller of the two subunits is the 30*S* (*left*). The larger is the 50*S*, seen in three different views characterized by a kidney shape (*second from left*), a "nose" (*third from left*) and a groove (*fourth from left*). The two subunits join to form a 70*S* ribosome (*right*).

somal particles in cold cesium chloride solution for 36 hours and then recovering the core particles from the band in the middle of the centrifuge cell [*see middle illustration on following page*]. This procedure was troublesome, however, and unsuitable for large-scale preparation of the particles. We therefore tried omitting the centrifuge step. Reasoning that it was surely the particular salt solution and not the physical centrifugation that disrupted the subunits, we simply kept 50S subunits in the solution for 36 hours, expecting that irreversible splitting of the protein would take place, yielding the 40S core particles we wanted to study.

When we removed the cesium chloride and examined the products, however, we found to our surprise that the recovered particles behaved just like the original 50S ribosomal subunits. Why had there been no splitting of the proteins? We immediately realized the important implication of this experimental observation: The splitting of the 50S unit is reversible; the reaction is pushed in the direction of dissociation only by separation in the centrifuge. To test this supposition we prepared core particles and split proteins by the usual centrifugation method. Then we mixed them together and removed the cesium chloride. We found complete conversion of the core particles to intact ribosomal particles. In this way we succeeded in reconstituting the 50S ribosomal subunit from the 40S core and split proteins that had been derived from the 50S, and also in reconstituting the 30S subunit from the 23S core and the homologous split proteins.

In order to prove that the reconstituted ribosomal particles really had the

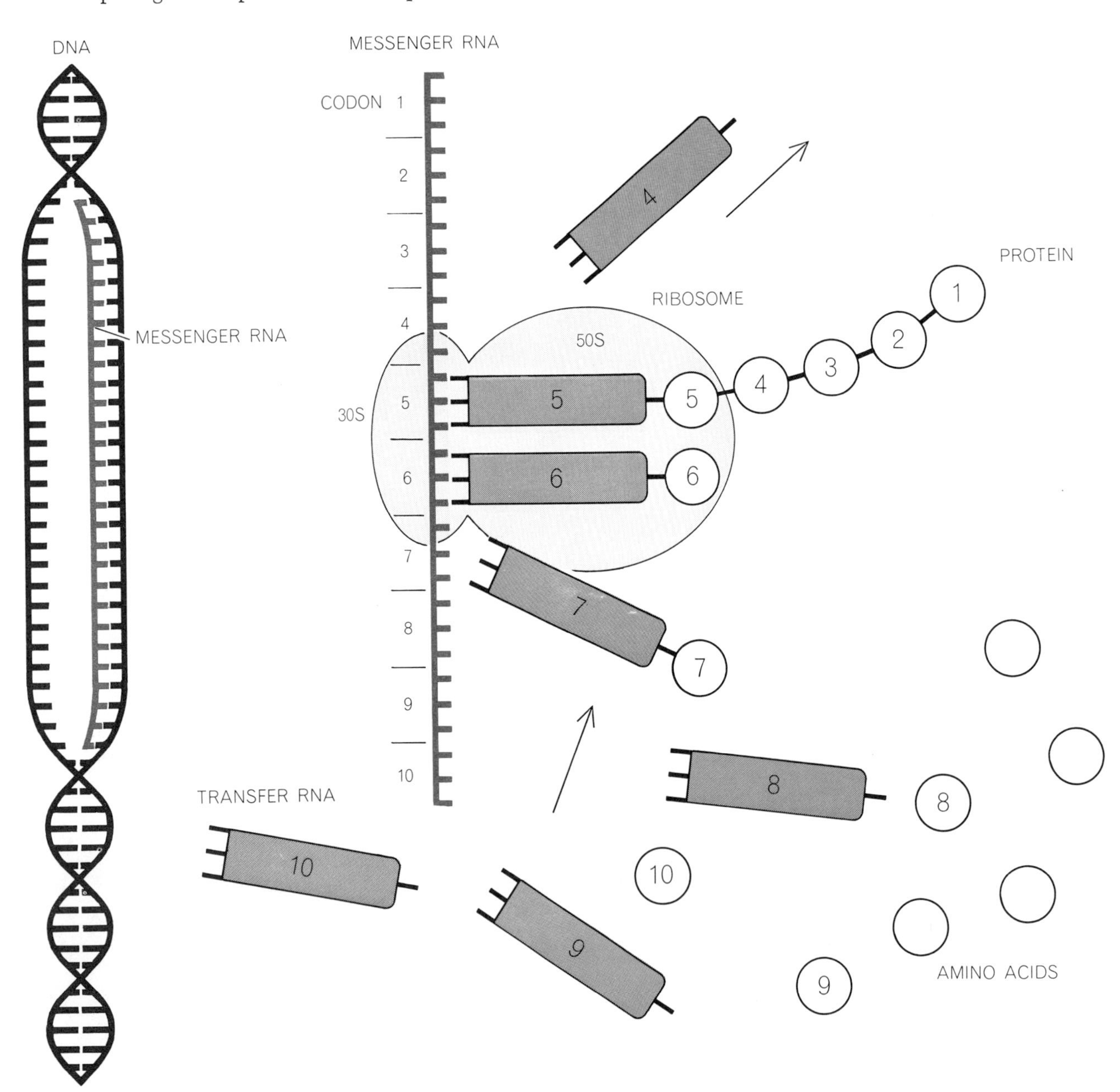

RIBOSOMES conduct protein synthesis. Genetic information is encoded in the sequence of bases (*horizontal elements*) in the double helix of DNA (*left*). This information is transcribed into a complementary sequence of RNA bases to form messenger RNA (*dark color*). Each group of three bases in the messenger RNA constitutes a codon, which specifies a particular amino acid and is recognized by a complementary anticodon on a transfer-RNA molecule (*lighter color*) that has previously been charged with that amino acid. Here amino acid No. 6, specified by the sixth codon, has just been bound to its site on the ribosome by the corresponding transfer RNA. It will bond to amino acid No. 5, thus extending the growing peptide chain. Then the ribosome will move along the messenger RNA the length of one codon and so come into position to bind transfer RNA No. 7 with its amino acid.

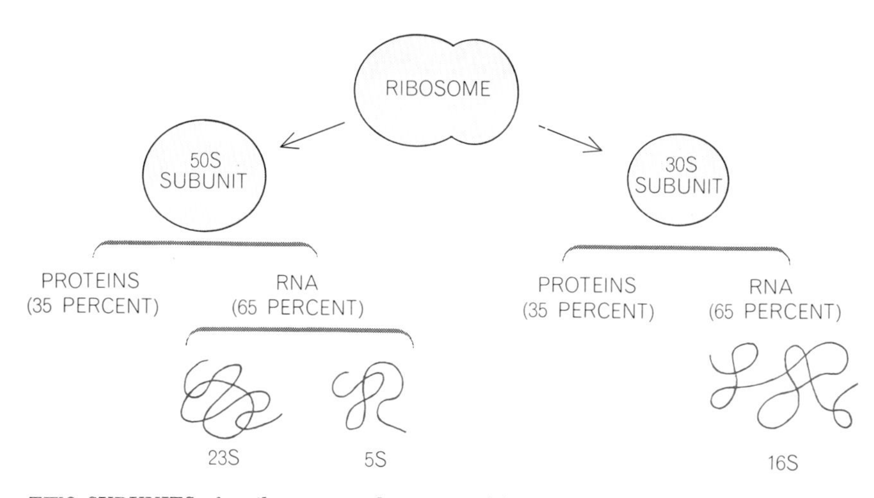

TWO SUBUNITS of a ribosome can be separated by spinning ribosomes in a centrifuge because the subunits are different sizes and move through the centrifuge cell at different rates. Both subunits are about 35 percent protein and 65 percent RNA. The 50*S* subunit contains a 23*S* and a 5*S* RNA molecule and the smaller 30*S* subunit has a 16*S* RNA molecule.

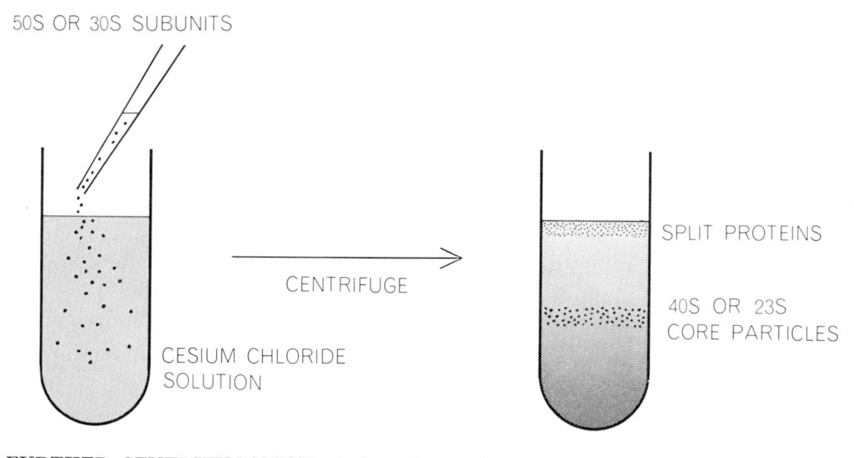

FURTHER CENTRIFUGATION of the subunits breaks them down. The subunits are added to a cesium chloride solution (*left*). Centrifugation establishes a stable density gradient in the solution (*right*), within which the subunit components form layers according to density. Some proteins split off, leaving "core" particles of RNA and other proteins.

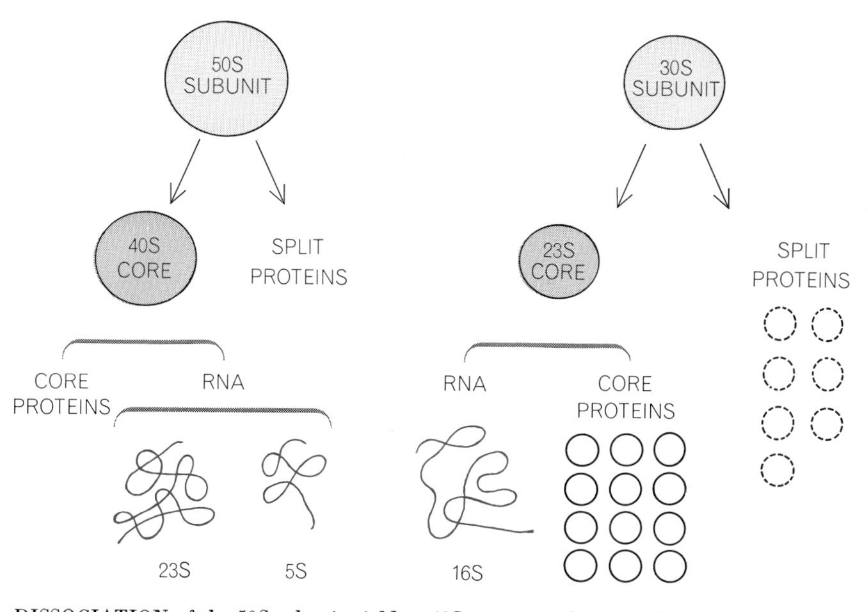

DISSOCIATION of the 50*S* subunit yields a 40*S* core particle and split proteins, dissociation of the 30*S* subunit a 23*S* core particle and split proteins. In the case of the 30*S* subunit there appear to be seven split proteins, with about 12 proteins remaining in the core.

same specific structure as the original ones, we had to demonstrate the functional integrity of the reconstituted particles. Before we could succeed in such experiments I left Osaka and moved to the University of Wisconsin. There Keiichi Hosokawa and I were able to show in 1965 that, whereas neither the 23*S* nor the 40*S* cores have any activity in a cell-free protein-synthesizing system, reconstituted 30*S* and 50*S* particles have activities comparable to the original intact 30*S* and 50*S* ribosomal subunits. At the same time Theophil Staehelin and Meselson, who were taking a similar approach at Harvard, independently succeeded in demonstrating the reconstitution of the ribosomal subunits from core particles and split proteins. The functional capabilities of the reconstituted particles can be assayed in various ways. For example, the function of 30*S* particles is usually assayed by measuring the rate of protein synthesis directed by messenger RNA in the presence of intact 50*S* subunits and other necessary components. One can also test the subunits' ability to bind several different transfer RNA's in the presence of various messenger RNA's and the ability to bind messenger RNA itself.

The success in reconstitution, although it involved the dissociation and reassociation of only some of the ribosomal proteins (the split proteins), had several important implications. First, the experiment showed that at least part of the ribosome assembly in the test tube is spontaneous; no extraribosomal template or enzyme is required. Second, it provided a system in which the functional roles of individual split proteins could be analyzed. The 30*S* split-protein fraction consists of seven proteins. By column chromatography we separated the proteins into five pure protein components and one fraction containing two proteins, and we showed that the five proteins differ from one another in amino acid composition. Then we determined the functional need for each of the purified proteins by omitting one of them at a time in reconstitution experiments. We found that three of the five purified proteins are essential for reconstitution, and that the omission of either one of the others has only a partial effect. From this type of experiment we could conclude that all five of the purified proteins are chemically and functionally distinct, and that some of them are absolutely required whereas others are not (although they are required for full activity in protein synthesis).

The partial-reconstitution system was a first step toward the functional analysis of the ribosome, but the information that could be provided by such a system was limited. To accomplish complete analysis we needed a way to reconstitute ribosomes entirely from free RNA and completely separated proteins. In 1967 Peter Traub and I began systematic attempts at complete reconstitution with the 30S subunit.

We assumed that we must do the reconstitution in two steps, first making 23S cores from 16S RNA and proteins and then making complete 30S subunits from the 23S cores and the split proteins. We therefore prepared 16S RNA by treating 23S cores with phenol. We separated proteins from other 23S cores by treating the core particles with urea and a high concentration of lithium chloride. We mixed these core proteins with the 16S RNA under several different conditions, hoping to obtain 23S core particles. Then we added the split proteins to the reaction mixture, recovered the particles by centrifugation and assayed the activity of the recovered particles in a cell-free protein-synthesizing system. As typical enzyme chemists, we felt it was essential to protect the ribosomal proteins and any sensitive intermediates from inactivation by heat, and so we performed all these operations in a cold room and kept everything on ice.

Our initial attempts were failures. We could find only very slight protein-synthesizing activity. It seemed that the reaction might not be possible. Then we realized that living *E. coli* cells multiply most rapidly at 37 degrees Celsius (body temperature) and not at all at freezing temperatures, and that multiplying cells must certainly be assembling ribosomes quite efficiently. We also recognized that the cytoplasm of living *E. coli* contains a rather high concentration of salts; the salts might discourage nonspecific RNA-protein aggregation and thereby promote the specific RNA-protein assembly reaction. We therefore attempted the reconstitution at 37 degrees and with a high concentration of a salt, potassium chloride. Success! The 30S ribosome could be self-assembled [*see illustration on page 44, bottom*]. We found, indeed, that the reconstitution of 30S subunits from RNA and proteins is independent of the order of addition of proteins, whether they are core proteins or split proteins. By simply mixing all the proteins prepared directly from the 30S subunit with 16S RNA and incubating them in an optimal ionic environment at about 40 degrees C. for 10 minutes we were able to convert almost all the 16S RNA in the mixture into 30S particles [*see top illustration on page 44*]. In protein-synthesizing activity, in protein content and in sedimentation behavior the reconstituted particles were almost identical with the original 30S particles.

From this point one could proceed in many directions. One basic goal was to determine the functional role of the 16S RNA and of each 30S ribosomal protein. With regard to the RNA, we first considered the specificity requirements. For example, is an RNA molecule

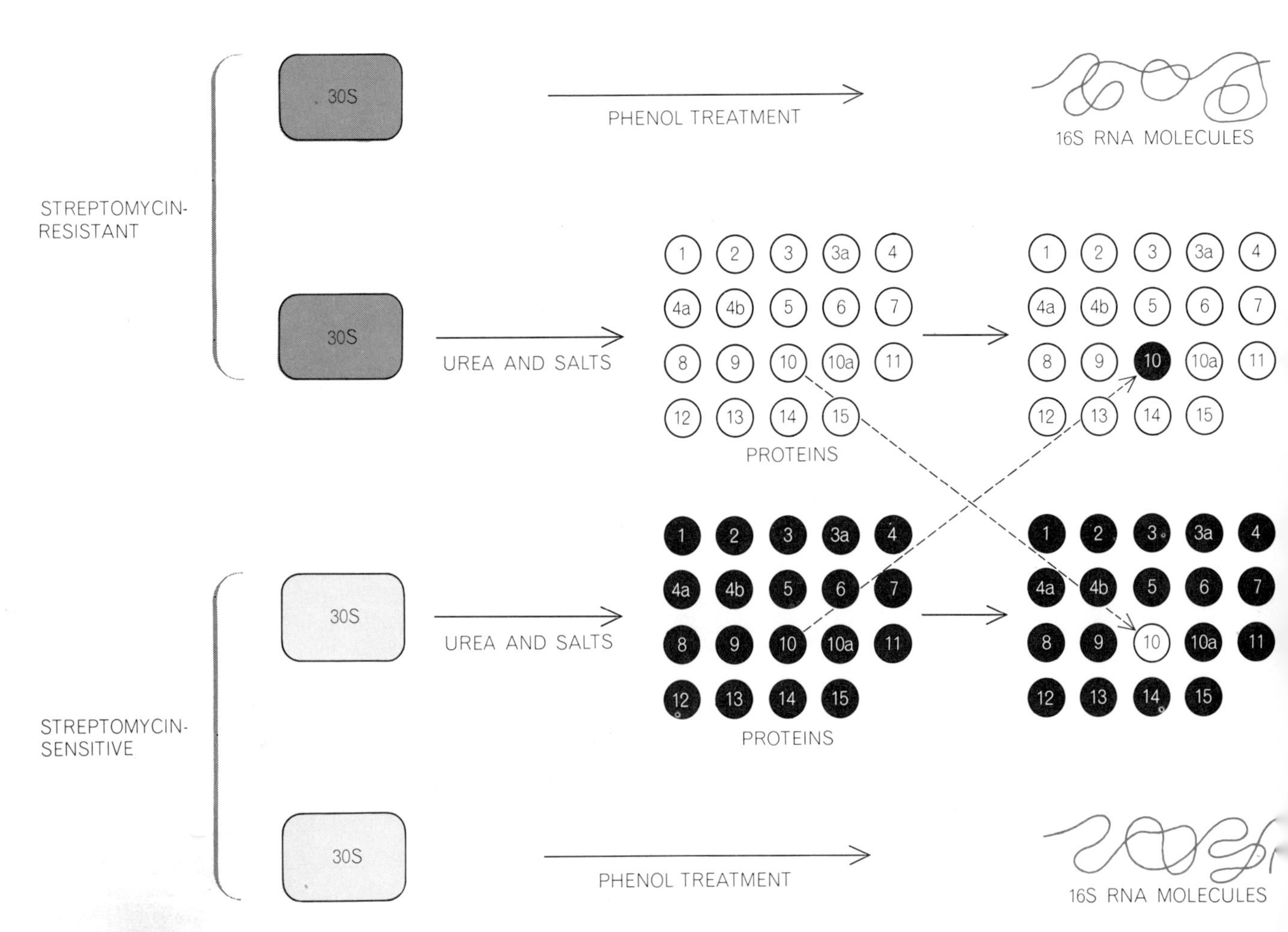

ONE PROTEIN in the 30S subunit is responsible for the effect of streptomycin on ribosome functioning. Proteins and RNA are separated from subunits derived from bacterial cells that are resistant to streptomycin and from other cells that are susceptible to its effect. The proteins are isolated. When the protein designated *P*10 from susceptible cells is combined with all the other proteins from resistant cells and with the RNA from resistant cells, the 30S ribosomes that result turn out to be susceptible to streptomycin. On the that is merely similar in size to *E. coli* 16S RNA competent to reconstitute a physically and functionally intact 30S subunit? No. Neither 16S ribosomal RNA from yeast cells nor *E. coli* 23S RNA degraded to a 16S-sized fragment was active in reconstitutions with *E. coli* 30S proteins. In fact, the inactive products of such combinations did not even resemble 30S ribosomes physically, judging by their sedimentation behavior. This finding certainly came as no surprise; one would expect the requirements for a functional ribosome to be stringent. How stringent? When we performed the reconstitution with 16S RNA from one bacterial species and 30S ribosomal proteins from another species that is distantly related, we found that in many cases such "artificial" ribosomes were as active as the respective homologous RNA-protein combinations. We conclude from this that although there is definitely a specificity requirement for the RNA, the requirement is not absolute.

In determining the roles of the various ribosomal proteins our basic approach was to perform the reconstitution with one component omitted or specifically modified and then see if physically intact 30S particles were formed and, if so, whether or not they were functionally active. We first had to separate the 30S protein mixture into each of its 19 components. The fractionation of 30S ribosomal proteins had already been achieved by research groups at the University of Geneva, the Max Planck Institute for Molecular Genetics in Berlin, the University of Wisconsin and the University of Illinois. We employed methods similar to theirs, relying mainly on various types of column chromatography.

We then did the reconstitution with 16S RNA and 19 purified proteins rather than with the unfractionated protein mixture used in our earlier experiments. The extent of the reconstitution was not as good as it was with unfractionated proteins (and we therefore could not exclude the possibility that there are some ribosomal components other than the RNA and 19 protein molecules), but the reasonably high efficiency of the reconstitution made it possible to undertake the functional analysis of the separated protein components.

The first protein we studied in detail was the one responsible for sensitivity or resistance to the antibiotic streptomycin. Earlier studies had indicated that the drug's primary site of action is the bacterial ribosome, specifically the 30S subunit. When streptomycin is added to a cell-free system containing ribosomes from a streptomycin-sensitive strain, it inhibits protein synthesis. Streptomycin also causes the misreading of certain synthetic messenger RNA's, that is, it induces the incorporation into proteins of amino acids other than the ones dictated by the genetic code. This misreading effect of streptomycin was discovered first by Julian E. Davies, Walter Gilbert and Luigi Gorini at Harvard [see "Antibiotics and the Genetic Code," by Luigi Gorini; Scientific American Offprint 1041].

Bacteria can become resistant to streptomycin through mutation, and streptomycin does not inhibit synthesis or cause misreading of messages in a cell-free system if 30S ribosomal particles from streptomycin-resistant mutants are used. Traub and I showed that the component altered by the mutation was not the RNA but had to be in the protein fraction; 30S particles reconstituted from the protein of a resistant mutant and the RNA of a susceptible strain were resistant to streptomycin in cell-free protein-synthesizing systems, whereas the reverse combination produced 30S particles that were susceptible to streptomycin. Makoto Ozaki and Shoji Mizushima took over the job of identifying the altered protein. We purified 30S ribosomal proteins from both susceptible and resistant bacteria, systematically substituted single proteins from a resistant strain in a mixture of proteins from a susceptible strain [*see bottom illustration at left*] and assayed the reconstituted ribosomes for their response to streptomycin. In this way we established that a single protein, one we had designated *P*10, determines the susceptibility of the entire 30S ribosomal particle to the inhibitory action of streptomycin, its susceptibility to streptomycin-induced misreading and its ability to bind the antibiotic.

Having learned how an alteration in a given protein can affect the function of the ribosome, we investigated what happens to the ribosome when this protein is simply left out. Is the ribosome still able to assemble itself and, if so, how are its functional capabilities altered? We found that in the absence of *P*10, RNA and the other ribosomal proteins can still assemble into particles that sediment at 30S. These *P*10-deficient particles have several interesting properties, however. Under the conditions of the assays these particles show high activity when a synthetic messenger RNA is used as a template, but their activity is weak when directed by RNA from a natural source. It is known that a special mechanism for initiating protein synthesis is needed in the system directed by natural messenger RNA's but not in the system directed by synthetic messenger RNA's [see "How Proteins Start," by Brian F. C. Clark and Kjeld A. Marcker, Offprint 1093]; the *P*10-deficient particles cannot carry out this special initiating function.

The other interesting finding is that the frequency of translation errors with the *P*10-deficient particles is much reduced not only in the presence of streptomycin but also in the presence of certain other antibiotics, of ethyl alcohol or of high concentrations of magnesium ions, all of which are known to induce translation errors. In fact, the deficient particles read synthetic messenger RNA more accurately in the presence of error-inducing agents than normal ribosomes do even in the absence of such agents. In other words, it appears that protein *P*10 plays a role in increasing the frequency of errors in the translation of the genetic message. The inherent ability of ribosomes to make mistakes may be ad-

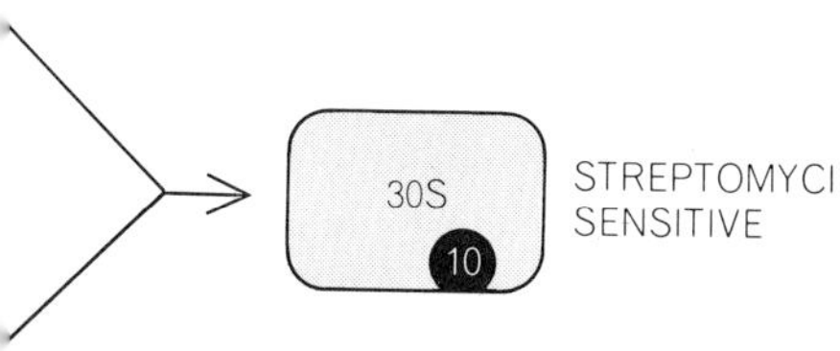

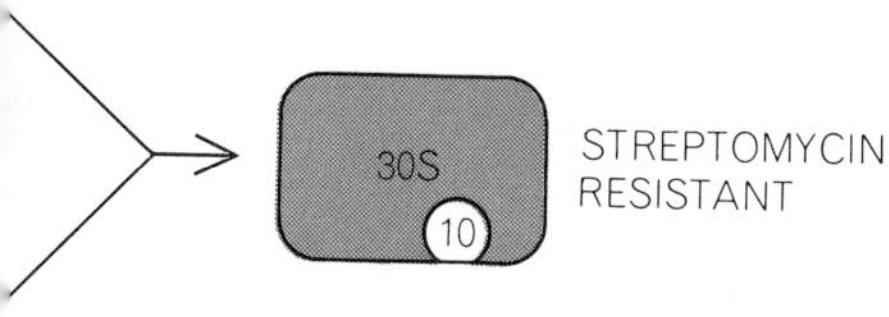

30S 10 STREPTOMYCIN-SENSITIVE

30S 10 STREPTOMYCIN-RESISTANT

other hand, when *P*10 is derived from resistant cells and all the other components are taken from susceptible cells, the resulting subunit is resistant to streptomycin.

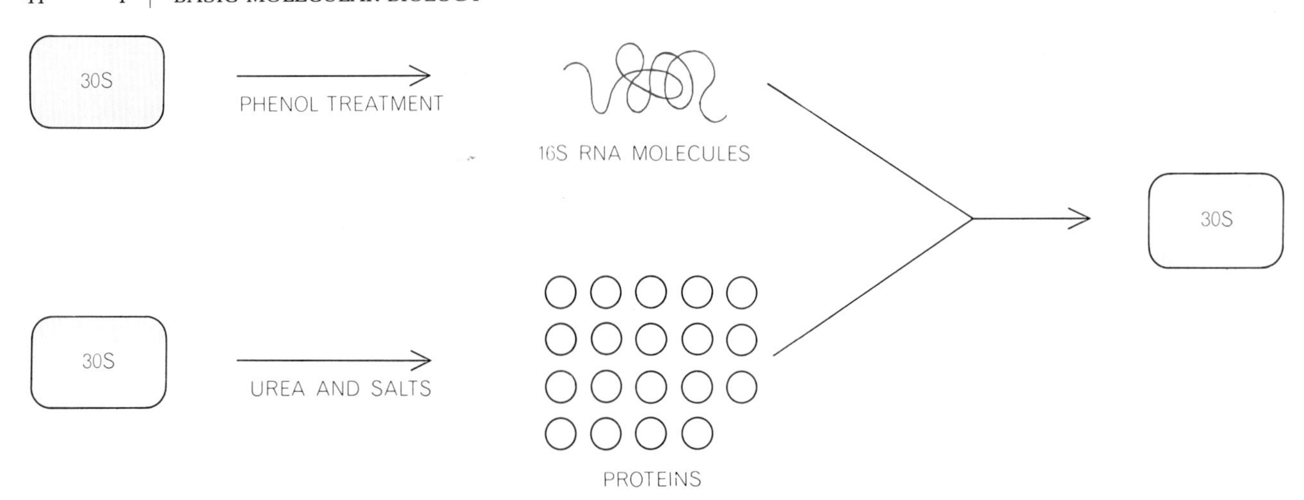

CORE AND SPLIT PROTEINS need not be assembled separately. Reconstitution can be accomplished by mixing all ribosomal proteins from 30S subunits with 16S RNA in the correct ionic environment and incubating at 37 to 40 degrees C. for 10 to 20 minutes.

vantageous to bacterial cells, as Gorini suggested, since it can suppress the effects of harmful mutations. On the other hand, the property may simply be an unavoidable consequence of the complexity of the machinery.

Although we have not yet completed similar detailed analyses of all the proteins, preliminary experiments with them lead to certain general conclusions. Omitting any one of several proteins affects a number of known 30S ribosomal functions. Conversely, several different 30S ribosomal functions are affected by the omission of any of a number of proteins. That is, these functions seem to require the presence of more than one protein component, and so one can say that the 30S ribosomal proteins function cooperatively. We have also found that the omission of some proteins drastically affects physical assembly. Particles formed in the absence of one of these proteins are deficient in several other proteins, including some that were present in the reconstitution mixture. In other words, the presence of certain proteins is essential for other proteins to be bound. In this sense the assembly process itself is cooperative.

One of the most effective tools that are available for the study of a reaction is chemical kinetics, the study of the rate at which a reaction proceeds. Most chemical reactions have a distinct kinetic mode, or "order," which is determined experimentally from the reaction rate's dependence on the concentration of the reactants. Since the 30S reconstitution reaction involves at least 20 components, one might expect the rate to have a very high order of dependence on their concentration—so that doubling the concentration would increase the rate by as much as 2^{20}, or more than a million times! To be sure, this would be an absurdity, since it would mean that it was necessary for all the components to collide simultaneously in order to form a complete subunit, and subunit formation would be an incredibly rare event. As a matter of fact, most chemical reactions, even those involving a number of reactants, turn out to have a first-order or second-order dependence on reactant concentration (that is, doubling the concentration doubles or quadruples the reaction rate). The reason is that most complex reactions proceed in steps, with one unimolecular rearrangement or bimolecular interaction being slower than the others and hence determining the overall rate.

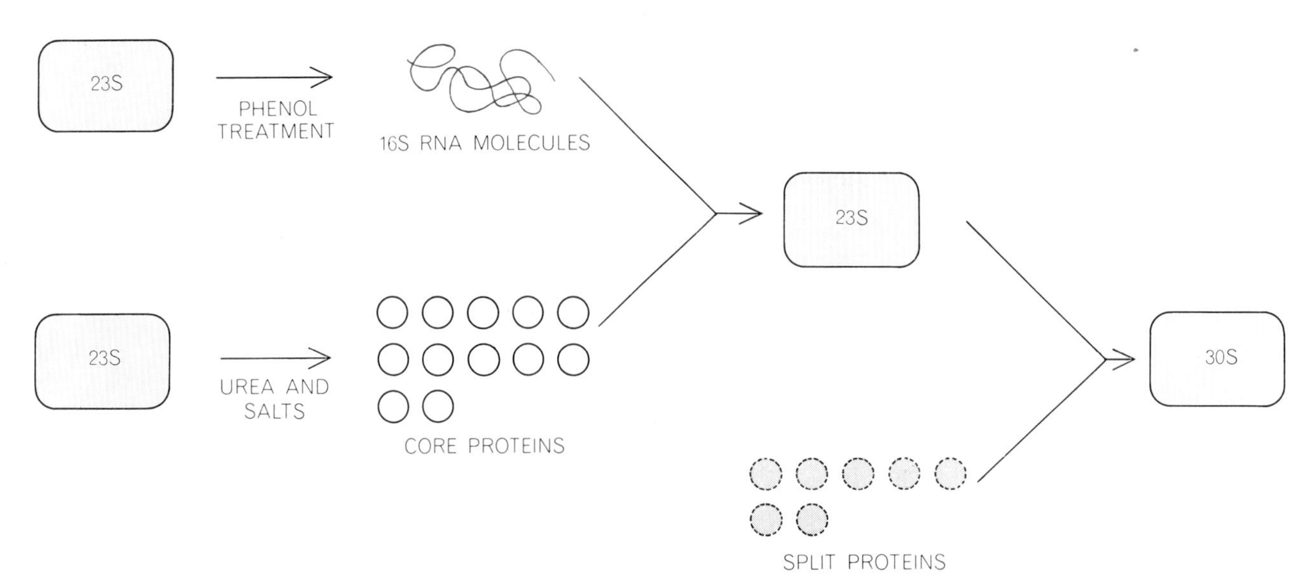

RECONSTITUTION of subunits was first accomplished as shown here. Core particles were treated with phenol to prepare 16S RNA and with urea and salts to yield the core proteins. The RNA and proteins were combined at 37 degrees Celsius to form 23S core particles. Then the reaction was completed by the addition of the split proteins, and in this way the 30S subunits were formed.

This generalization, however, has in the past applied to reactions involving far fewer components than are needed for reconstitution, and so we expected to find a somewhat higher order of reaction. We were rather surprised to observe that our assembly was in fact a first-order reaction; whether we double or halve the concentration of the reactants, the time it takes for all the components to assemble themselves into completed ribosomes is roughly the same—about five minutes under optimal conditions in the test tube. When one observes first-order kinetics in a reaction involving more than a single component, one can conclude that there is an intermediate step, involving a relatively slow rearrangement of a single component, that must take place before the reaction can be completed. This is exactly what we observe in the reconstitution reaction. The slow step in the assembly process may be occurring at any time before the binding of the 20 proteins to the RNA molecule is complete—after none of them or only some of them are bound—or after all are bound; in any case, our observation would be the same.

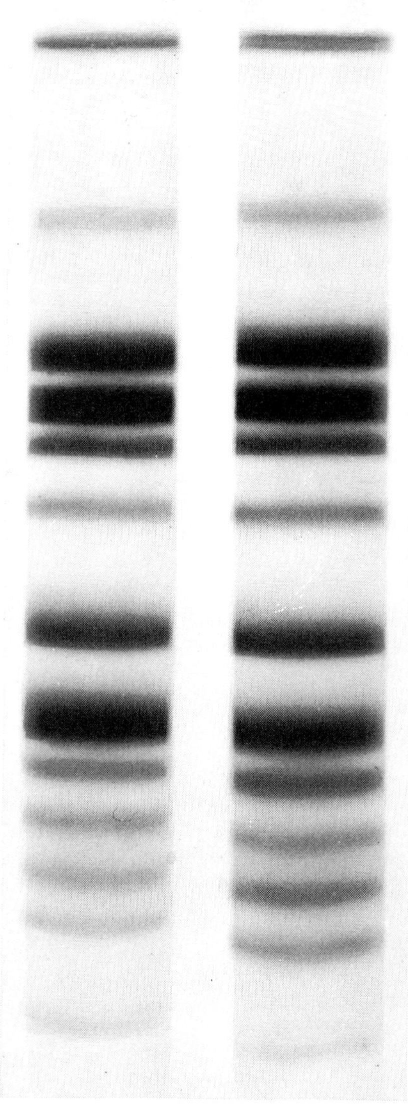

ALL THE PROTEINS normally found in a 30S subunit are found in a reconstituted 30S subunit, as shown by a comparison of electrophoresis results for the natural (*left*) and the reconstituted (*right*) particles. The protein mixtures are layered onto the top of a polyacrylamide gel column. When an electric current is applied, the proteins migrate down the column, each protein forming a band that moves at a different rate depending on the charge and the size of the molecule. Staining visualizes each protein.

We had noted with interest that the reaction is extremely dependent on temperature. It turned out that a considerable amount of heat energy is required to effect the rearrangement of the unknown intermediate product—about 40,000 calories per mole of ribosomes. On the other hand, many of the proteins attach themselves in an ice-cold solution. We were therefore able to isolate the intermediate, activate it by warming the solution and then observe the almost instantaneous binding of the rest of the proteins to form a completed ribosomal subunit. And so we can describe the general nature of the pathway of self-assembly: a rapid binding of some of the proteins to the RNA, a slow structural rearrangement of this intermediate that requires thermal energy and then a rapid binding of the rest of the proteins.

After obtaining all this information on assembly in the test tube, one comes to an obvious question: Do the same principles operate in the living cell? The problem of ribosome synthesis in living cells was being attacked long before test-tube assembly was even seriously considered. The early work was done by the group including .Britten, Brian J. McCarthy and Roberts at the Carnegie Institution, and they were followed by a number of groups, notably Shozo Osawa's in Japan and David Schlessinger's at Washington University. In a series of intricate experiments they delivered short pulses of radioactive components to growing bacterial cells and monitored the flow of the labeled components into ribosomes. It was possible in this way to postulate the presence of several classes of precursor particles, but it was difficult to isolate and analyze them. The obvious limitations of such an experimental approach encouraged investigators to seek other directions. One was genetics. Genetics has been a powerful tool for identifying the flow of intermediates in numerous biosynthetic pathways because one of the easiest ways to find out how something works is to see what happens when it does not work. For example, it should be possible to isolate mutants that are defective in a specific step in the biosynthesis of ribosomes; the step reveals itself by the accumulation of the precursor whose conversion is blocked by the mutational defect.

What was needed was a systematic method of isolating mutants defective in ribosome assembly. The trouble is that since ribosomes are essential for growth such mutants are ordinarily inviable and cannot be cultured. They can be isolated only as "conditionally lethal mutants": cells with defects such that the organism is inviable under one condition but functions normally under some other condition. As I have mentioned, in our detailed study of the test-tube reaction we had been struck by the remarkable dependence of the reaction rate on temperature. If the same principle operated in living cells, we reasoned, then many mutational defects in ribosomes or in related components should manifest themselves more severely at lower temperatures, and so some assembly-defective mutants should be conditionally lethal—viable at high temperatures but inviable at lower temperatures. They could therefore be isolated as cold-sensitive mutants.

Our reasoning proved to be correct. Christine Guthrie, Hiroko Nashimoto and I have isolated a large number of cold-sensitive mutants of *E. coli*, a significant fraction of which appear to be defective specifically in ribosome assembly. (Independently John L. Ingraham and his co-workers at the University of California at Davis have found abnormal ribosome biosynthesis at low temperatures in cold-sensitive mutants of the related bacterium *Salmonella typhimurium*.) By sucrose-gradient sedimentation we have already identified three distinct classes of particles, from three different mutants, that accumulate in cells grown at 20 degrees C. Two of these particles appear to be precursors of 50S subunits; the third appears to be a 30S precursor. While proceeding with the biochemical characterization of these particles we are also conducting genetic analyses of the various mutants in the hope of obtaining information on the genetic organization and genetic control of the ribosome and of ribosome assembly. We hope that through the coupling of genetic techniques with the biochemical techniques of test-tube reconstitution this sophisticated and complex cellular instrument will soon be understood on a truly molecular level.

II

THE BIOLOGICAL BASIS OF GENETIC MANIPULATION

II THE BIOLOGICAL BASIS OF GENETIC MANIPULATION

INTRODUCTION

In order to introduce new DNA permanently into a cell, several conditions must be satisfied: (1) the DNA of interest must be isolated, (2) a carrier DNA molecule or *vector* must be provided, (3) there must be a means of splicing together the two DNA molecules, and (4) there must be a means of getting the spliced molecule into a *host* cell. All four conditions are now attainable, and it is interesting to note that the original technology derived from investigations unrelated to the study of recombinant DNA.

In 1952 it was observed in Salvador E. Luria's laboratory that when *E. coli* bacteriophage T2 is grown on a particular strain of *E. coli*, the phages produced are modified in that they lose the ability to grow on certain bacteria (*restricting* hosts), which are adequate for phages that are not grown on the *modifying* host. This phenomenon is called *host modification and restriction;* it was later shown by Matthew Meselson and Robert Yuan that restricting hosts contained an enzyme, a *restriction endonuclease,* which makes cuts in the DNA molecules at particular base sequences. These susceptible sequences are normally protected by *methylation* (the introduction of a methyl group onto an adenine or cytosine in the sequence), and bacteria lacking the methylating ability yield phages that are restricted. Later many restriction endonucleases were isolated from a variety of microorganisms; these endonucleases apparently destroy foreign DNA, as explained in the article by Luria.

Each restriction endonuclease recognizes only a single base sequence and makes, in opposing DNA strands, two nearby breaks (one in each strand) which are displaced by several bases (see article by Stanley Cohen at the end of this section). Therefore, all fragments produced are terminated by short complementary single strands, a fact first recognized by Ronald Davis and Janet Mertz at Stanford University when they observed that the fragments are capable of circularization. Since the single strands can be joined permanently under appropriate conditions, and since all fragments produced from the DNA of any cell by a particular restriction endonuclease are terminated with the same single strands, it thus becomes possible to join, or splice, any pair of DNA fragments.

In another useful splicing technique, one uses *terminal transferase,* an enzyme that can synthesize an extension of one of the two DNA strands. With this enzyme it is possible to attach a single strand containing only adenine (A) to the end of one fragment and a strand containing only thymine (T) to the end of the other. Because of the base pairing between A and T, described in the first article by Crick in Section I, the A-containing strand and the T-containing strand form a short double-stranded segment which couples the two fragments. This effect is also described in the articles by Stanley Cohen in this section and by Clifford Grobstein in Section III.

Studies of infectious drug resistance, as described in the article by Royston C. Clowes, provided the first DNA carrier, or *vector*. The drug-resistance factors, or *R* factors, carry genes for resistance to certain antibiotics; thus their presence in a bacterium is easily detected by the capability of the bacterium to grow on agar containing the antibiotic. The *R* factors are circular DNA molecules that are easily isolated. The important property making them effective vectors for carrying foreign genes is that purified *R* factor DNA can be taken up by a drug-sensitive bacterium and permanently establish itself as a functioning *R* factor in the cell. Thus, if DNA from a cow, for instance, is treated with a restriction endonuclease and the fragments are spliced with fragments of *R* factor DNA, an antibiotic-sensitive bacterium can be infected with the spliced molecules, and the presence of cow DNA in the bacterium is indicated by the antibiotic resistance.

Another suitable vector is the DNA of *E. coli* bacteriophage λ, described in the article by Allan M. Campbell. This DNA is also infective and can be made a permanent part of the cell by means of a phage-mediated insertion process called lysogeny. Certain regions of the phage DNA can be dispensed with, and standard genetic techniques employed to replace those regions with foreign DNA (Section I). Thus a foreign DNA molecule can be made part of *E. coli* DNA by lysogenization and its presence can be detected by simple genetic tests for the λ phage genes. A λ phage containing foreign DNA can also be propagated by infection of bacteria. This has two applications. First, large amounts of phage DNA are easily purified, thus yielding an ample supply of a particular foreign DNA, if that is desired. Second, if the phage infects a sensitive bacterium, the phage DNA replicates, producing 50 to 100 copies per cell; if the foreign DNA is transcribed and translated, the cell becomes a rich source of the protein coded for by the gene. This development has been a tremendous boon in that it permits the large-scale purification of proteins that are normally produced only in small quantities or that are not easily assayed.

Isolation of all of the DNA from an organism is reasonably simple. Although the entire DNA molecule can sometimes be used for the preparation of some recombinant organisms, it is frequently necessary to work with only a single gene or group of genes. If the property determined by the gene is easily detected (e.g., by the ability of a bacterium to form a colony on a particular kind of agar or by production of a colony with a characteristic size, shape, or color), the "shotgun" approach described by Clifford Grobstein in Section III can be used. But if it is not, that part of the DNA molecule corresponding to the particular gene must be isolated. This isolation is often a formidable task: in fact, in most cases, it is not possible at the present time. However, for some genes it is possible to isolate the messenger RNA transcribed from that gene. Once the messenger RNA has been isolated, the particular DNA can be obtained by one of two methods. In the first, described in the article by Donald D. Brown, the investigator makes use of the fact that a single strand of DNA can be combined with an RNA molecule having the complementary base sequence. Thus the total DNA from an organism is isolated and converted to single strands, and these strands are mixed with the messenger RNA. From this mixture, a DNA-RNA complex (one strand of DNA and one strand of RNA) can be isolated, and the DNA strand can then be separated from the RNA. The DNA strand has the same base sequence as the original DNA from which the messenger RNA was transcribed. The isolated DNA strand can then be converted to double-stranded DNA by the polymerization reaction catalyzed by the enzyme *E. coli* DNA polymerase I, which Arthur Kornberg discovered and has described in his article. This method is probably useful only for genes that exist in the cell in multiple copies.

In the other method, which is the one presently preferred, the enzyme *reverse transcriptase* is used. Several years ago, two investigators who were working independently, Howard M. Temin and David Baltimore, isolated the reverse transcriptase from certain RNA tumor viruses. This enzyme copies an RNA molecule and synthesizes a single-stranded DNA molecule whose base sequence is complementary to that of the RNA model, or template—as described in the article by Temin. One can then convert this single strand to double-stranded DNA, again using the *E. coli* DNA polymerase I. The double-stranded DNA can then be easily purified from the polymerase reaction mixture and coupled to other DNA vectors by means of the terminal transferase procedure. The reverse transcriptase procedure has recently been used to clone the gene responsible for synthesis of insulin.

In the final article of this section, Stanley N. Cohen describes these procedures in detail and discusses the advantages and drawbacks of each.

SUGGESTED ADDITIONAL READING FOR PART II

GENETIC MANIPULATION OF MICROORGANISMS: POTENTIAL BENEFITS AND HAZARDS. R. Curtis *in Annual Reviews of Microbiology* **30,** 1976, pages 507–533.

RECOMBINANT DNA. R. L. Sinsheimer in *Annual Reviews of Biochemistry* **46,** 1977, pages 415–438. This article has excellent figures showing how recombinant DNA is made. It also has an extensive reference list.

The Recognition of DNA in Bacteria

6

by Salvador E. Luria
January 1970

Some bacteria have enzyme systems that scan invading DNA molecules injected by viruses and break them unless they are chemically marked at specific recognition sites

The genetic code closely resembles a universal language. As far as anyone knows every word in this language (that is, every triplet of nucleotides) means the same thing for all forms of life: it specifies a particular amino acid. By the same token, organization of a sequence of the words into a sentence, written in the form of a molecule of deoxyribonucleic acid (DNA), carries a similarly unvarying meaning: it specifies the construction of a particular protein. Hence the genetic script, like the script of a book printed in a universal language, is read in the same way by all organisms.

There are situations, however, where the organism examines the DNA script not as a linguist would but as a bibliophile would. A bibliophile may find in the structure of a book's script little signs or marks that identify the printer. It turns out that DNA often has just such identifying marks, and that these decisively affect the behavior of the organisms that recognize them. This unexpected finding has begun a fascinating chapter in the study of molecular biology and evolution.

The story begins with a curious discovery Mary L. Human and I made nearly 20 years ago in experiments with the bacteriophage, or bacterial virus, known as T2 [see "The T2 Mystery," by Salvador E. Luria; SCIENTIFIC AMERICAN Offprint 24]. Ordinarily T2 readily invades and multiplies in the bacterium *Escherichia coli*, but we found that when the virus was grown in a certain *E. coli* strain (called *B*/4), almost all the daughter phages that came out were altered in such a way that they could no longer multiply in the usual *E. coli* hosts of T2. It developed in further experiments that the altered T2 could multiply perfectly well in dysentery bacilli, and this breeding had the effect of transforming the phage back to full ability to multiply in *E. coli* bacteria.

Since the change T2 had undergone was completely reversible, it was obviously not a genetic mutation; an alteration of a gene or genes would be expected to persist in the progeny that inherited it. What, then, could account for the modification of T2's character? The phenomenon we had observed was not a freak; this was soon shown in other experiments with phages conducted by Giuseppe Bertani in my laboratory at the University of Illinois and independently by Jean J. Weigle at the California Institute of Technology. They found, for example, that when a certain phage called lambda (λ) was grown in the cells of an *E. coli* strain called *C*, only one in 10,000 of the daughter λ phages coming out of *C* could reproduce in a different *E. coli* strain called *K*. The descendants of the few phage particles that did manage to grow in *K* were then able to multiply in *K* cells as well as in *C* cells. A single cycle of growth in *C* cells, however, would restore these phages to the original form: capable of growing in *C* cells but not in *K*. Similarly, it was found that certain other strains of *E. coli* bacteria could restrict the growth of specific phages or modify their form [*see illustrations on page 52 and 53*]. A given phage could be modified to a series of different forms by shifting it from one host strain to another; each time the daughter phages became adapted to the new host, and the descendants could be returned to the original form by shifting them back to the original host.

The enigma of the changeable phages did not begin to clear up until 1961, when a Swiss investigator, Werner Arber of the University of Geneva, discovered a clue to the reason for the different reception of phages by different strains of a bacterium. Arber and his colleague Daisy Roulland-Dussoix found that when phage lambda injects its genetic material into an "alien" bacterium (one that will restrict the phage's multiplication), the cell breaks up most of the phage's DNA molecules into small fragments. This happens, for example, when phages grown in *E. coli* of strain *K* invade bacterial cells of strain *B*. The broken DNA of course cannot reproduce new phage. A few of the DNA molecules injected into the *B* bacterium manage, however, to remain intact, and they multiply, producing about 100 daughter phages. These now have the ability to multiply in *B* cells. It turns out that two of them (that is, about 2 percent of modified phages) can also still multiply in *K* bacteria as well. Arber performed a pretty experiment that revealed why this is so [*see top illustration on page 55*].

He used tracer isotopes—heavy nitrogen (^{15}N) and heavy hydrogen (^{2}H, or deuterium)—to label the DNA of phages bred in *K* cells. This was done by growing the bacteria and phages in a medium containing the heavy isotopes. The "heavy" phages were then employed to infect *B* bacteria growing in a medium of ordinary (light) nitrogen and hydrogen. When the new crop of daughter phages emerged from the lysed bacteria, Arber separated the phages by weight by means of a centrifuge that layered the particles (in a cesium salt solution) according to their relative buoyancy, based on their density. Most of the daughter phages proved to be "light," indicating that they were composed of material newly synthesized in the infected bacteria. A few particles, however, contained DNA that was half-heavy; that is, their weight indicated that half of the DNA was made up of new material and the other half of material from the heavy parent phages that had infected the bac-

PROBABILITY OF SUCCESSFUL INFECTION IN

	E. COLI B/4	*E. COLI* B/4	*S. DYSENTERIAE* SH
T2 • B	1	1	1
T2 • SH	1	1	1
T2 • B/4 = T*2	10^{-5}	10^{-5}	1
T2 • B/4 • SH	1	1	1

PROBABILITY OF SUCCESSFUL INFECTION IN

	E. COLI K	*E. COLI* B	*E. COLI* C
λ • K	1	10^{-4}	1
λ • B	4×10^{-4}	1	1
λ • C	4×10^{-4}	10^{-4}	1

MODIFICATION OF PHAGES can be demonstrated by growing phages in one strain of bacteria and observing how successfully the progeny grow when another strain serves as the host. In these tables the original host strain of T2 or lambda (λ) phages is indicated, in the first column, by a suffix such as *B* or *B*/4, which represents *E. coli* strains *B* or *B*/4 respectively. The other columns indicate the probability of successful growth when the phage is obliged to grow again in the same strain or in another one. Fifteen years ago the author and Mary L. Human discovered that when phage T4 was grown in *E. coli B*/4, yielding T2 · *B*/4 (originally called T*2), the phage was so altered that it would scarcely multiply in its usual hosts. It would grow freely, however, in dysentery bacilli (*Shigella dysenteriae Sh*). Daughter phages from this cycle, designated T2 · *B*/4 · *Sh*, were completely normal.

teria. Since the phage DNA is a typical double-strand molecule, it could be deduced that in these phages one strand of their DNA was light and the other strand was heavy—contributed intact from the parent phage. Arber found that the phage particles containing this hybrid DNA retained the ability to multiply in *K* cells as well as in *B* cells. There was also a small fraction of fully heavy daughter phages (both DNA strands being heavy), and these too proved able to grow in both *K* and *B*.

The result of Arber's experiment made clear that the lambda phage's DNA strands carried some kind of marking that identified specific phages for the bacteria. In a phage generated in *K* cells the DNA somehow acquired a *K* marking, and a phage grown in *B* cells had its DNA marked *B*. When phage DNA that lacked the right marking was injected into a bacterium, the DNA was almost sure to be broken down. The experimental results also said something more: they said that the specific marking on the DNA molecule was attached by a stable, covalent chemical bond, since the mark was retained during the chemical events attending the construction of the daughter phage. Arber confirmed the stability of the bond by using highly purified phage DNA (instead of the intact phage) to infect bacteria; the DNA retained its specific marking even after going through the chemical purification treatment.

Arber's findings raised a number of interesting questions. What was the nature of the DNA markings? How did a bacterium recognize the absence of the right markings in phage DNA? What mechanism in the bacterium was responsible for destroying DNA with the "wrong" markings? A number of laboratories have looked into these questions and have worked out much of the answer. Actually two sets of answers have come out, one for phages of the lambda type (with *K*, *B* or *C* markings) and one for the T2 and other "T even" phages. Let us look first at the lambda markings.

It was known even before Arber's discoveries that every kind of DNA can occasionally have a seemingly spurious methyl group (CH_3) attached to some of the nucleotides. For example, the methyl group can be tacked on at a certain position in adenine; the base is then called methylaminopurine, or MAP, instead of simply A. Similarly, the base cytosine (C) can be methylated, so that it becomes methylcytosine (MC). In each case the methyl group is added (with the help of a special enzyme) only after the DNA molecule has been built [*see bottom illustration on page 55*]. And it plays no observable part in the molecule's functioning: MAP behaves just like A, and MC just like C. In short, the methyl group does not alter the genetic spelling.

Could one now suppose the addition does change the style of the script? Is the methyl group perhaps a kind of serif tacked onto the letters? Arber tried the experiment of cutting down the amount of methylation of the phage DNA. He did this by growing *K* bacteria in a medium in which they were deprived of the amino acid methionine, the precursor of the substance that donates methyl groups to DNA. It turned out that phages grown in these bacteria generally did not acquire the *K* marking. Apparently they did not have the serifs (methyl groups) that would identify them as *K*.

Arber found further that the methyl groups serve as markers not just anywhere on the DNA molecule but only at certain strategic spots. This was clearly shown in experiments with a very tiny phage known as fd. Attached to each DNA molecule of this phage are a few MAP groups. Arber and his colleague Urs Kühnlein found that when the fd phage was grown in *K* bacteria, it had two fewer MAP groups than when it was grown in *B* bacteria. The *K*-marked phages failed to grow in *B* cells; these cells broke down the phage DNA.

This suggests that the fd phage's DNA has two sensitive sites, located on adenine bases. If the two sites are methylated (that is, marked MAP), the phage can grow equally well in both *K* and *B* bacteria. If, however, these sites are unmethylated, the *B* bacterium recognizes the phage as being the "wrong" kind and breaks down its DNA. Presumably the recognition is effected by a special enzyme that can split the DNA molecule. The *B* bacterium also has a methylating enzyme that can convert the two critical A bases to MAP (that is, attach a serif) and thus make the phage DNA acceptable so that it multiplies in *B* cells [*see top illustration on page 56*].

This interpretation has now been confirmed by various experiments. Some of the specific enzymes that break the vulnerable DNA sites and some of those that can methylate them have been isolated and identified. For example, Arber's group in Geneva and Matthew S. Meselson and Robert T.-Y. Yuan at

Harvard University have identified the enzymes that break the DNA molecules marked *K* and those that break the DNA molecules marked *B*. As expected from Arber's vulnerable-site hypothesis, it turns out that in each case the enzyme makes just a few breaks in the DNA molecule. Similarly, Arber and his colleagues have isolated a marking enzyme: from bacteria of the *B* strain they extracted the enzyme that can transfer methyl groups from a suitable methyl-donor substance (S-adenosyl methionine) to the *K*-marked (unmethylated) DNA of fd phages.

Thus we see that the bacteria possess a well-defined system for marking and recognizing phages. Shifting from our printing metaphor, we might say that each strain of *E. coli* bacteria stamps its own trademark on the DNA of the phages it produces, just as a factory brands its commercial product.

Work in several laboratories has elicited further details of the bacteria's branding and recognition system. It has been learned that the system in *E. coli* involves three closely linked genes. One directs the synthesis of the DNA-breaking enzyme, another controls the synthesis of the methylating enzyme, and the third gene generates a mechanism that is responsible for the recognition of the critical DNA sites that are to be either broken or methylated. It seems likely that this mechanism, or component, is a protein chain that associates itself with both the breaking and the methylating enzymes, and that can recognize the specific sequence of nucleotides that represents the critical DNA sites [*see bottom illustration on page 56*].

The marking and restricting system is widespread among the *E. coli* and related strains of bacteria; it is possessed not only by the *K* and *B* strains but also by others (except for the *C* strain and certain others that have lost the recognition system, perhaps through mutation). Also Arber and his co-workers found that DNA incompatibility is not confined to the case of invasion by a phage. It also applies to exchanges of DNA among the bacteria themselves. When a female *E. coli* cell mates with a male cell carrying the wrong DNA brand (for example, if the female is *B* and the male *K*, or vice versa), the female on receiving the male's DNA will break it down, just as if it belonged to a phage.

The answer to the T2 mystery turned out to be a different story. The investigation of this problem, carried out in our laboratory at the Massachusetts Institute of Technology by my colleagues Toshio Fukasawa, Costa P. Georgopoulos, Stanley Hattman and Helen R. Revel, demonstrated that the DNA in the T2 phage is not branded in the same way as that in the lambda and fd phages.

The DNA of T2 (and of its even-numbered relatives T4 and T6) carries an oddity: the base cytosine always has a hydroxymethyl group (CH_2OH) attached to it, so that the nucleotide contains hydroxymethylcytosine (HMC) instead of cytosine as its base. The discovery of the HMC base by Gerard R. Wyatt of Yale University and Seymour S. Cohen of the University of Pennsylvania played a major role in phage biochemistry. In most of the HMC nucleotides of these phages one or more sugar molecules (glucose) normally are attached to the hydroxymethyl group [*see illustration on page 57*]. What our laboratory team discovered was that when an *E. coli* bacterium of the *B*/4 strain

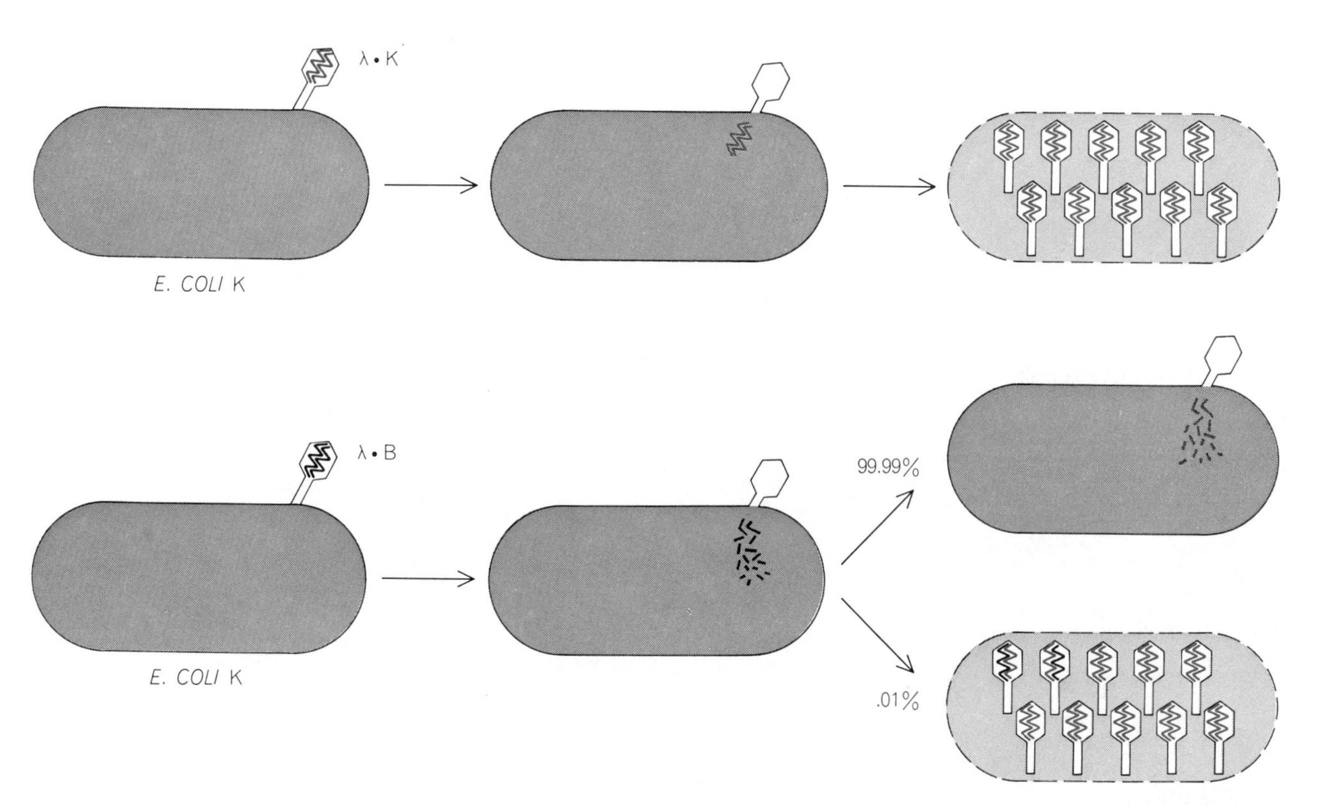

GROWTH RESTRICTION is demonstrated by cells of *E. coli* strain *K*. No restriction occurs when *K* cells are infected by phage $\lambda \cdot K$ (*top*). The deoxyribonucleic acid (DNA) of the phage (*colored zigzag shape*) enters the cell and exploits the cell's chemical machinery to produce about 100 new phage particles, which are released when the cell lyses, or dissolves. Restriction occurs (*bottom*) when *K* cells are infected by $\lambda \cdot B$ particles: lambda phages previously grown on *E. coli* strain *B*. The DNA of the phage (*black zigzag*) enters the cell but is broken down. In about one cell in 10,000, however, the phage DNA manages to multiply and give rise to phage progeny. About 2 percent of the progeny can grow in both *B* and *K* cells because their DNA is a hybrid molecule consisting of one strand of DNA (*black*) from the original $\lambda \cdot B$ and one strand (*color*) newly made inside the *K* cell. The remaining progeny are now modified so that they will grow normally in *K* cells but with a probability of only 10^{-4} in *B* cells.

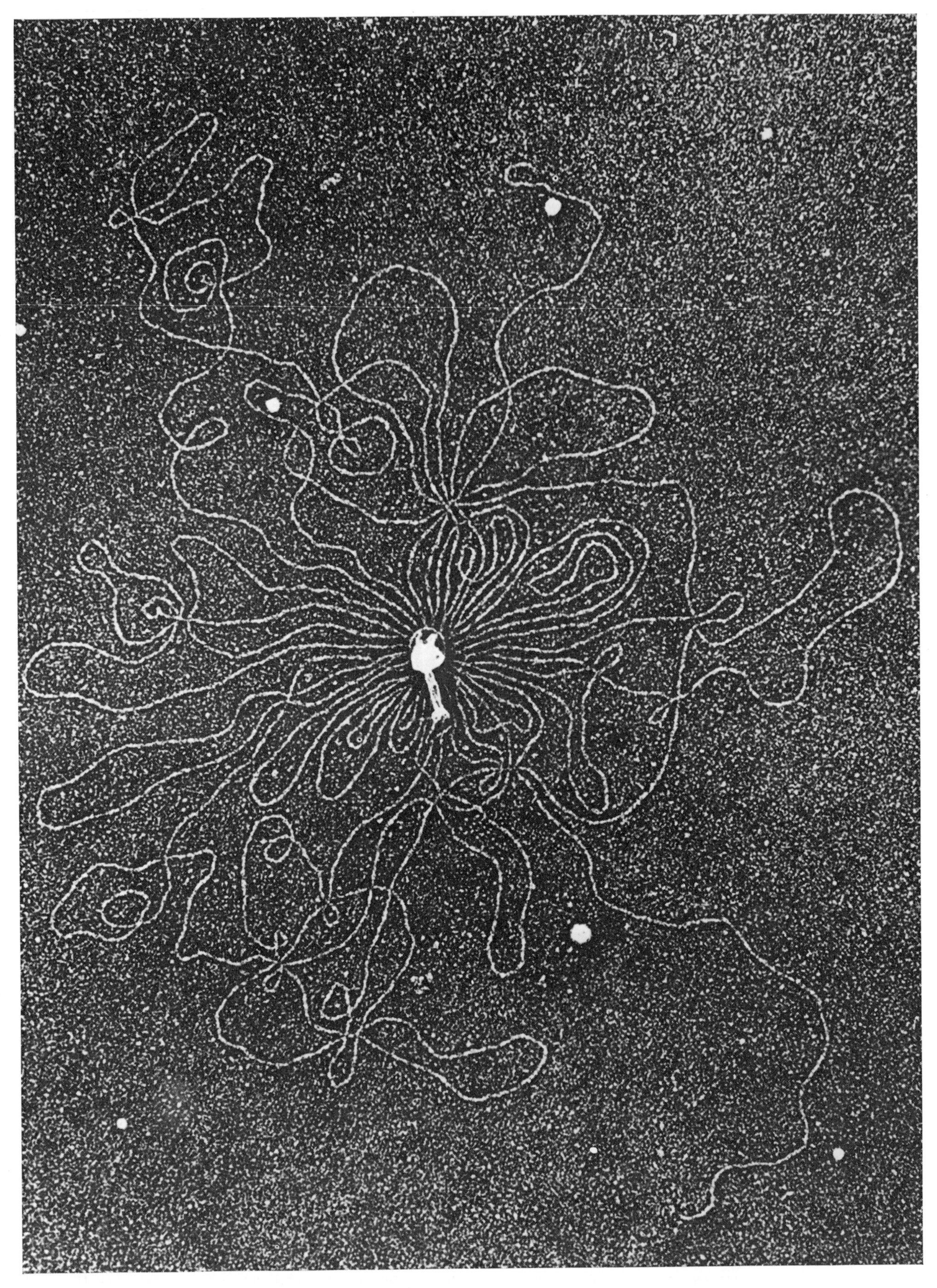

MOLECULE OF BACTERIOPHAGE DNA is shown here as a long, tangled thread after being released from the head of the T2 phage particle. The magnification is 100,000 diameters. The T2 bacteriophage is one of several "T even" phages that infect cells of *Escherichia coli*. When the phage DNA molecule is modified in a certain way, it can be "recognized" as an invader by the cell and destroyed. This electron micrograph was made by Albrecht K. Kleinschmidt of the New York University School of Medicine.

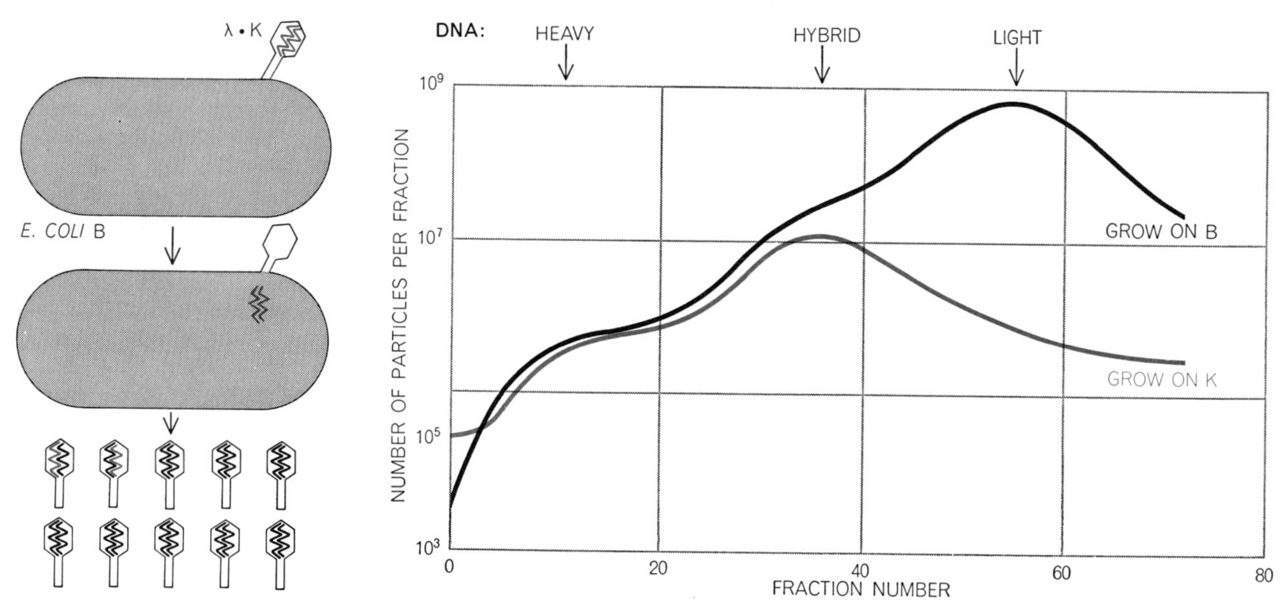

LABELING OF DNA demonstrates that phage particles possessing the ability to grow on both *B* and *K* strains of *E. coli* are predominantly those with hybrid DNA molecules. The experiment was conducted at the University of Geneva by Werner Arber and Daisy Roulland-Dussoix. They infected cells of strain *B* made up of atoms of ordinary weight with phage λ · *K* that had been grown on *E. coli K* cells incorporating heavy hydrogen (deuterium) and heavy nitrogen (^{15}N). Thus the DNA molecules in particles of λ · *K* were labeled with heavy atoms. The progeny produced in the light *B* cells were spun in a centrifuge containing a gradient of cesium chloride. The phage particles become distributed in the centrifuge tube according to their weight: heaviest particles at the bottom, lightest at the top. The phage particles in different fractions are then tested for their ability to grow on *B* and *K* cells, with the result plotted at the right. The particles containing hybrid DNA, about 2 percent of the total, grow in both kinds of cell. A small fraction of phage inherits two heavy strands of DNA from its λ · *K* parent and can also grow in both *B* and *K*. For this experiment Arber used a special strain of *E. coli B* that can impress the *B* modification on the phage DNA but lacks the *B* restriction function.

modifies, or brands, a T2 phage, the way it does so is to fail to attach the glucose to the HMC at the sensitive sites in the DNA molecule. In short, it changes the DNA from "sweet" to "sour"!

The soured phage can no longer multiply in *E. coli* cells; the cells detect the absence of glucose at certain critical sites and break down the DNA. Desugared T2 can, however, grow in the cells of dysentery bacilli, and these cells mark the phage DNA by attaching glucose at the necessary sites, thus restoring the phage's ability to grow in *E. coli.*

The special strains of *E. coli* that transform the phage from sweet to sour lack a substance (uridine diphosphoglucose, or UDPG) that is needed to donate glucose to the phage's DNA. This is one way for a T2 phage to acquire the sour branding. The brand can also be put on

"MARKED" BASES found in phage DNA that has been modified are 5-methylcytosine (MC) and 6-methylaminopurine (MAP). They are formed from the standard bases cytosine and adenine by the addition of a methyl group (CH_3), supplied by S-adenosyl methionine. Adenine (A) and cytosine (C) supply two of the four "letters" that spell out the genetic message in DNA molecules.

PHAGE	CHARACTERISTIC	PROBABILITY OF GROWTH IN *E. COLI* B
fd • K	2 MAP PER DNA MOLECULE	.001
fd • B	4 MAP PER DNA MOLECULE	1.0
fd m1 • K	3 MAP PER DNA MOLECULE	.03
fd m2 • K	2 MAP PER DNA MOLECULE	1.0

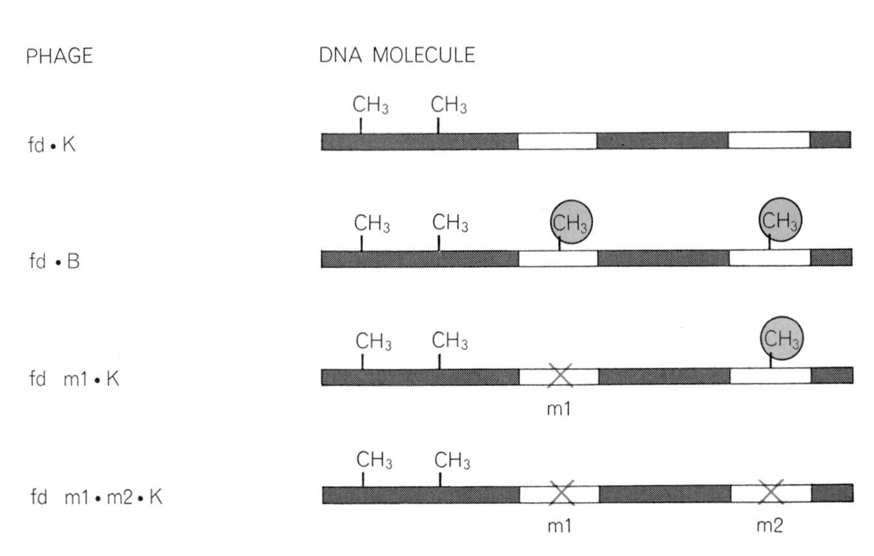

VULNERABLE SITES in the DNA molecule of the phage fd are represented by the white regions in the lower part of the illustration. These sites, or sequences of bases, are broken with a probability of 97 percent by the "restricting" enzyme of *E. coli B* unless an A (adenine) base in that sequence has been converted to MAP by the modifying (that is, methylating) enzyme present in *B* cells. There are also two irrelevant CH_3 groups elsewhere in the fd DNA molecule. The table (*top*) shows for various strains of fd the probability of growth in *E. coli B*. Strains of fd · *K* with certain mutations, *m*1 and *m*2 (*indicated by* X *marks*), are insensitive to the restricting enzyme in *E. coli B* even if they are not methylated.

BACTERIAL GENE	FUNCTION	MUTANTS
r	RESTRICTION MECHANISM ("NUCLEASE" COMPONENT)	r⁻: RESTRICTING FUNCTION LOST
m	MODIFICATION MECHANISM ("METHYLASE" COMPONENT)	m⁻: MODIFYING FUNCTION LOST
s	SPECIFICITY (SITE-RECOGNITION COMPONENT)	s⁻: BOTH RESTRICTING AND MODIFYING FUNCTIONS LOST

GENETIC CONTROL OF RECOGNITION SYSTEM seems to involve three closely linked genes in *E. coli*. Gene *r* makes the restricting enzyme that breaks the phage DNA molecule unless it carries the proper MAP marks. Gene *m* makes the methylating enzymes that provide the marks. Gene *s* makes a component responsible for specific recognition of sites where the marks may or may not be present. Presumably this recognition component is part of both enzymes: the one that can break DNA and the one capable of methylating it.

it, however, by a genetic mutation in the phage itself, that is, by an alteration of the phage genes that normally are responsible for production of the enzymes that attach glucose to the HMC [*see illustration on page 58*]. The *E. coli* bacteria, for their part, possess special genes that endow them with the ability to recognize and destroy sour DNA. These genes produce two enzymes, apparently different, that recognize and attack specific unsweetened sites in the DNA molecule. Experiments have shown that mutation of these two genes eliminates the rejection mechanism, so that unsweetened phages can grow in the mutated *E. coli* cells.

The two enzymes responsible for distinguishing between sweetened and unsweetened DNA have not yet been isolated. All we know about them so far (from genetic analysis) is that they are not the same as the enzymes that detect the difference between the methylated and unmethylated brands on the DNA of lambda and fd phages and on the DNA of different strains of *E. coli* bacteria.

How does the enzyme, in each case, recognize the brand? By what mechanism does the enzyme (protein) molecule find and identify the significant chemical markings on the DNA molecule? This is an intriguing question that applies to many other phenomena in molecular biology. For example, the enzymes that bring about the replication of DNA and those that build RNA molecules using DNA as a template must find the right place on the DNA at which to start the construction of the new molecule. How do they recognize these starting points?

We must suppose that the identifying marker on the DNA molecule in each case is a certain sequence of nucleotides, and that the enzyme establishes recognition by temporary attachment to this sequence—by fit and "feel," so to speak. There are good reasons to believe the identifying sequences are generally rather short; the length of a globular protein would not span many nucleotides in their linear array in the DNA molecule. Arber has suggested that the identifying sequences in branded DNA are probably no more than six to eight nucleotides long. It may soon be possible to work out the exact sequence of bases that constitutes the recognition site for at least one of the restrictive enzymes. This would be a big step forward in the study of protein-DNA interaction.

What kind of search do the enzymes carry on? Are the sensitive sites recognized and attacked only when enzymes

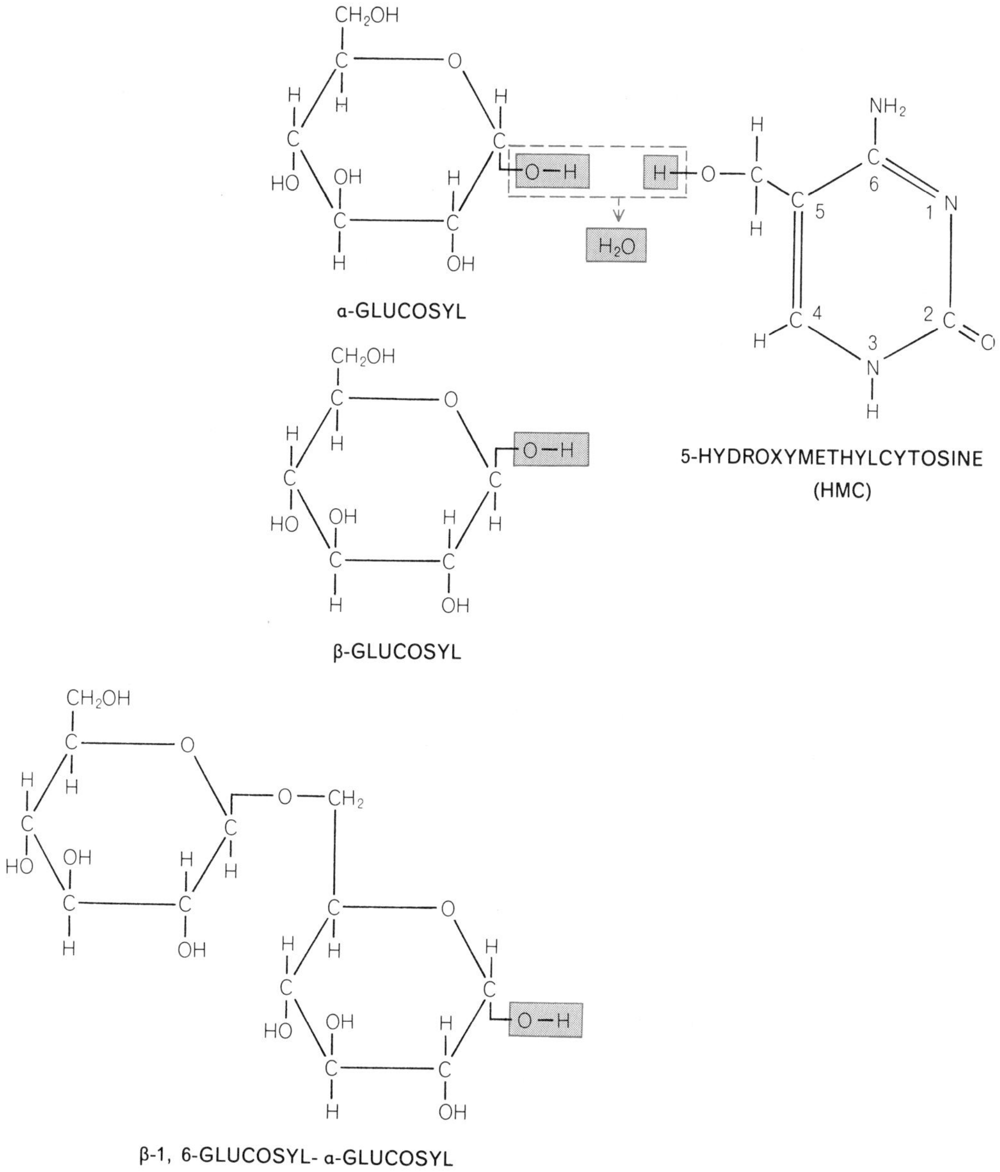

α-GLUCOSYL

5-HYDROXYMETHYLCYTOSINE (HMC)

β-GLUCOSYL

β-1, 6-GLUCOSYL- α-GLUCOSYL

"SWEET" AND "SOUR" LABELS provide another recognition system that enables *E. coli* cells to discriminate among invading molecules of phage DNA. In this system, which applies to the phages T2, T4 and T6, the base cytosine (C) is replaced by 5-hydroxymethylcytosine (HMC), a base formed when a hydroxymethyl group (CH_2OH) replaces hydrogen on the No. 5 carbon atom of cytosine. The normal phage has 70 percent or more of its HMC bases linked to one of the three sugar structures shown at the left. Such DNA molecules can be regarded as sweet. When the sugar units are missing, the DNA is sour and is readily broken down by *E. coli* cells.

happen to fall on them directly? Or do the enzymes browse along the phage DNA molecules, possibly as the latter nose into the bacterium, until they find the telltale markers? A finding by Georgopoulos and Revel in our laboratory suggests that the second alternative may be correct. They found that even a trace of "sweetness" (that is, glucosylation of comparatively few of the HMC bases, such as occurs in certain mutant T2 phages) was sufficient to protect the DNA of these phages from an enzyme that attacked only one particular site on the molecule and then only when it was unsweetened. It seems reasonable to assume that the small amount of glucose in these phages attaches itself to HMC groups at random, and therefore that the HMC groups at the particular site this enzyme attacks may often be unsweetened. We can guess the reason this site usually escapes attack is that the enzyme explores the DNA molecule and stops its exploration or falls off as soon as it encounters any glucose.

Another puzzling question is posed by the finding that even if only one (either one) of the two strands of a phage's DNA carries the protective marking, an enzyme will not break the molecule. This indicates that in the case of a vulnerable molecule the enzyme scans both strands separately before breaking the molecule. How does it manage to scan the strands individually?

The entire set of phenomena I have described in this article raises a more general question. What is the evolutionary significance of the DNA-branding system? What function does it serve for the bacteria? One obvious suggestion is that the system gives a bacterium a defense against certain phages. Evolutionary development could, however, have provided the bacteria with more effective defense mechanisms, such as eliminating the surface sites on the bacterial cell to which phages attach themselves. Furthermore, the bacteria that can recognize and destroy branded DNA have no defense against phages that are not branded. It is hard to believe the branding system evolved simply to protect bacteria of one strain against phages coming from another strain. A more plausible speculation is that the system serves primarily to prevent "undesirable" mixing of the bacterial genes (DNA) between the strains of bacteria themselves. Presumably in nature the *E. coli* bacteria do mate or otherwise exchange genetic material. If this is so, the ability of a *B*-strain bacterium to reject *K*-marked DNA (and vice versa) is an effective device for keeping each strain "pure." Thus the branding-and-rejection system facilitates the evolution of bacterial strains in diverging directions, in the same way that isolation mechanisms in cross-fertilization play a role in the evolution of plant and animal species.

The case of sweet and sour DNA seems to tell a clearer and more dramatic story. The T2 phages have the ability to break down the DNA of the bacteria they attack, which contains cytosine instead of HMC. Hence the hydroxymethylation of cytosine in the phages' own DNA may have been an important step in the evolution of their ability to multiply in the bacteria. We can surmise, then, that in reply some strains of *E. coli* by natural selection evolved the ability to break down phage DNA marked with HMC. The phage in turn evolved the capacity to tack glucose onto HMC to protect the vulnerable sites of its DNA. Here, then, the evidence of the branded DNA may be unfolding for us a scene in the dynamic acting out of evolution at the most elementary level: the chemistry of the genetic material itself.

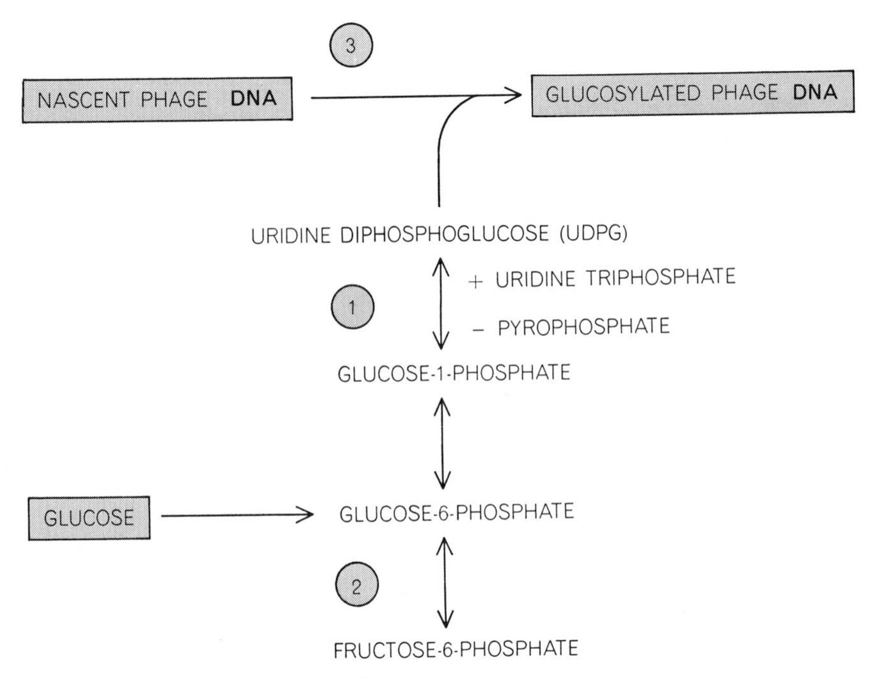

SYNTHESIS OF SWEET DNA requires a series of biosynthetic steps culminating in the addition of glucose units to the nascent DNA of the T-even phages. Mutants of *E. coli* can have blockages in the enzyme systems that carry out either or both of the first two numbered steps. Mutants with a blockage in enzyme No. 1 produce T-even phage with unglucosylated, or sour, DNA (T* phage). Mutants with a block in enzyme No. 2 can make the glucose donor substance, uridine diphosphoglucose, only if glucose is present in the growth medium. The last step, No. 3, is carried out by enzymes not normally present in *E. coli* cells; they are made from instructions coded in the DNA molecules of T-even phages.

The Molecule of Infectious Drug Resistance

7

by Royston C. Clowes
April 1973

R factors, which transmit resistance to antibiotics from one bacterial strain to another, are carried on extrachromosomal genetic elements called plasmids, some of which have now been isolated and measured

In 1971, only a few months after the antibiotic gentamycin had been introduced for the treatment of bacterial infections, strains of bacteria had developed resistance to the antibiotic. This development was not too surprising. Ever since sulfonamides were first administered to treat bacterial infection it had been known that repeated exposure of bacteria to an antibiotic results in the development of a resistant strain through the processes of mutation and selection. A bacterium that has spontaneously undergone a gene mutation that makes it no longer susceptible to the antibiotic's effect will live and multiply in the presence of the drug, and its similarly resistant progeny constitute the resistant strain. Mutations are rare, however; usually only one bacterial cell in a population of several hundred million will have mutated, so that the physician has a chance to treat the infection successfully by administering antibiotics before resistant cells develop in large numbers.

The resistance to gentamycin was different. The bacteria from the first patient in whom a gentamycin-resistant strain was detected were resistant not only to gentamycin but also to three other common kinds of antibacterial agent: chloramphenicol, ampicillin and sulfonamide. Then, from a second patient, investigators isolated four different bacterial strains, each of which carried resistance to the same four antibiotics. Moreover, when they grew any of these bacteria in laboratory glassware with a bacterial strain that was susceptible to the antibiotic's effects, the susceptible strain also became resistant to all four antibiotics in a matter of hours. From the investigators' knowledge of many similar cases reported during the past decade they realized that they were dealing with a case of infectious antibiotic resistance. Each strain isolated from the patients carried an "infectious antibiotic resistance factor," or *R* factor: a complex consisting of a gene producing gentamycin resistance, genes determining resistance to the other three antibiotics and genes that facilitate the transfer of the *R* factor.

Intestinal bacteria that give rise to such diseases as enteritis or typhoid fever and carry *R*-factor-controlled resistance to all known antibiotics have now been isolated in most parts of the world. Because the genes of these *R* factors are transferred independently of the genes on the bacterial chromosome, and because they are "cured," or lost, if bacteria carrying them are grown in the presence of certain dyes or other agents, they are known to exist independently of the chromosome; they are carried on the extrachromosomal genetic elements called plasmids. A strain carrying an *R* factor is able to conjugate, or pair, with other bacterial cells. When that happens, a copy of the *R* plasmid is transmitted to the second cell, which thereby acquires all the drug resistances carried on the plasmid and also the potential of transmitting those resistances to still more cells.

During the 1960's genetic experiments in a number of laboratories established the mode of transference of *R* factors and revealed something of their composition and structure [see "Infectious Drug Resistance," by Tsutomu Watanabe; SCIENTIFIC AMERICAN, December, 1967]. More recently the effort in my laboratory at the University of Texas at Dallas and in other laboratories has been to learn more about the physical nature of the plasmids that constitute the *R* factor: their size and structure, the ways in which they multiply and how they may evolve.

It seemed likely that *R* factors are composed of the genetic material DNA, and that had been confirmed by transferring *R* factors to *Proteus mirabilis,* a strain of bacteria that has DNA of unusually low density. When DNA extracted from a *Proteus* culture is suspended in a solution of the salt cesium chloride and spun at high speed in a centrifuge, the DNA molecules move to positions in the centrifuge tube where their density is equal to the density of the salt surrounding them. The contents of the tube are separated into a number of fractions, each from a different level in the tube. Fractions containing DNA are then identified either by measuring the extent to which they absorb ultraviolet radiation of a certain wavelength or by first incorporating radioactive subunits in the DNA of the live culture and then measuring the radioactivity of each fraction [*see illustrations on page 61*]. If the

→

R-FACTOR PLASMIDS, shadowed with platinum vapor, are enlarged 62,000 diameters in an electron micrograph made by Michiko Mitani. The plasmids are molecules of looped DNA. The extended one (*bottom*) is a "nicked duplex loop"; the distance around it can be measured to establish its molecular weight. The other one (*top*) is a "supercoiled" form.

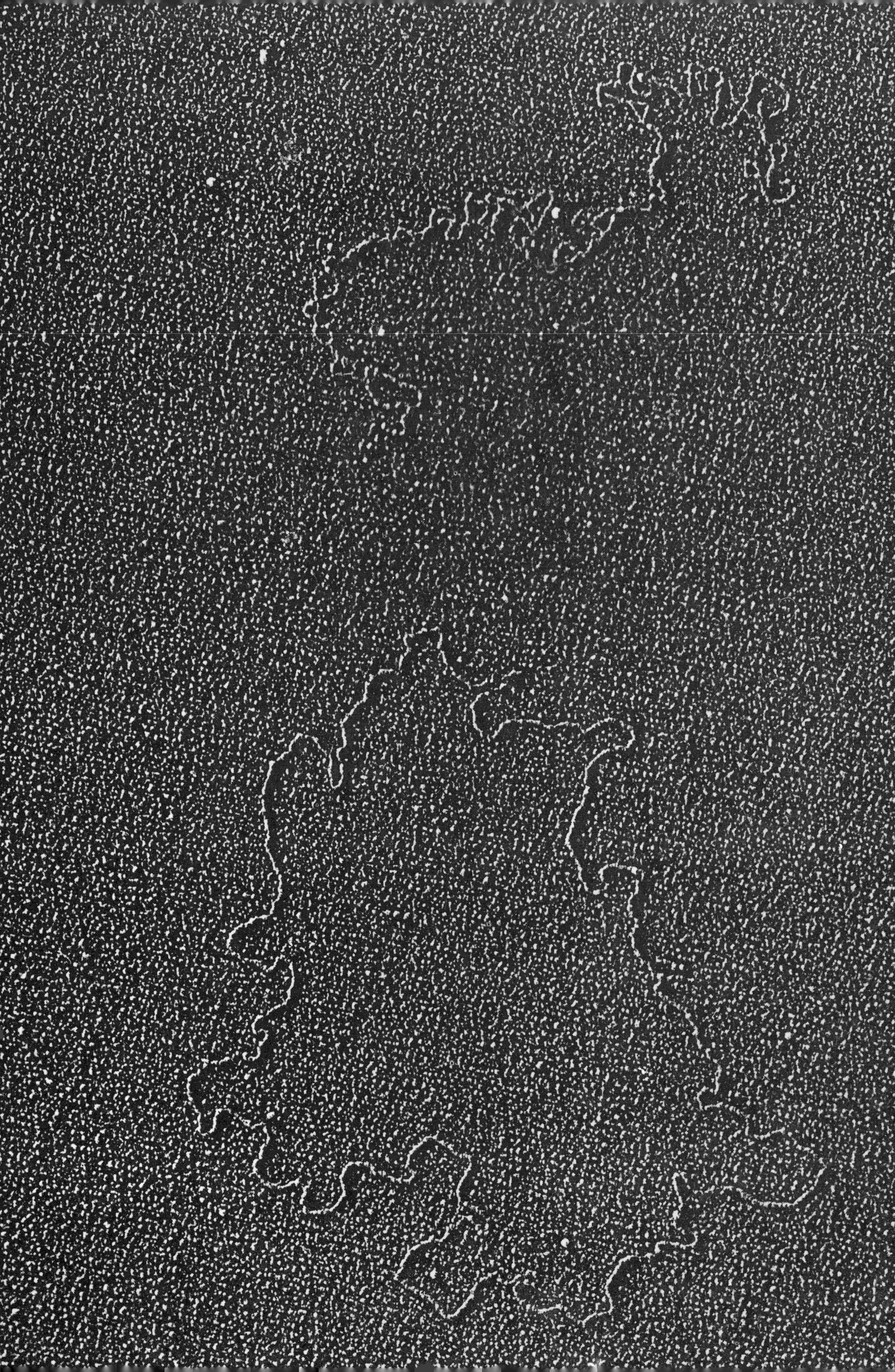

DNA is extracted from a *Proteus* strain to which an *R* factor has been transferred, one finds in addition to the main band of DNA characteristic of the *Proteus* chromosome a smaller "satellite" band at a greater density. If the strain of *Proteus* is then "cured" of its *R* factor, this denser satellite band is no longer seen, and so it is clear that the satellite band must be the DNA of the *R* plasmid.

The bacterial chromosome is a single giant DNA molecule in the form of a closed loop about a millimeter long (about 1,000 times as long as the cell). After DNA isolated from a culture of *Proteus* harboring an *R* factor had been centrifuged in cesium chloride, my colleagues Taizo Nisioka and Michiko Mitani and I took samples from the fractions containing the *R*-plasmid DNA and prepared them for examination in the electron microscope by what is called the microdrop technique. In this method the DNA samples are mixed with a solution of a basic protein such as cytochrome *c*. When small drops of the mixture are put on a material such as Teflon, the protein molecules spread over the surface of the drops as oil does on water, in a film one molecule thick. The acidic groups of the DNA become bound to the basic groups of the protein, so that the long DNA molecules, teased out gently without breaking, become extended at the surface of the film [*see top illustration on page 62*]. A sample is taken from this surface to an electron-microscope specimen grid and dried, and the DNA molecules can be viewed in the microscope.

We found that the DNA of the *R* factor consists of closed looped molecules, some of them extended and some of them tightly coiled [*see illustration on opposite page*]. The distance around all the extended loops was about the same. After storage over many months the proportion of tightly coiled molecules decreased, but the number of extended loops, which were the same size as the original ones, increased. It was clear, therefore, that the DNA of the *R* factor is made up of looped (or what are often called "circular") molecules of a specific size.

Knowing the length of other DNA molecules of known molecular weight, and thus the molecular weight per unit length of DNA, we could calculate the molecular weight of the *R*-plasmid molecules. By this method and others all *R* factors so far investigated have been shown to be looped DNA molecules, each with a characteristic size. Their molecular weights range from about one

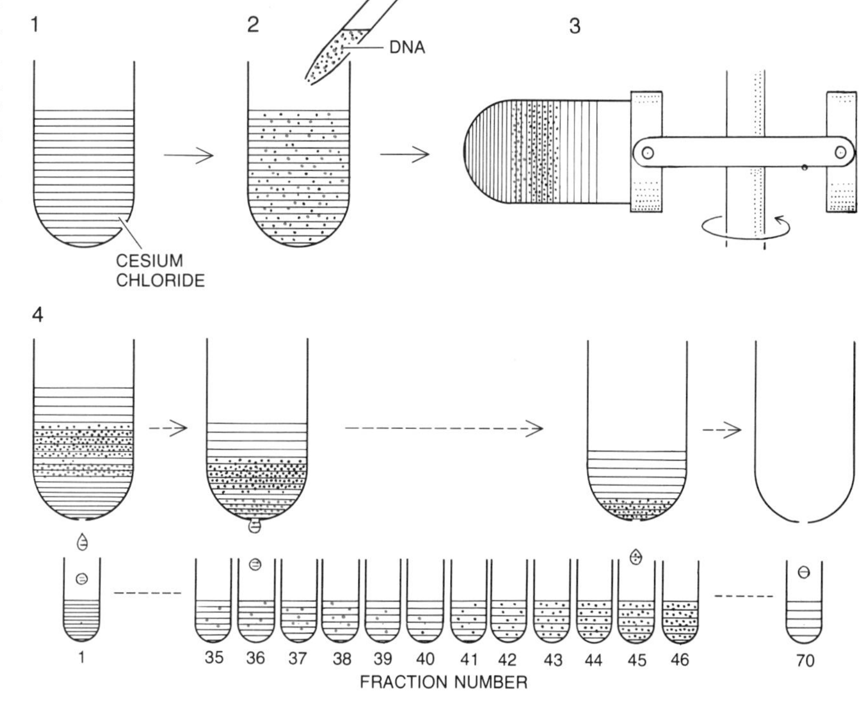

R-FACTOR DNA is separated from bacterial DNA by density-gradient centrifugation. A solution of cesium chloride (*1*) to which DNA from a culture of *Proteus mirabilis* carrying an *R* factor has been added (*2*) is spun in an ultracentrifuge for 48 hours (*3*). The DNA's move to points in the centrifuge tubes where their density is equivalent to that of the cesium chloride. The contents of a centrifuge tube are collected drop by drop in smaller tubes (*4*); each fraction contains material from a different level and hence of a different density.

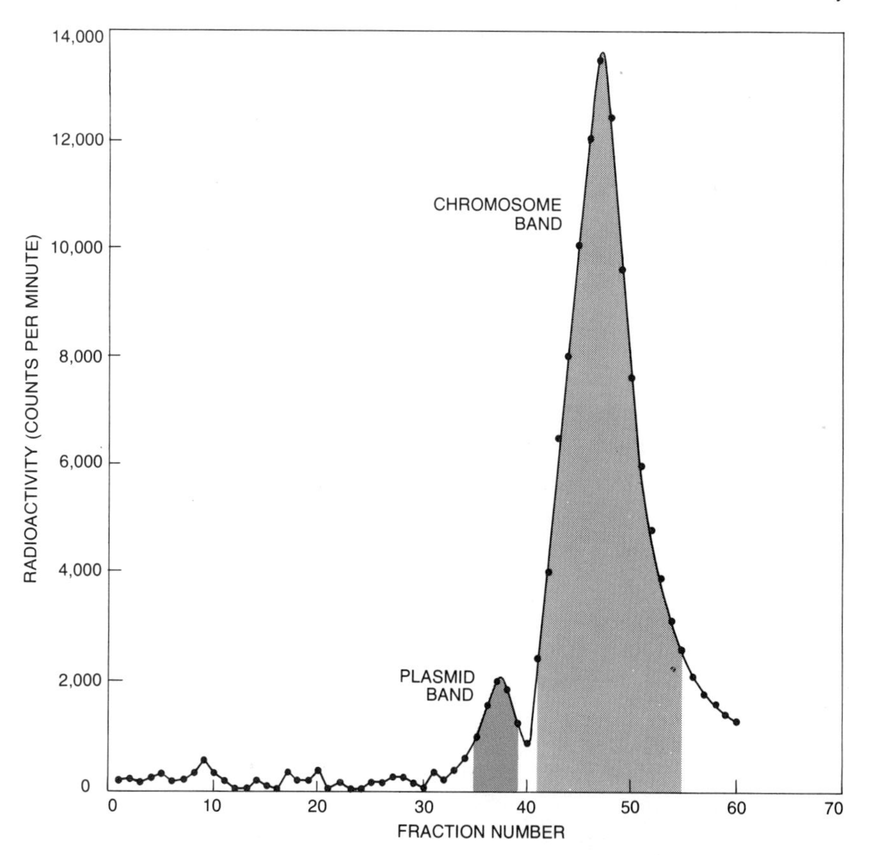

DNA-CONTAINING FRACTIONS can be identified if radioactive subunits were previously incorporated in the bacteria; the amount of DNA in each fraction is proportional to the radioactivity. In the case of *Proteus mirabilis* carrying an *R* factor there are two DNA peaks. The major, less dense band (*gray*) is the DNA of the *Proteus* chromosome. The satellite band (*color*), which comes from earlier fractions, represents the denser DNA of the *R* factor.

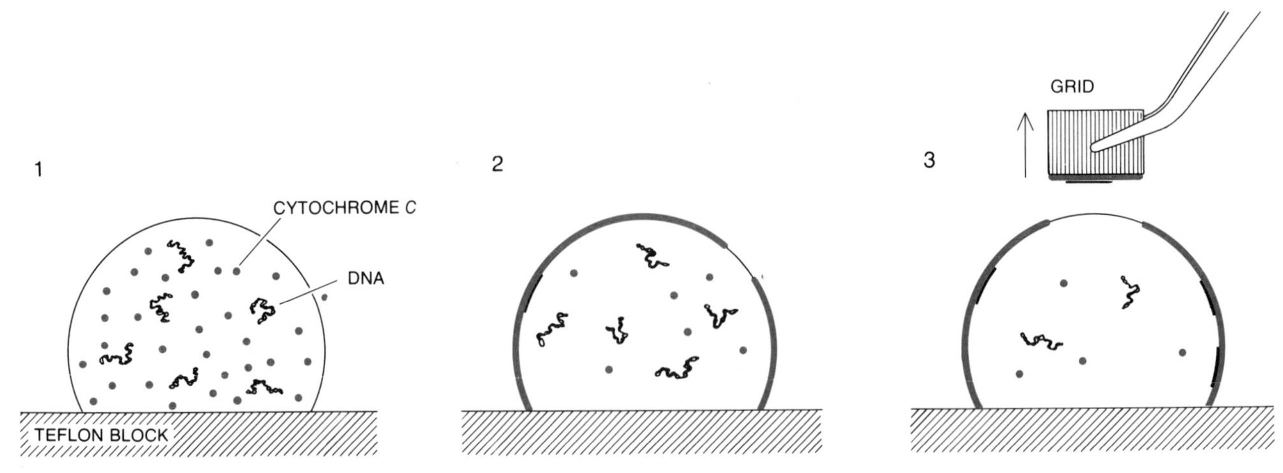

MICRODROP TECHNIQUE developed by Dimitrij Lang and Michiko Mitani at the University of Texas at Dallas mounts DNA molecules for electron microscopy without distorting them. A drop of water containing DNA (*black*) from a plasmid-rich centrifuge fraction, salt, formaldehyde and the protein cytochrome *c* (*color*) is placed on a water-repellent surface (*1*). The cytochrome *c* molecules form a monomolecular film at the surface of the drop and the positively charged protein attracts the negatively charged DNA molecules, which unfold against the film (*2*). A part of the film, with the DNA, is picked up on a microscope specimen grid (*3*).

megadalton to more than 100 megadaltons. (A dalton, the unit of molecular weight, is the weight of one hydrogen atom. One megadalton is a molecular weight of one million.) This spectrum of sizes is about the same as that found for the DNA of bacterial viruses and is at most only a small fraction of the size of the bacterial chromosome.

The fact that *R*-plasmid molecules are circular and tightly coiled was of great value in our subsequent work. When DNA molecules are extracted from cells, they are subjected to shearing forces that may break them. The small size and tight coiling of plasmid molecules make them resistant to these forces. More important, a circular molecule must by its nature have remained unbroken during extraction, and so its size and weight are true measures of what the native molecule was inside the cell. The special properties of various species of circular DNA have been investigated intensively by two groups headed by Robert L. Sinsheimer and Jerome Vinograd at the California Institute of Technology. The Cal Tech groups found circular DNA molecules of two kinds. In one the two strands of the DNA double helix are joined to form a duplex loop; these structures are "supercoiled," presumably so that the long DNA molecules can be fitted inside a cell. In the other kind only one of the DNA strands is joined; these molecules are not supercoiled. If a sin-

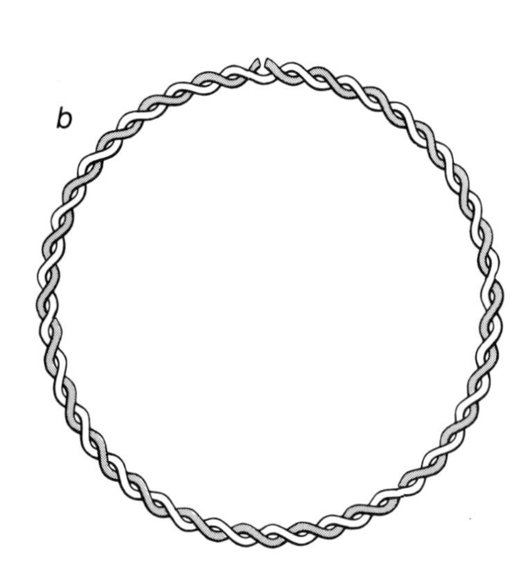

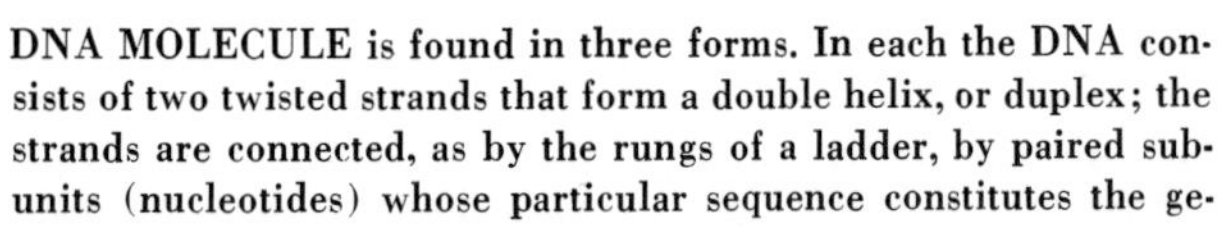

DNA MOLECULE is found in three forms. In each the DNA consists of two twisted strands that form a double helix, or duplex; the strands are connected, as by the rungs of a ladder, by paired subunits (nucleotides) whose particular sequence constitutes the genetic code. The basic form is the linear duplex (*a*). If the ends of the duplex are joined through one strand, the structure is a nicked duplex loop (*b*). If both strands are joined, the structure is an intact duplex loop, usually found in the supercoiled form (*c*).

gle break is made in one strand of a supercoiled molecule, the molecule loses its coils and is converted to the uncoiled structure, which is therefore called a "nicked duplex loop" [*see bottom illustration on page 62*]. The two kinds of circular DNA we had found in *R* factors were just such nicked duplex loops and supercoiled molecules.

There is an effective method of distinguishing supercoiled DNA from other DNA, which depends on the insertion of molecules of ethidium bromide, one of the dyes that can "cure" plasmids, between adjacent DNA subunits. This intercalation, as it is called, results in two changes: the length of the DNA molecule is increased, so that it becomes less dense, and the helix becomes slightly less twisted. Large numbers of ethidium bromide molecules can intercalate in a nicked duplex loop or a linear duplex, without the untwisting effect leading to coiling. In contrast, the number of dye molecules that can be inserted in a supercoiled loop is strictly limited because there is a limit to the further coiling that is produced as a result of untwisting. The limitation on intercalation limits the associated extension in length and thus leads to a smaller decrease in density for supercoiled loops than for nicked duplex loops or linear duplex DNA. Therefore if ethidium bromide is added to a mixture of these three types of DNA molecule, the supercoiled loops can be separated from the other forms as a denser band after centrifugation.

Since, as we have seen, a major portion of *R*-factor DNA exists in the cell as supercoiled loops, it can be isolated and separated by this procedure from the chromosomal DNA. That DNA, being about 100 times longer, is generally broken into linear fragments when it is extracted from the cell. The procedure provides a means of separating plasmid DNA even in host bacteria such as *Escherichia coli*, in which the chromosomal DNA does not differ in density from the *R* plasmid as it does so conveniently in *Proteus*. Since *E. coli* and related bacteria are the natural hosts of many *R* factors, the procedure enables one to study the plasmids as they are found in nature.

The same experiments can be used to measure the number of *R*-plasmid molecules in each cell. When bacteria are grown in the presence of a radioactive subunit of DNA such as thymine, the DNA in the cell takes up this radioactive label. If we assume that the uptake of radioactive thymine into plasmid DNA occurs to the same extent as it does into chromosomal DNA, and if most of the plasmid DNA exists as supercoiled loops, then the relative amounts of DNA in the plasmid and in the chromosome can be measured by the relative amounts of label in the denser satellite peak and in the main peak. Since we know the molecular weights of a number of *R* plasmids, and other experiments have shown that the molecular weight of the *E. coli* chromosome is 2,500 megadaltons, we can estimate the relative number of *R*-plasmid molecules per chromosome from their relative DNA content in the cell. By this method and others many *R*-plasmid molecules have been found to exist in an approximate one-to-one relation with the chromosome. Although the DNA's of the *R* plasmid and of the chromosome are physically separated, then, there appears to be some regulatory mechanism in the bacterium such that the *R* plasmid replicates only once for every chromosomal replication. How

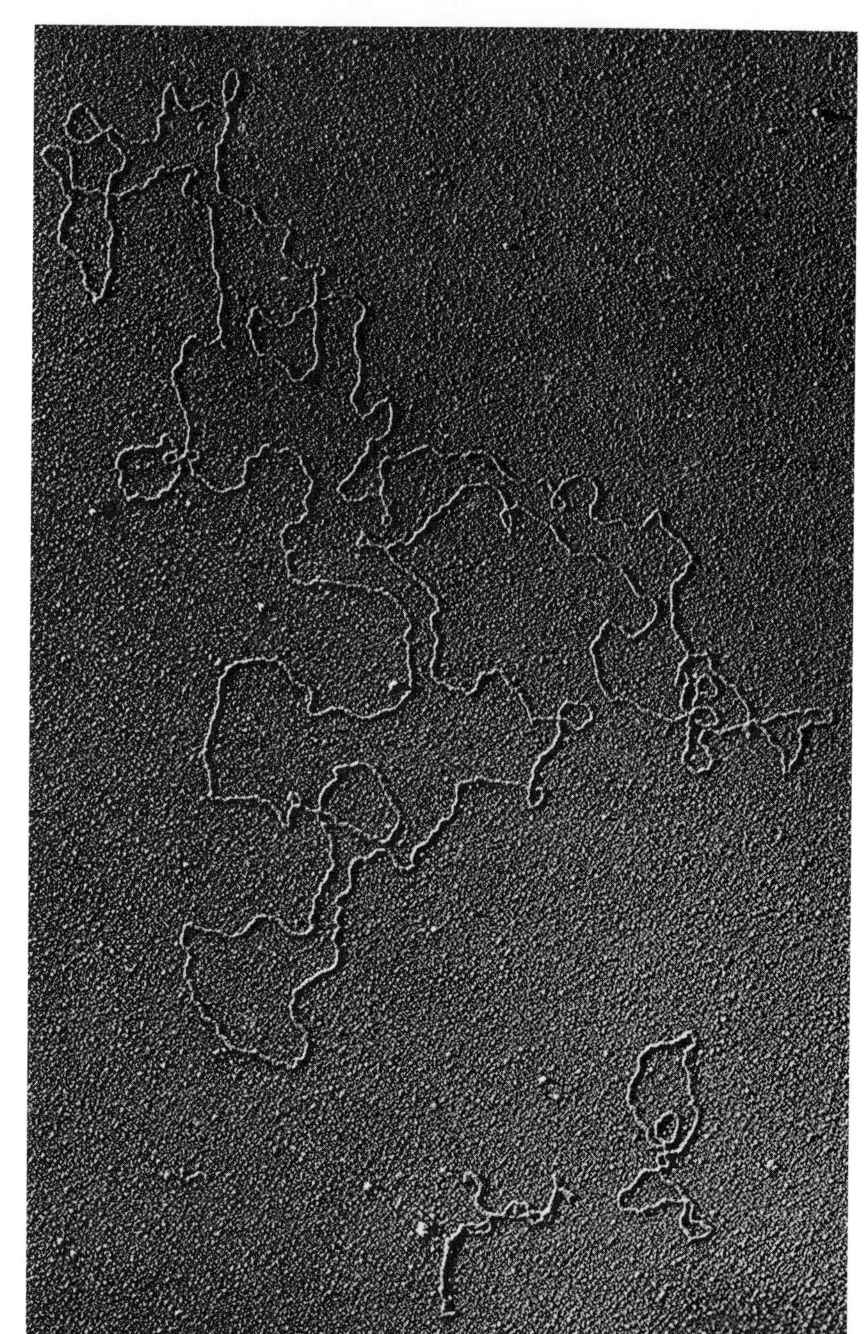

DNA MOLECULES of the *R* factor delta (*S*), a plasmid aggregate, are enlarged 40,000 diameters in an electron micrograph made by Christine Milliken. The large nicked duplex loop (*top*) is the delta plasmid, the transfer factor. The small molecules (*bottom*) are *S* plasmids, carrying streptomycin resistance. One is supercoiled, one is a nicked duplex loop.

this regulation is effected is not known and is the subject of intensive work in a number of laboratories.

Before 1955 infectious antibiotic resistance factors were rare. By 1965 it was found that up to 60 to 70 percent of all common intestinal bacteria from hospitals and other clinical sources carried *R* factors, usually determining resistance to three or more antibiotics. *R* factors can also be transferred to bacterial species responsible for cholera and plague, and similar elements are now common in staphylococci. The epidemic spread of these factors among pathogenic bacterial strains has made an understanding of how they have evolved—and how they evolve so quickly—an important problem in clinical bacteriology today.

One of the original hypotheses for *R*-factor evolution suggests that it depends on a process known as gene pickup. To explain this concept we must consider another bacterial plasmid, the *F* factor, which determines sexuality and conjugation in *E. coli.* Like the *R* factors, the *F* factor is infectious, can transmit itself sequentially by conjugation and is an intact duplex DNA loop (of about 62 megadaltons), and there is about one of these *F* plasmids for every chromosome. The infectious transfer of the *F* factor and of a number of *R* factors has recently been elegantly defined in molecular terms by Daniel Vapnek and W. Dean Rupp of the Yale University School of Medicine. They have shown that conjugation leads to a break in one of the two strands of the intact *F*-factor duplex loop and the transfer of that strand to the recipient cell. There it synthesizes its complementary strand and subsequently forms an intact duplex loop; meanwhile the strand retained in the donor also forms its complementary strand and is converted into an intact duplex loop [*see top illustration on these two pages*].

From donor cells that harbor an *F* factor (called F^+ donors) rare mutant cells can be selected in which the *F*-plasmid DNA has been inserted within the bacterial chromosome without causing a break in the chromosomal loop [*see bottom illustration on these two pages*]. Conjugation by one of these mutant strains results in a similar break within the DNA of the inserted *F* plasmid. Thereafter a strand of DNA consisting of a part of the *F* plasmid is transferred to the recipient cell, taking with it the attached chromosomal strand. This chromosomal transfer occurs very frequently, so that the mutant strains are called *Hfr* (for high-frequency recombination). Different *Hfr* cells are formed after the insertion of the *F* plasmid at different sites along the chromosome.

Still a third type of donor is derived from *Hfr* donors by a reversal of the insertion process, leading to the release of the integrated *F* factor from the chromosome. This release usually occurs at a point slightly different from the point of insertion, in which case a segment of the bacterial chromosome is incorporated within the DNA of the released *F* plasmid and some of the plasmid DNA remains in the chromosome. These released *F* factors that have "picked up" chromosomal DNA are known as *F*-prime factors, and bacterial strains in which they are present are called *F*-prime donors. A similar mechanism of gene pickup was suggested as the basis for the evolution of *R* plasmids. The idea was that a "transfer factor" resembling *F* was integrated in the chromosome and then released, together with a segment of adjacent chromosomal DNA on which a gene controlling drug resistance was located.

Some recent findings make the idea of evolution of *R* plasmids through gene

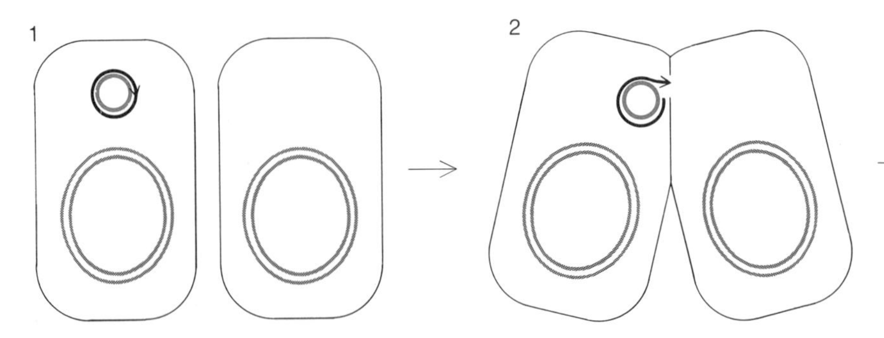

INFECTIOUS TRANSFER of a bacterial plasmid such as an *F* factor or one of the *R* plasmids is diagrammed. Two bacteria are shown about to conjugate, or pair (*1*); each of them has a circular chromosome (*gray*), and one of them, the donor cell, also has a

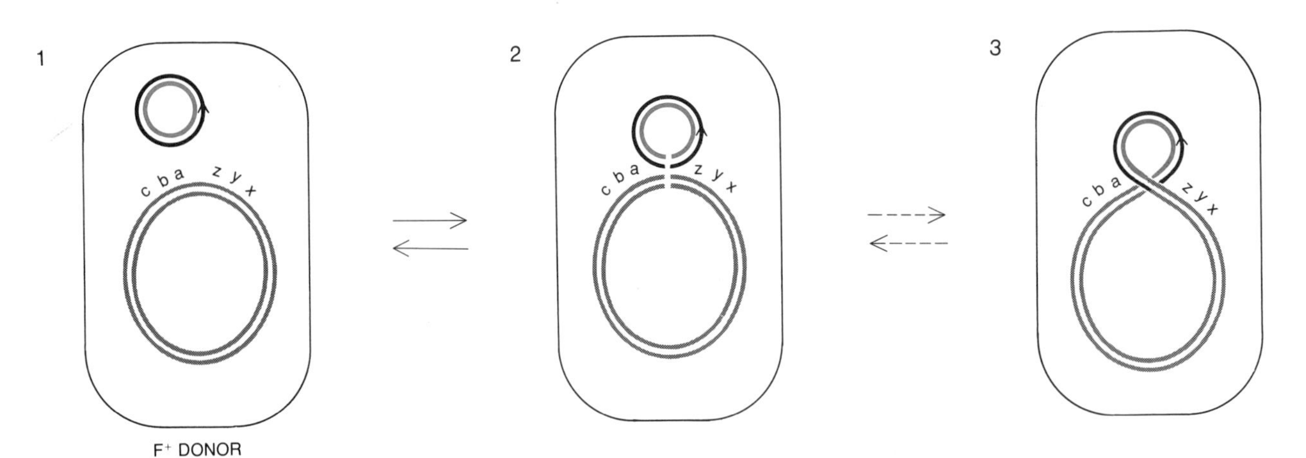

DONOR CELLS can be of three types, the relations among which are illustrated here with the *F* factor as a model. An F^+ donor (*1*) usually transfers *F* as in the illustration at the top of the page. Sometimes, however, a break occurs in both strands of the *F* plasmid and of the chromosome (*2*) in a region where the nucleotide sequences of plasmid and chromosome are similar enough so that the two sets of ends join (*3*) and the plasmid is inserted into the chromosome (in this case between genes *x, y, z* and *a, b, c*) to form

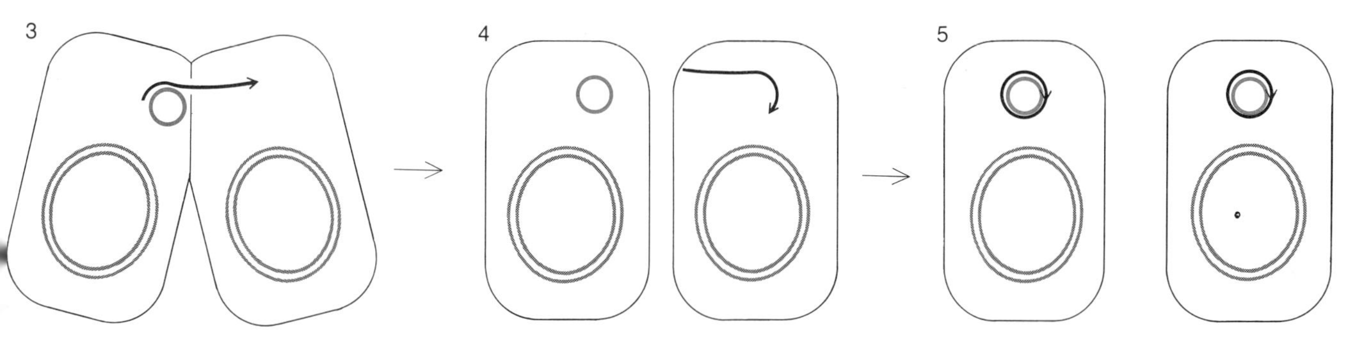

plasmid, an intact duplex loop (*color and black*). On pair formation (*2*) one strand of the plasmid DNA breaks at a particular point (*arrow*) and passes into the recipient cell (*3, 4*). After transfer each strand of DNA synthesizes its complementary strand, so that two intact duplex loops are formed. Each cell now has a copy of the original plasmid and is therefore a potential donor cell (*5*).

pickup seem less likely. The most telling argument comes from discoveries about the mode of action of antibiotic resistance mediated by *R* factors. Many antibiotic molecules exert their lethal effect by becoming attached to bacterial ribosomes: the cellular particles where proteins are synthesized. In this way they interfere with protein synthesis. In all bacteria where they have been analyzed resistance to these antibiotics that arises from mutations of chromosomal genes is effected by changes in the structure of the ribosomes such that the antibiotic can no longer interfere with protein synthesis [*see illustration on page 66*]. In contrast, *R*-factor resistance to these same antibiotics involves the direct modification of the antibiotic molecule in such a way that it can no longer interact with the ribosome, which remains unchanged. Since the mode of action of *R*-factor resistance appears to be so different from that of chromosomal resistance, *R*-factor genes (unless they arise from chromosomal genes as yet unidentified) must be extrachromosomal in origin.

Although the DNA of any one of a number of *R* factors isolated from either *E. coli* or *Proteus mirabilis* consists of

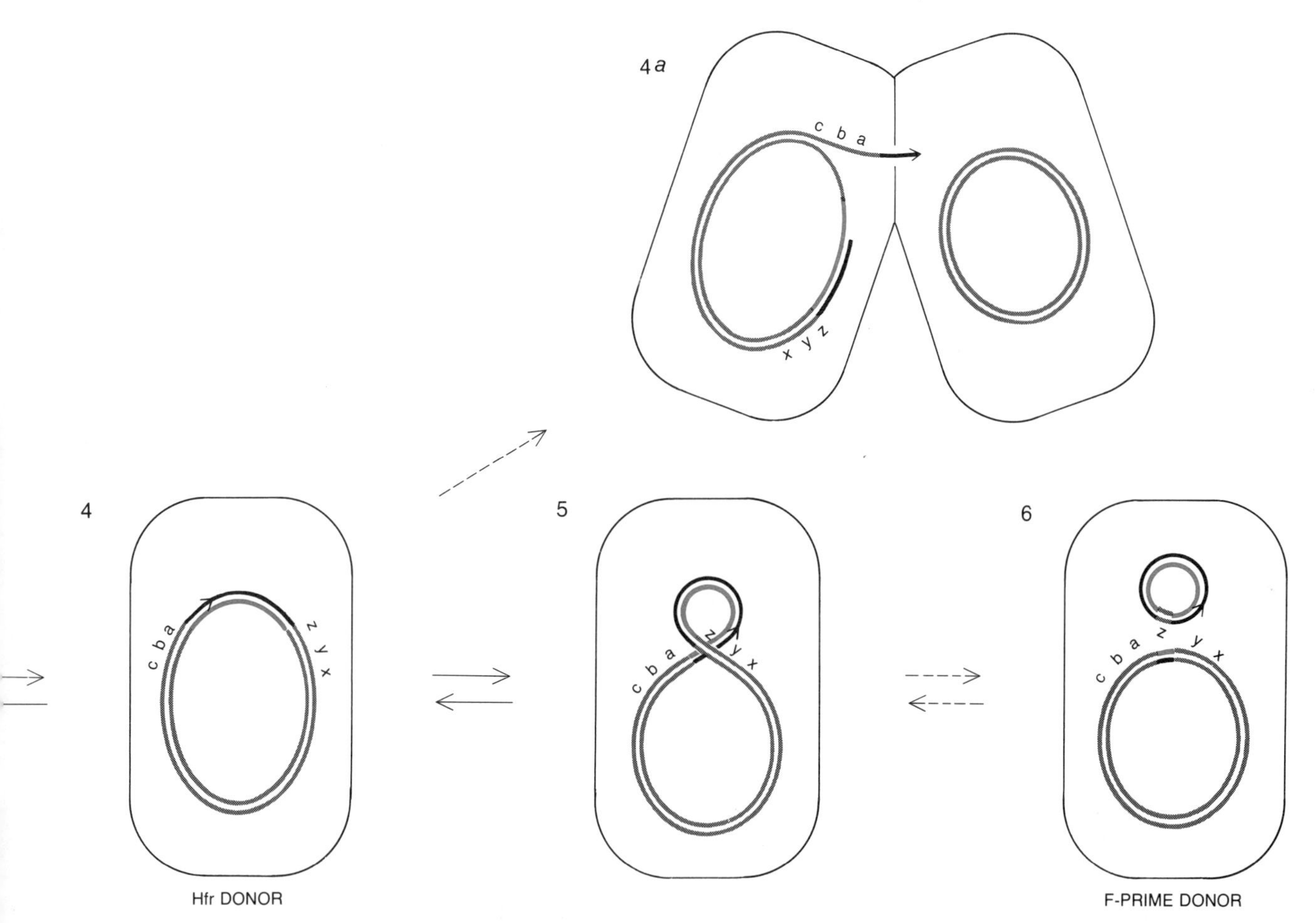

an *Hfr* donor cell (*4*). When an *Hfr* cell conjugates, one strand of the *F* plasmid breaks and the first part of the plasmid strand is transferred to the recipient along with the chromosome (*4a*). Occasionally the *F* plasmid is released from the chromosome of an *Hfr* donor. If this occurs by breakage and reunion at a point different from where the plasmid was incorporated, segments of chromosomal and plasmid DNA are interchanged. The plasmid, having picked up a chromosomal region, becomes an *F*-prime factor (*5, 6*).

molecules of the same size, certain exceptions are found. One of the first reported was in the case of the 222/*R* plasmid, which determines resistance to tetracycline, chloramphenicol, streptomycin and sulfonamide. DNA from a *Proteus* host harboring this plasmid produces a satellite band with three distinct peaks [*see illustration on opposite page*]. When we examined a sample from each peak in the electron microscope, we found molecules of 12 megadaltons in the peak of greatest density, molecules of 70 megadaltons in the middle peak and molecules of 58 megadaltons in the lowest-density peak. The sizes and densities of these molecules suggested that some of the 222/*R*-factor plasmids (which in *E. coli* were all 70 megadaltons in size) were divided into two smaller molecules of 12 and 58 megadaltons in *P. mirabilis*. Since the peak of greatest density is about the same size as the other two peaks (that is, it has the same amount of DNA) in spite of the fact that the 12-megadalton molecules in this peak are much smaller than the molecules in the other peaks, there must be many more of the 12-megadalton molecules than of the larger molecules. It is therefore likely that after the 70-megadalton molecule subdivides into two smaller ones the 12-megadalton molecule can replicate independently. Other experiments show that a modified plasmid similar to the 58-megadalton molecule can be derived from the 70-megadalton molecule, and this plasmid can also replicate normally.

The natural replication of the bacterial chromosome (or any other DNA molecule) appears to be possible only beginning at a unique point of origin on the chromosome, and this point appears to be essential for replication. If an *R* factor such as 222/*R* had in fact evolved by sequential chromosomal pickup, it would be unlikely that enough of the chromosome would have been picked up to have much chance of including the unique point of origin. The *R* factor would therefore be expected to have only the one point of origin for DNA replication that was present in the original transfer-factor molecule. Since in the *Proteus* host the 222 factor segregates into two molecules, both of which may replicate independently, this again does not fit the idea of evolution by sequential chromosomal pickup. On the other hand, these facts can be reconciled with an alternative idea: that the genes on *R* factors arise from molecules that were originally extrachromosomal and capable of independent replication.

The *R* factors first studied by genetic and physical methods gave results similar to those of 222/*R* in *E. coli*, that is, all the antibiotic-resistance genes were transmitted together with the genes determining transfer properties as though they were all part of the same genetic structure. The physical analysis of the DNA of these plasmids confirmed this unitary structure, disclosing only a single molecular species for each *R* factor, as shown in the case of 222/*R* in *E. coli*. Since all the antibiotic-resistance genes and the genes determining transfer are integrated in a single structure, these *R* factors are called plasmid cointegrates. A number of other factors have recently been shown to have different properties, in particular an *R* factor studied in *E. coli* by E. S. Anderson and Malcolm Lewis at the Central Enteric Reference Laboratory in London. That factor controls resistance to streptomycin, ampicillin and tetracycline but transfers only some of its properties to some recipient cells [*see illustration on page 68*]. After conjugation some of the recipient cells carry infectious resistance to only one of the three antibiotics; others carry either streptomycin resistance or ampicillin resistance that is no longer infectious. Still other cells have lost all antibiotic resistance but, when they are incubated with the noninfectious streptomycin- or ampicillin-resistant cells, can potentiate the transfer of this noninfectious resistance. Anderson and Lewis therefore suggested

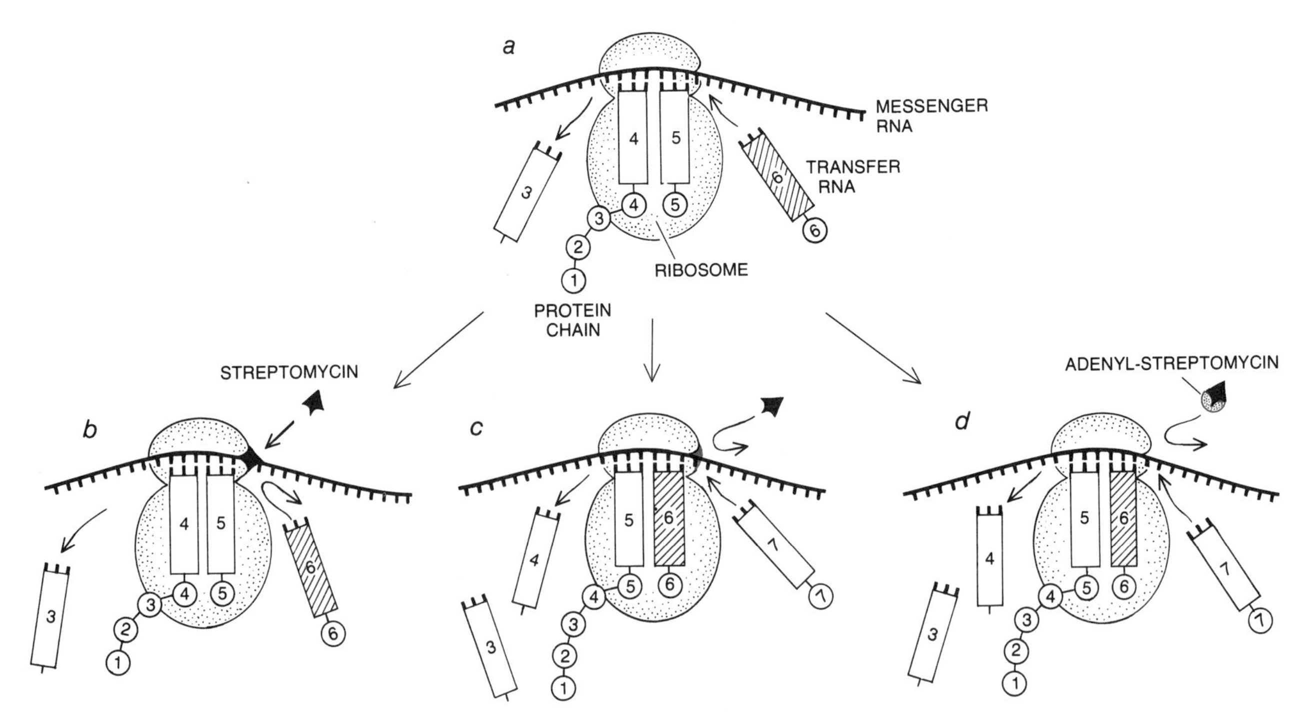

MODE OF ACTION is different in chromosomal and in plasmid-mediated antibiotic resistance. Streptomycin inhibits protein synthesis in bacteria. Ordinarily transfer RNA assembles amino acids on ribosomes according to instructions recorded in messenger RNA (*a*). Streptomycin attaches itself to a site on the ribosome and prevents protein synthesis, perhaps by blocking transfer-RNA attachment (*b*). Mutation of a chromosomal gene changes the ribosome's structure so that the streptomycin cannot bind (*c*), but *R*-factor genes make an enzyme that attaches an adenine molecule to the streptomycin, which no longer fits the unchanged ribosome (*d*).

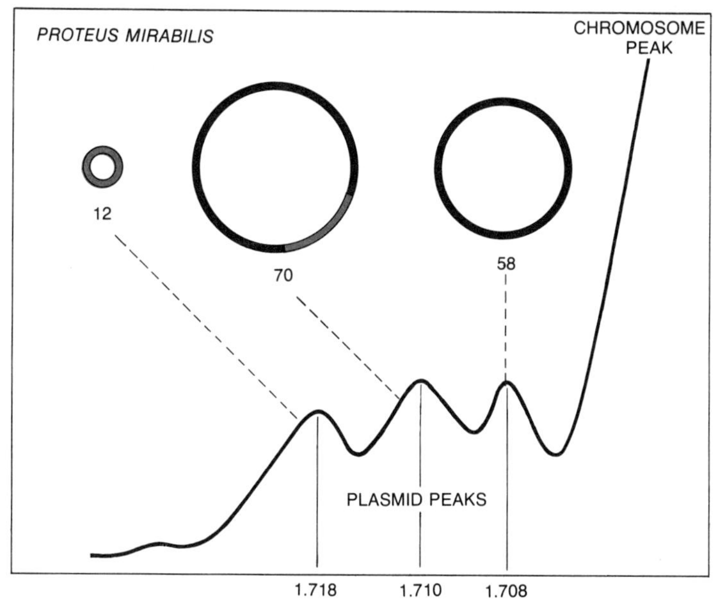

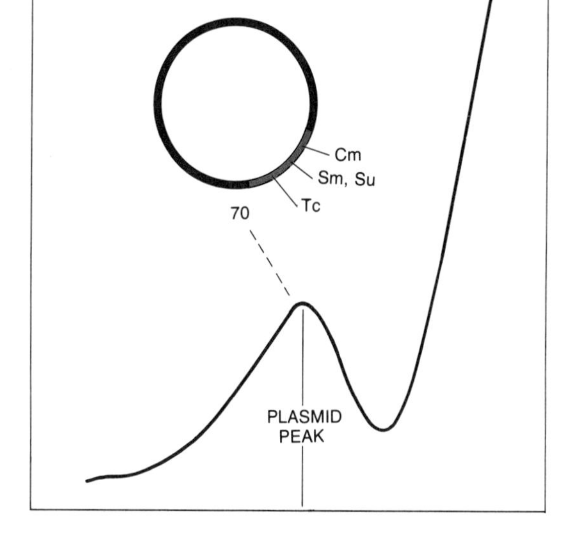

PLASMIDS representing the 222/*R* factor in *Proteus mirabilis* and *Escherichia coli* are compared after centrifugation and fractionation, with ethidium bromide present in the case of *E. coli*. In *Proteus* there are three peaks in the satellite band. In each peak a different plasmid, measuring respectively 12, 70 and 58 megadaltons, is found. In *Escherichia*, on the other hand, there is only one plasmid peak, containing 70-megadalton molecules. The genes for resistance to four antibiotics are on the 12-megadalton plasmid, those for the transfer factor on the 58-megadalton plasmid. The two plasmids are integrated in the 70-megadalton plasmid cointegrate.

that these antibiotic-sensitive cells contain a plasmid (called delta) that controls only transfer-factor properties.

In my laboratory I undertook a biophysical study of the DNA molecules isolated from these cultures in collaboration with Christine Milliken. This study has fully confirmed the conclusion, reached by Anderson and Lewis from their genetic studies, that the original strain harbors several independent plasmids. The delta plasmid controls transfer-factor properties, a second plasmid (called *S*) controls noninfectious streptomycin resistance and a third (called *A*) controls noninfectious ampicillin resistance. The delta plasmid was found to be joined with a tetracycline factor (*T*) to form a composite delta-*T* plasmid controlling infectious tetracycline resistance. When *A* or *S* is present in the same cell as a transfer factor (delta or delta-*T*), it too can be transferred to recipient cells.

Satellite DNA separated from *E. coli* host cells by centrifugation showed the following features: DNA from delta^{+} cells (cells with transfer-factor properties but no drug resistance) contained a single species of duplex looped DNA molecules of 60 megadaltons; DNA from *S* or *A* cells (respectively carrying noninfectious streptomycin or ampicillin resistance) was also a single species but was a tenth as large (six megadaltons); DNA from delta (*S*) or delta (*A*) cells (respectively carrying infectious streptomycin or ampicillin resistance) was found to contain two distinct species of molecules, 60 megadaltons and six megadaltons in size.

It appears, then, that an *R* factor can be composed of an aggregate of two or more plasmid molecules. That is, delta (*S*) has the properties of an *R* factor, since it can transfer infectious streptomycin resistance, but those properties are due to two independent molecules, one (delta) controlling transfer-factor properties and the other (*S*) controlling noninfectious streptomycin resistance. As a result of conjugation brought about by the presence of the delta factor, delta itself is transferred and the probability of simultaneous transfer of *S* is very high. The majority of recipient cells therefore inherit both delta and *S* molecules and show the characteristics of infectious streptomycin resistance. Occasionally some of the recipient cells acquire only one or the other of the two molecules, and then they acquire either transfer-factor properties or noninfectious streptomycin resistance. Similar properties are proposed for delta (*A*) cells.

In distinction to delta and delta-*T*, which like all other *R* plasmids were found in a one-to-one molecular ratio with the chromosome, there were more than 10 copies of either the *A*- or the *S*-plasmid for each chromosome. This new type of *R* factor, delta (*S*) or delta (*A*), is termed a "plasmid aggregate." Its evolution is difficult to explain by gene pickup and is clearly more consistent with extrachromosomal evolution. The plasmid aggregate may be a more primitive evolutionary type that can evolve, through a joining of the separate molecules, into a plasmid cointegrate such as delta-*T*.

E. coli is a normal inhabitant of the human alimentary tract; it seems unlikely that during its evolutionary history in the preantibiotic era such an organism would ever have come into contact with an antibiotic such as streptomycin, which is produced by a very different genus of microorganism usually found only in soil. How then can the bacterium produce an enzyme that recognizes streptomycin and inactivates it? (The evolution of chromosomal, as opposed to *R*-factor, resistance is easier to explain, since chromosomal resistance stems from a change in the structure of a normal bacterial product, the ribosome, through mutation.)

William Shaw of the University of Miami has pointed out that many *R*-factor resistances depend on the inactivation of antibiotics through such enzymic processes as acetylation, phosphorylation or adenylation, processes that normal metabolic compounds undergo in the normal bacterial cell. Shaw suggests that enzymes coded for by *R*-factor genes and operating to inactivate antibiotics may have evolved from enzymes that originally acted on normal bacterial molecules. He has found that resistance to chloramphenicol by a number of plasmids of different origin is due in all cases to its acetylation by enzymes that are similar in size and constitution and appear to differ only in the efficiency with which they can bind to chloramphenicol. He suggests that these enzymes may have evolved from one that originally acetylated a normal cell component. Mutations of the gene determining this enzyme might have modified the enzyme so that it bound to chloramphenicol with increasing efficiency, until now it binds to chloramphenicol rather than to the original molecule and can inactivate it effectively enough to prevent its antibiotic activity.

The rapid proliferation of *R* factors and the identification of *R*-plasmid aggregates suggest that similar entities could have importance in the normal regulatory control of bacteria. Genes on the bacterial chromosome determine the activities that enable the bacterial cell to grow and multiply in many very different environments. Under most circumstances most of these genes are not active; they are "switched off" except in the particular environment in which their function is needed [see "The Control of Biochemical Reactions," by Jean-Pierre Changeux; SCIENTIFIC AMERICAN Offprint 1008]. Now, if a certain function determined by several sequentially acting enzymes is required very infrequently, the continual replication of the genes that determine it may impose more of an evolutionary burden than would be compensated for by the rare need for the function. If, however, these determinant genes were carried on a plasmid present in only a few cells of a population, they would need to replicate only in those rare cells. Any time there arose a need for the function in question, these special genes could be acquired rapidly by all the cells in a progeny population if the plasmid also had transfer-factor properties or (even more economically) if other rare cells contained transfer factors that could interact to establish a plasmid aggregate.

The control of a number of activities in this way by their distribution to a number of different plasmids carried by different cells in the population would enormously increase the total gene pool and extend the overall metabolic potential of the species. This would appear to represent a logical extension of genetic

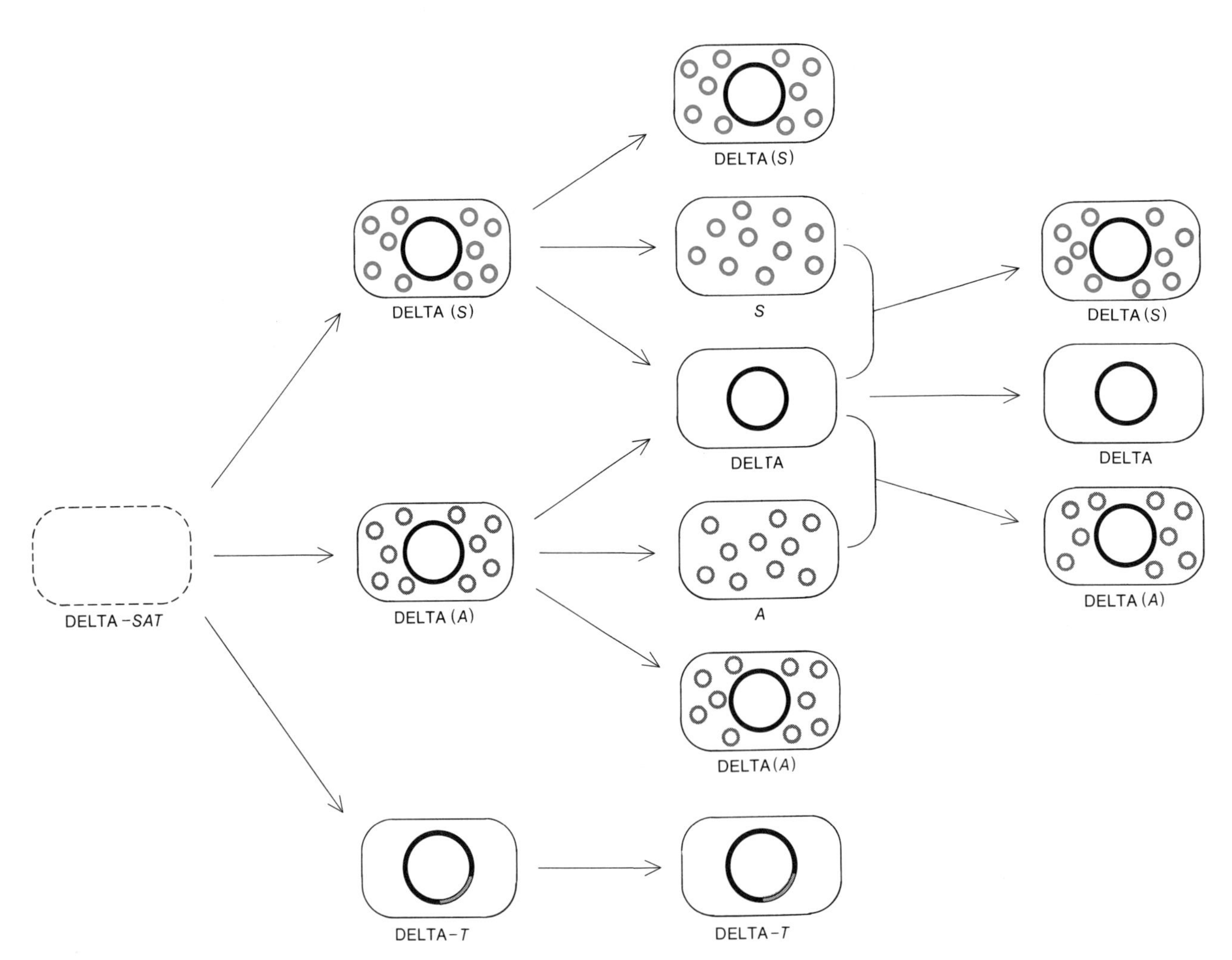

PLASMID AGGREGATE is an *R* factor composed of several independent plasmids. Here the original donor is resistant to streptomycin (*S*), ampicillin (*A*) and tetracycline (*T*) and has a transfer factor (delta). On conjugating it gives rise to three kinds of new donor, each carrying infectious resistance to one antibiotic. When these donor cells conjugate, the strains that result either have the same characteristics as the donor or have noninfectious resistance (either *S* or *A*) or have only the transfer factor (delta); tetracycline resistance, however, is always infectious. These properties have now been explained by the presence of four distinct plasmids: delta, measuring 60 megadaltons; *S* and *A*, each of six megadaltons, and delta-*T*, a plasmid cointegrate of 67 megadaltons.

regulation: from the present concept of individual cellular control by the switching on and switching off of chromosomal genes to a species-wide control of gene-pool synthesis.

As a matter of fact, entities that are similar to *R* factors but that control enzymes involved in metabolism have recently been recognized. For example, genes for enzymes catalyzing the breakdown of complex organic molecules to enable them to be used as a source of carbon have recently been shown to be situated on plasmids in *Pseudomonas,* a genus of bacteria well known for its ability to grow on almost any kind of organic substance. Certain strains of *Pseudomonas* are enabled to grow on camphor by a number of enzymes determined by one plasmid, and a plasmid controlling a sequence of genes that allow another strain to grow on salicylic acid has been identified. Since *R* factors have recently been found in *Pseudomonas* strains (which are also being seen more often in postoperative infections), it would not be too surprising if much of the nutritional versatility of *Pseudomonas* turns out to depend on a series of plasmids, each controlling enzymes that make it possible for the organism to grow on unusual organic compounds.

Plasmids, which are being found in an increasing number of bacterial species, share many properties with bacterial viruses, particularly the "temperate" type of bacterial virus, which in the latent state establishes a stable symbiotic relation with its host bacterium. Bacterial plasmids can be thought of as symbionts without even the limited lethal effects of temperate viruses, whose spread depends on the eventual destruction of their host cells. Since it has been recognized that certain components of plant and animal cells, such as chloroplasts and mitochondria, also rely for their maintenance partly on extrachromosomal DNA, the possibility exists that plasmids not only may be common in many bacterial species and be concerned with activities other than antibiotic resistance but also may be closely related to other extrachromosomal elements in all cells.

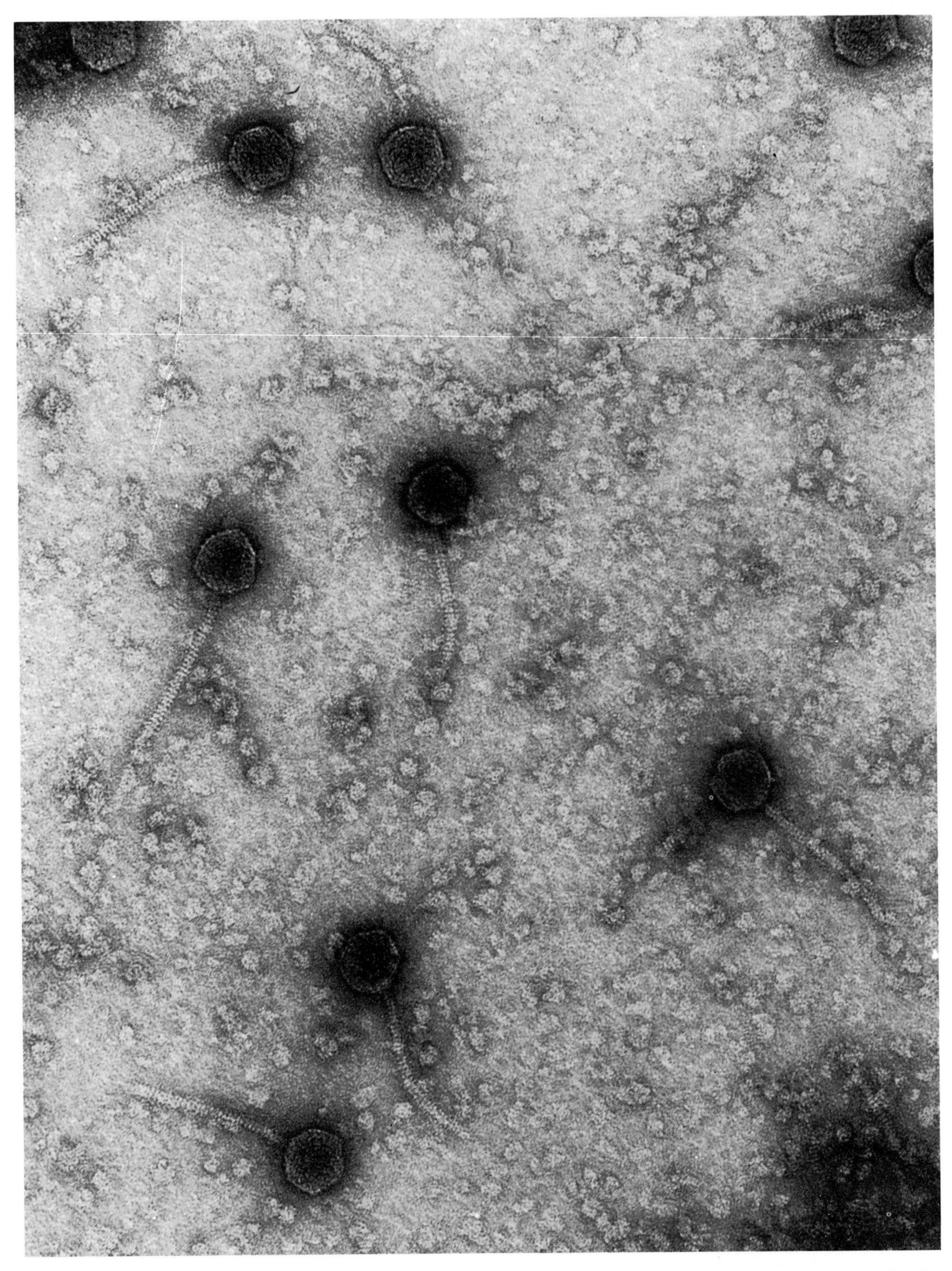

BACTERIAL VIRUSES, or phages, of the strain designated lambda are magnified 200,000 diameters in this electron micrograph made by William C. Earnshaw and Philip A. Youdarian of the Massachusetts Institute of Technology. Each virus particle has an icosahedral head containing a molecule of infectious viral DNA, and a flexible hollow tail with which the phage attaches itself to the outer membrane of its bacterial-cell host and injects its DNA into the interior of the cell. Negative staining with uranyl acetate clearly reveals the subunits that make up the virus's protein shell. The phage-lambda particles shown here were extracted from an infected strain of the colon bacillus *Escherichia coli*, together with a large number of bacterial ribosomes (*lightly stained particles scattered over background*).

How Viruses Insert Their DNA into the DNA of the Host Cell

8

by Allan M. Campbell
December 1976

Some viruses are able to coexist peaceably with their host cell for long periods, incorporating their genes into the host chromosome. The details of the insertion process are now fairly well understood

The imagery of medicine often seems to portray viruses as aggressive organisms bent on human destruction. In actuality their role in disease is merely a by-product of their parasitic existence. Unable to reproduce by themselves, viruses must invade living cells and redirect some of the cellular machinery to the production of new virus particles. In the process many of the host cells are destroyed, leaving the tissue of which they are a part damaged and giving rise to disease.

Although virus production and cell destruction are the most dramatic outcomes of virus infection, they are by no means the only outcomes possible. Like other parasites, many viruses find it advantageous to persist innocuously within their host cell for an indefinite period, actively multiplying only when the host weakens or stops growing. The existence of such latent viruses was first suspected in the 1920's, but it was not until the early 1950's that André Lwoff and his colleagues at the Pasteur Institute established unequivocally that a latent virus, or "provirus," can be transmitted from one cell generation to the next without external reinfection. Working with bacteriophages, the viruses that infect bacteria, they showed that the provirus exists in a "lysogenic" state. The term lysogenic refers to the fact that the provirus can come out of the dormant state and give rise to mature virus particles that lyse, or dissolve, the bacterial cells. Among the agents that can induce the provirus to resume its former mode of multiplication are ultraviolet radiation, X rays and carcinogenic chemical compounds. In the intervening 20 years the further study of bacteriophages has revealed much about the mechanism of lysogeny.

A typical bacteriophage consists of a single linear molecule of nucleic acid enclosed in a protein coat. Resembling a minute hypodermic syringe, the phage attaches itself to a bacterial cell and injects its strand of DNA into the interior. Once inside the cell, the viral DNA may begin directing the manufacture of new virus particles or become a provirus by incorporating itself into the DNA of the bacterial cell's long, threadlike chromosome.

What then switches the provirus from the lysogenic state to the active production of virus particles? A mechanism was put forward by François Jacob and Jacques Monod of the Pasteur Institute in 1961 and later demonstrated by Mark Ptashne of Harvard University. In the provirus only a few genes are expressed, and the product of one of them is a "repressor" protein that combines with the viral DNA and prevents the expression of the other viral genes, particularly those responsible for the independent replication of the viral chromosome. Inducing agents such as X rays alter the metabolism of the bacterium in such a way that a substance is produced that inactivates the repressor. If all the repressor molecules in a given lysogenic bacterium are simultaneously inactivated, the viral genes are sequentially expressed. Proteins needed for the replication of the viral DNA are made first, followed by the head and tail proteins of the virus particle, which spontaneously assemble inside the bacterium. Unit segments of viral DNA are then packaged into the phage heads. Finally, some 60 minutes after the cell was exposed to the inducing agent, the cell bursts, releasing about 100 virus particles that are capable of infecting other cells.

When a single lysogenic bacterium multiplies to form a colony, every cell of that colony is potentially capable of manufacturing virus particles and can express that potentiality when its repressor is destroyed. The viral genes have therefore been added to the bacterial genes in such a way that both sets of genetic information are inherited by the cell's descendants. In order for this to happen the viral chromosome must replicate like any normal cellular component, and it must be partitioned at cell division so that each daughter cell receives at least one copy of it.

Such a result can be achieved in at least three possible ways. One is for the viral chromosome to insert itself into the host chromosome, after which it can be passively replicated and distributed at cell division as part of the host DNA. Another way is for the viral chromosome to establish itself independently; it replicates and is distributed like the host chromosome but is separate from it. A third way is for the viral chromosome to replicate separately from the host chromosome and in multiple copies. In that case the viral chromosome might not be distributed regularly when the bacterial cell divides, but if the number of copies is large enough, the chance of a daughter cell's receiving no copies whatever is small. If we survey all known viruses, we find that the first two mechanisms exist in nature, and the third is observed in certain mutant viruses produced in the laboratory. One property all three mechanisms have in common is that once the virus has become established as a hereditary component of the cell, there is nothing particularly novel in its mode of inheritance from then on.

Bacteriophage lambda, which was discovered by Esther M. Lederberg of Stanford University as a provirus carried by the K12 strain of the colon bacillus *Escherichia coli,* is the best-understood genetically of the bacteriophages and remains one of the favorite experimental organisms of molecular biologists. Soon after Lwoff's demonstration of heritable lysogeny several investigators were able to achieve genetic crosses between strains of *E. coli* that were lysogenic for phage lambda and strains that were nonlysogenic for it. They got the surprising result that the characteristic of being lysogenic or nonlysogenic was distributed among the progeny of the cross exactly as one would expect if the trait were being determined by a gene at a specific location on the bacterial chromosome. In fact, lambda behaved in crosses as if it was closely linked to the *gal* operon, a cluster of bacterial genes concerned with the metabolism of the sugar galactose.

There is now general agreement that the lambda provirus is inserted into the *E. coli* DNA at a specific site between the *gal* operon and the *bio* operon, which is responsible for the synthesis of the vitamin biotin. The significance of the insertion process is not confined to virology. It shows one way in which two genetic units, each capable of existing and replicating independently, can coalesce into

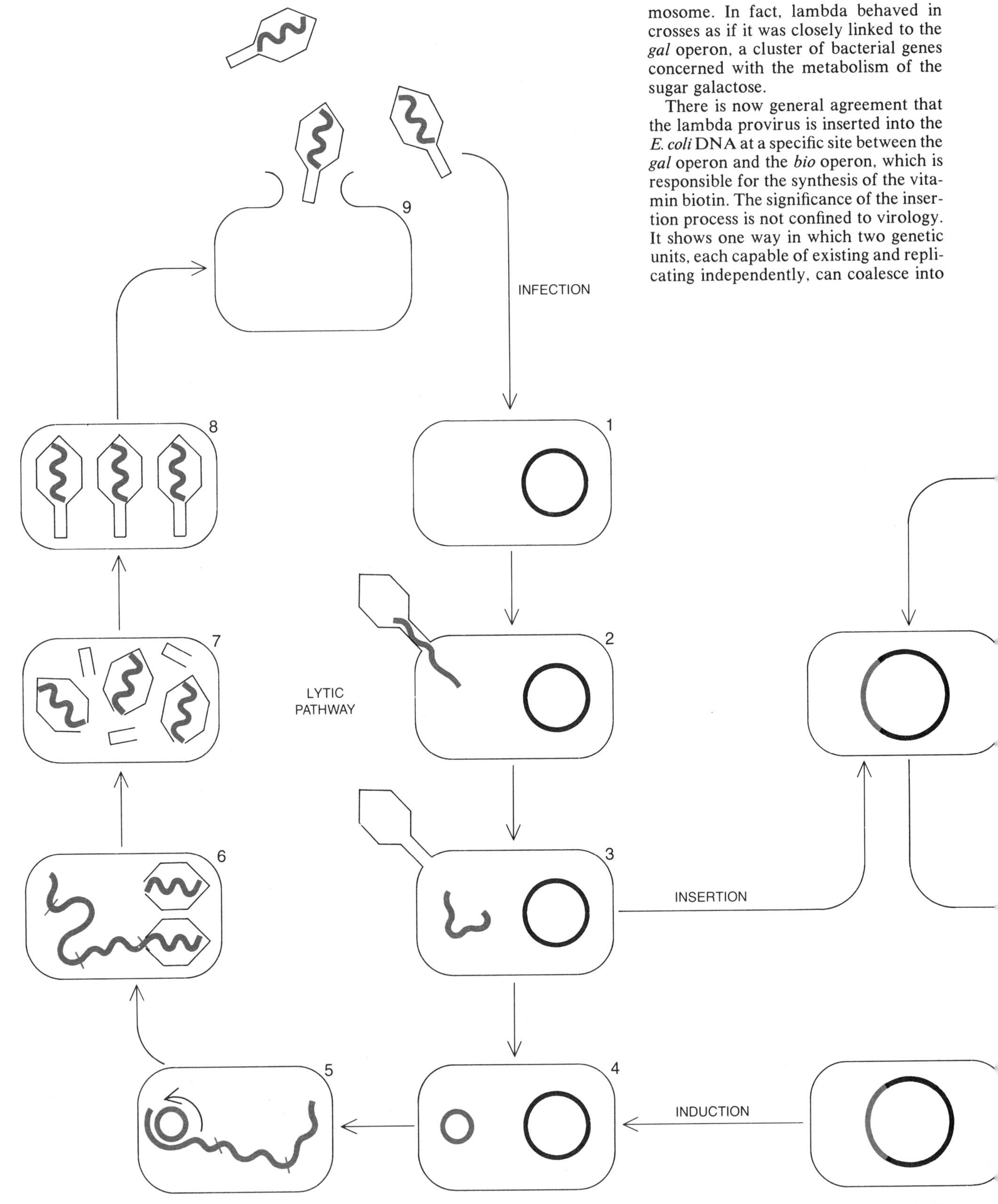

LIFE CYCLE OF PHAGE LAMBDA shows that lysis (dissolution) and death of the infected bacterial cell are not inevitable. After a molecule of viral DNA is injected into a healthy *E. coli* cell (*2*) the host may take either of two paths. In the productive, or lytic, pathway the viral DNA forms a circle (*3 and 4*) and replicates by the "rolling circle" process to give rise to a long "sausage string" of DNA containing multiple copies of the viral genes (*5*). Next the viral genes direct the synthesis and assembly of the head and tail proteins of the virus particles and the packaging of unit DNA segments into the heads (*6 and 7*). Heads and tails then assemble spontaneously inside bacterium, giving rise to mature phage particles (*8*). Finally, some 60 minutes after infection, the host cell bursts, releasing about 100 progeny virus particles that can then infect other *E. coli* cells (*9 and 1*). Alternatively, depending on the conditions of infection, the phage-lambda DNA can live quietly within the host, establishing itself as a semipermanent part of the bacterial chromosome (*10*). It is then replicated and segregated at cell division (*11 and 12*) and passed on to succeeding cell generations for an indefinite period. The latent virus,

a larger unit. Phage lambda is only one of a number of elements for which insertion is known or suspected. Most of these elements are viruses or the small pieces of bacterial DNA called plasmids, some of which can transfer resistance to antibiotics and other properties from one bacterial cell to another. In addition there are the "insertion sequences" of bacteria and the "transposable elements" of maize, which are known only for their ability to move occasionally from one chromosomal location to another and may well have no independent existence.

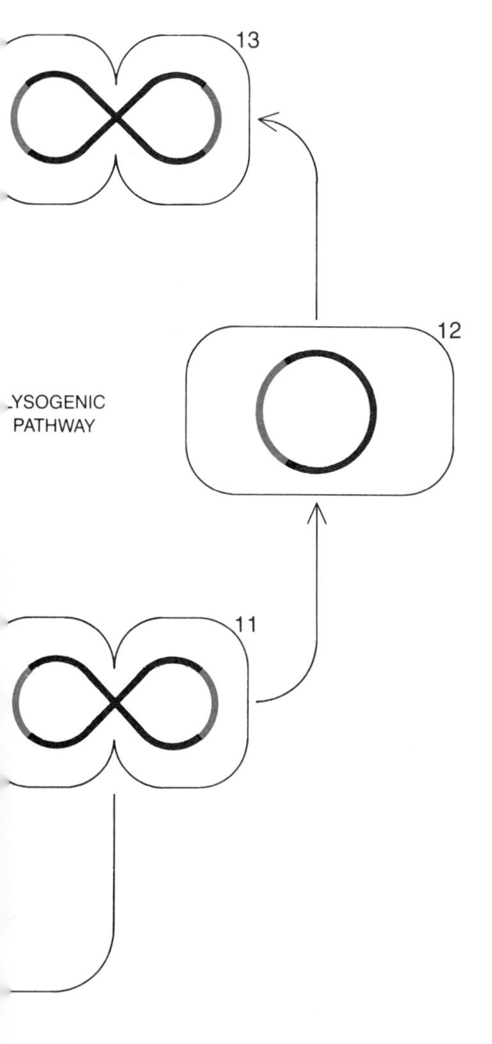

or "provirus," still retains the potential to grow lytically, however. Bacteria harboring a provirus are called lysogenic, meaning that they carry a property that can lead to viral multiplication and the fatal lysis of the host cell. Exposure of lysogenic bacterium to ultraviolet radiation, X rays or chemicals such as nitrogen mustard and organic peroxides can induce the provirus to return to lytic pathway.

Although the widespread occurrence of genetic elements that can insert themselves into preexisting chromosomes is incontestable, there is a considerable divergence of opinion about their importance in the normal life of the host or in evolution. One extreme viewpoint is that these elements are basically foreign to the cells that harbor them; they are invaders like viruses that have somehow got into cells and chromosomes and do not really belong there. The alternative view is that much of the DNA of present-day chromosomes is derived from elements that were originally foreign but that through a series of small evolutionary steps gradually became naturalized citizens of the intracellular community. A somewhat intermediate position is that the mobility of these elements plays some essential role in the normal development of multicellular organisms from the fertilized egg into the adult.

Studies of the mechanism by which new genetic elements are added to existing chromosomes cannot provide any direct information about the origin or function of such elements. The studies are, however, relevant in one respect. If it turned out that the DNA of the added elements was hooked onto the rest of the chromosome in some unusual way, by connections that were not normal features of chromosome structure, then a clear distinction between foreign DNA and indigenous DNA would be implied. It therefore seemed particularly important to firmly establish whether or not the foreign DNA of some prototypical examples such as the lambda provirus is directly inserted into the linear sequence of genes on the host cell's chromosome.

The concept of genes' being arranged in a linear sequence, and the experimental basis of that concept, have a long history. In 1913 A. H. Sturtevant of Columbia University crossed fruit flies differing from each other by several genetic traits and analyzed the frequencies of progeny exhibiting new combinations of the parental traits. He concluded that the determinants for these traits (all of which happened to be on the same chromosome) distributed themselves as though they were on some one-dimensional structure, with new combinations arising from the redistribution of connected segments of that structure. It became possible to extend this kind of formal linkage analysis in the 1950's, when Seymour Benzer of Purdue University, working with small segments of phage chromosome, proved that the linkage maps of mutational sites within genes were also one-dimensional. These purely genetic studies were followed by cytological and biochemical work showing that the one-dimensionality of linkage maps was associated with the linear arrangement of the genes along the chromosome and ultimately with the linear sequence of nucleotides along the double helix of DNA.

The question of whether all the linear DNA segments in a chromosome are joined end to end in one continuous double helix is still not completely settled, but the evidence is increasingly strong that they are. The uncertainty surrounding the question through the mid-1960's was such, however, that many bacterial geneticists found nothing particularly bizarre in the notion that the provirus might lie alongside the host chromosome rather than being inserted into it.

One way to investigate the question is to cross two lines of lysogenic bacteria, in which both the host genes and the viral genes are marked with mutations. In such bacterial crosses pieces of DNA from the donor cells are introduced into the recipient cells, where they pair with corresponding segments of the recipient cells' DNA and, in some fraction of those cells, replace them. Progeny in which a specific gene has been replaced can be recognized and selected for because they express a genetic trait characteristic of the donor, such as the ability to metabolize galactose. The recognition of the replacement of specific genes in the provirus has been made easier in recent years by the isolation of conditionally lethal mutant viruses, which fail to multiply in some condition under the control of the experimenter but which grow normally when that condition is changed. For example, temperature-sensitive mutants survive and multiply at one temperature (25 degrees Celsius) but not at another (42 degrees C.).

Such experiments throw some light on the provirus's mode of attachment. If, for example, the provirus is not linearly inserted into the host chromosome but instead projects from it sideways, one would expect that an exchange of genetic material occurring along the viral segment would not redistribute the genes lying on the main axis of the host chromosome. The fact is that redistribution of the host genes does occur, suggesting that insertion is indeed linear.

A second and more informative method of determining the topology of the lysogenic chromosome is deletion mapping, in which mutants marked by chromosomal deletions are crossed with strains that have other genetic markers. A deletion mutation involves the permanent loss from the chromosome of a string of neighboring nucleotides, numbering from one to many thousands. Since each deletion eliminates a continuous segment of chromosome, the order of the genes along the chromosome can be deduced by piecing together the information provided by the characteristics of the various deletion mutants observed. Deletion mapping provides a strict "betweenness" criterion for locating genes with respect to one another. The condition that every observed dele-

INSERTED ELEMENT	APPROXIMATE LENGTH IN NUCLEOTIDES	INSERTED INTO	SPECIAL PROPERTIES
BACTERIOPHAGE LAMBDA	50,000	*E. COLI* CHROMOSOME	SPECIFIC SITES ON VIRUS AND HOST CHROMOSOMES
BACTERIOPHAGE MU-1	37,000	*E. COLI* CHROMOSOME	SPECIFIC SITE ON VIRUS, ANY SITE ON HOST CHROMOSOME
BACTERIAL SEX FACTOR *F*	100,000	*E. COLI* CHROMOSOME	MANY SITES ON HOST CHROMOSOME
DRUG-RESISTANCE PLASMID	20,000	BACTERIAL FACTOR RESEMBLING *F*	TRANSMITS RESISTANCE TO ANTIBIOTICS BETWEEN BACTERIAL STRAINS
TUMOR VIRUS SV-40	5,000	HUMAN CHROMOSOME	DERIVED ORIGINALLY FROM MONKEY CELLS
TRANSPOSABLE ELEMENTS	(UNKNOWN)	MAIZE CHROMOSOMES	NO DETECTABLE EXTRA-CHROMOSOMAL PHASE
INSERTION SEQUENCE IS2	1,400	*E. COLI* CHROMOSOME	

ADDED GENETIC ELEMENTS, small pieces of DNA that can exist as part of the main chromosome or independently, have been observed in bacterial, maize and human cells. Some are viruses; others are not. When they are inserted, they introduce into the cell instructions governing additional biochemical reactions that may be superimposed on the cell's metabolism.

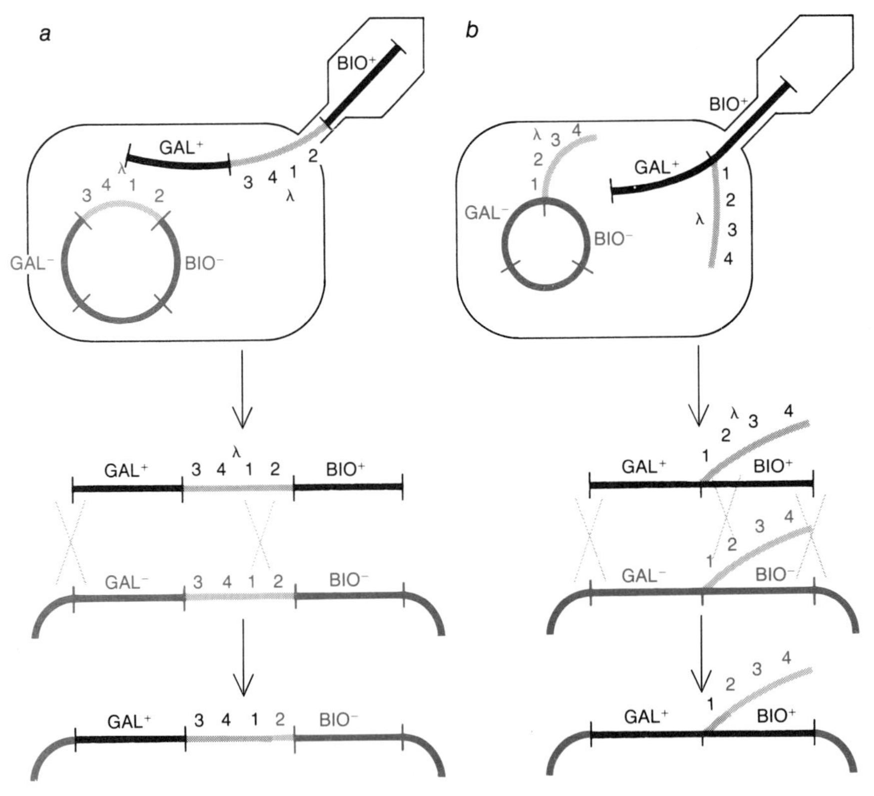

TYPES OF GENETIC EXCHANGE, or recombination, observed between two lysogenic *E. coli* chromosomes containing different genetic markers shed light on the mode of attachment of the phage-lambda genes. In the experiment depicted here a fragment of DNA from a bacterial cell capable of metabolizing galactose and synthesizing biotin ($gal^+ bio^+$) and containing a lambda provirus with mutations in genes *1, 2, 3* and *4* is introduced by means of an infective phage coat into a recipient bacterium that is unable to utilize galactose or synthesize biotin because of genetic mutations ($gal^- bio^-$) and that harbors a nonmutant provirus. Type of recombination that might occur if the viral genes were linearly inserted into each chromosome is shown in *a*. The mechanism of insertion proposed by the author requires the permutation of the order of genes along the viral chromosome, so that a genetic exchange between viral genes *1* and *2* would serve to recombine the flanking bacterial markers *gal* and *bio*. The resulting recombinant bacterium would be capable of utilizing galactose but incapable of synthesizing biotin, a prediction that agrees with observed results. For comparison, *b* indicates the expectation for one kind of nonlinear topology, in which the provirus joins to the chromosome as a branch.

tion should be representable as a linear segment is a highly restrictive one; a segment represented by two deletion markers must necessarily include all points that lie between them. For example, if the provirus is inserted between two identified host genes, then every deletion removing both of those genes must remove the provirus as well. This turns out to be the case in experiments, and it is again consistent with the model that the provirus is continuous with the host DNA.

Until the beginning of the 1970's genetic analysis of the type I have been describing provided the only precise information on the relation between the provirus and the chromosome; no direct physical information was available on the relevant nucleotide sequences within the DNA of the provirus. In principle the simplest approach would be to use direct methods of determining the sequence of nucleotides along the DNA chains of the virus and the lysogenic chromosome. Although such sequencing methods are improving rapidly, the identification of all 50,000 nucleotides in the lambda provirus would be a time-consuming and costly task. For many purposes adequate information can be obtained by exploiting the fact that single DNA chains with complementary nucleotide sequences can find each other and form double helixes in the test tube. Electron micrographs of DNA molecules formed this way, notably in the laboratory of Norman R. Davidson at the California Institute of Technology and that of Waclaw T. Szybalski of the University of Wisconsin, have demonstrated that the structures inferred from genetic results have a physical reality.

For these experiments double-strand DNA molecules extracted from virus particles are dissociated into single chains by heating. If the solution is then cooled slowly, double helixes of the complementary chains will re-form. When single-strand DNA's from two viruses with some nucleotide sequences in common (such as phage lambda and a deletion mutant of it) are mixed before the cooling step, new helixes can form not only between complementary chains from one virus but also between complementary chains from both viruses; the latter kind of chain is known as a heteroduplex. The nucleotides in the complementary segments of these hybrid chains pair up and form double helixes but the noncomplementary sequences do not, leaving single-strand loops that can be seen in electron micrographs.

By some ingenious manipulations Davidson and his colleagues Phillip A. Sharp and Ming-Ta Hsu were able to examine heteroduplexes between viral

DNA and the DNA of an inserted provirus. Although the most straightforward approach would be to make heteroduplexes from one DNA chain of phage lambda and the complementary chain of a chromosome from a lysogenic bacterium, that experiment is not yet feasible because of the difficulties in handling a DNA molecule the size of the bacterial chromosome. The same end was achieved by letting the phage-lambda DNA insert itself not into the entire bacterial chromosome but into a smaller DNA molecule: a derivative of the bacterial sex factor that had picked up from the bacterial chromosome the specific DNA segment into which lambda inserts. Strands of the sex factor with the lambda provirus inserted into them could be readily isolated intact and used to form heteroduplexes with DNA extracted from virus particles.

The combined results of genetic and physical studies make us quite confident

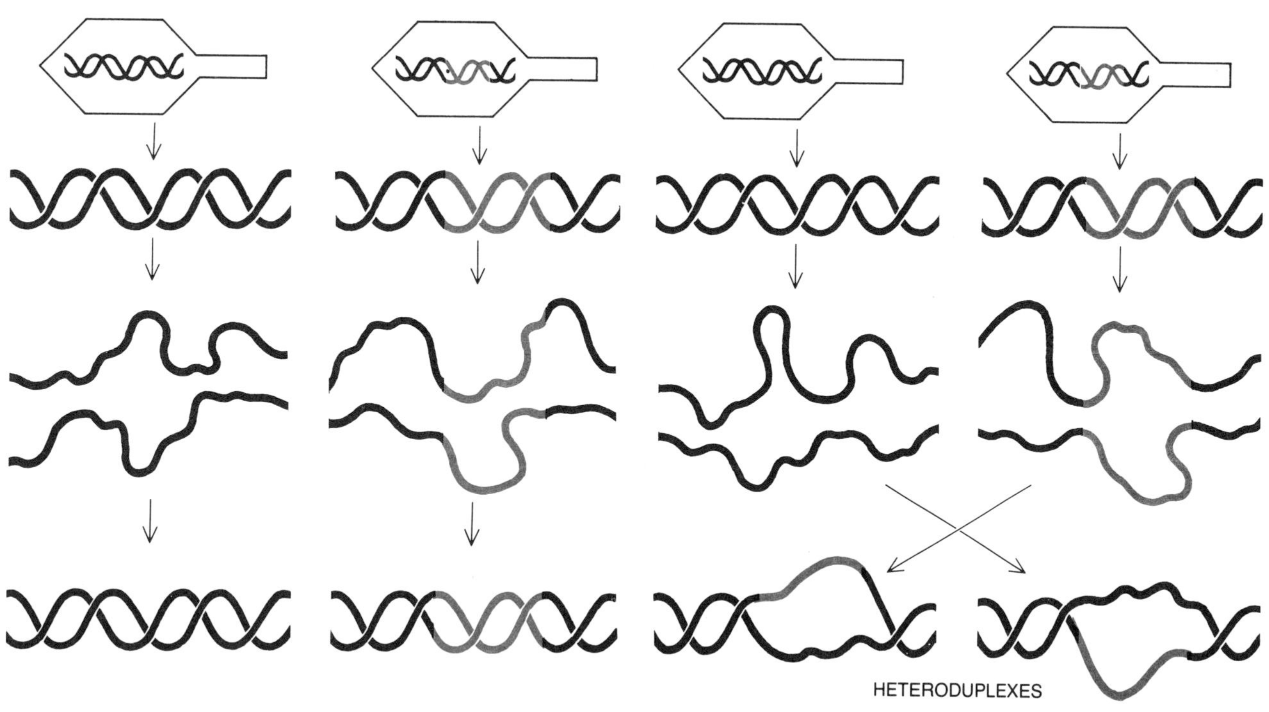

HYBRID DNA MOLECULES artificially formed in the test tube can be used to map viral genes physically. When the DNA double helix is heated, it unwinds, giving rise to two single-strand chains. If single chains having complementary and noncomplementary regions are mixed together at this stage and slowly cooled, some "duplexes" will form between the chains. When these heteroduplexes are viewed in the electron microscope, the two DNA chains will be double helixes where they have the same sequence of nucleotides and unpaired where they differ in sequence. With this method one can precisely map position of a given marker mutation along the DNA molecule.

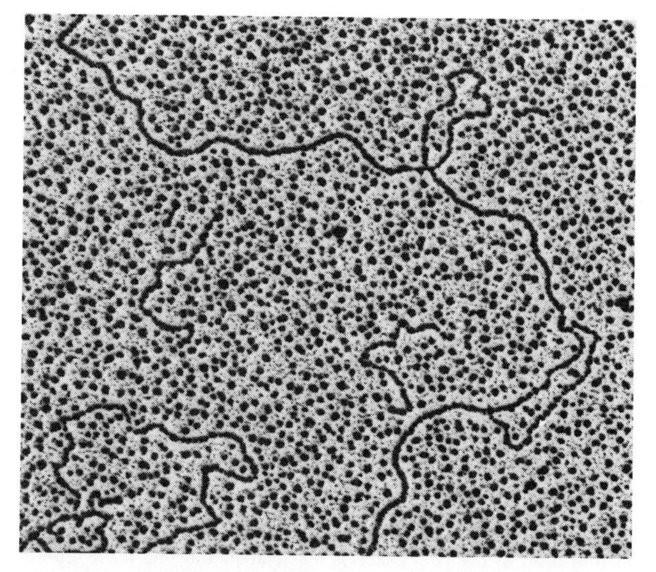

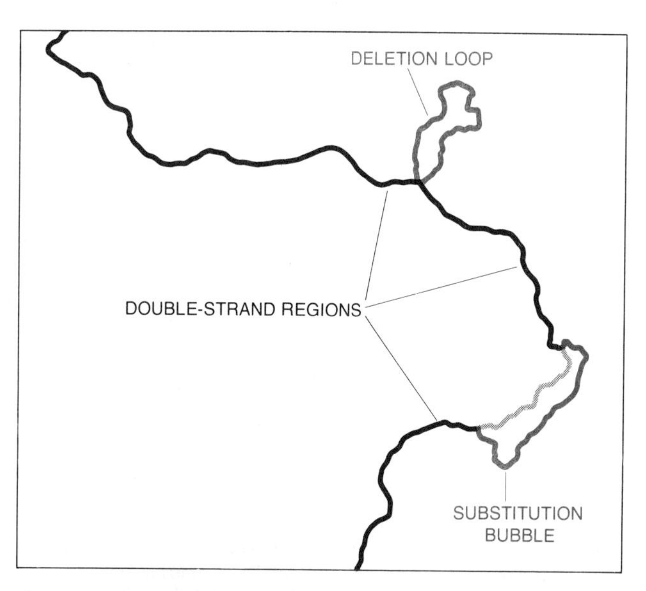

HETERODUPLEX between a strand of normal lambda DNA and a second strand incorporating two mutations is clearly visible in this electron micrograph made by Elizabeth A. Raleigh of M.I.T. (Only a small segment of the long viral DNA molecule is shown.) The loop of single-strand DNA at top right results from a deletion mutation that removed an entire segment of DNA from one strand. The "collapsed bubble" of single-strand DNA results from the substitution of several nucleotides in one strand by different ones, making that region of the two strands noncomplementary and hence unable to pair. The remaining portions of molecule shown here are double-helical.

that we now know the structure of the lysogenic chromosome. The steps by which this structure is formed and dissociated into its component parts are the subject of current research. During the life cycle of phage lambda, DNA must be cut and rejoined at the ends of the viral chromosome and at the ends of the provirus. The lambda DNA injected into the bacterial cell is in linear form, but before it is inserted into the bacterial chromosome its ends are joined so that it makes a circle. During insertion the circle is opened at a different point. As a result, although the provirus and virus chromosomes are both linear structures, the order of the genes along the two is not identical.

How the ends of the viral DNA are joined to form a circle is known, thanks largely to the work of A. D. Kaiser and

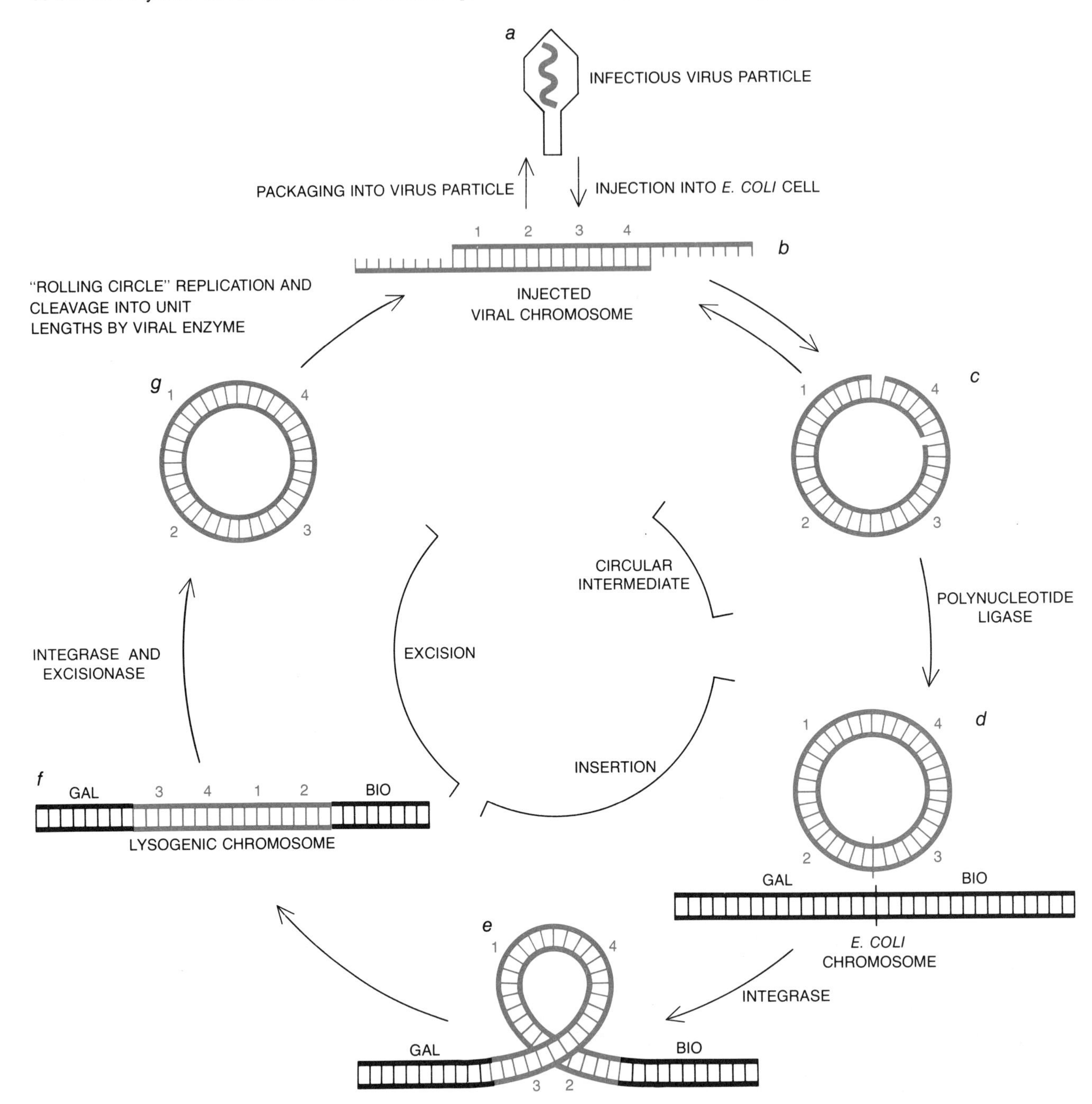

INSERTION AND EXCISION of the phage-lambda genes into the *E. coli* chromosome require the action of both bacterial and viral enzymes. As the DNA of phage lambda is packaged in the viral particle (*a*) it is linear and double-helical, except for complementary unpaired segments 12 nucleotides long at the ends of the two nucleotide chains (*b*). In solution this linear form comes to equilibrium with a circular form that has staggered "nicks" 12 nucleotides apart in the two complementary chains (*c*). When viral DNA is injected into the bacterial cell in the course of infection, the two nicks in the open circle are sealed by the bacterial enzyme polynucleotide ligase, so that both chains of the circle are now closed throughout their length (*d*). This circular intermediate then interacts with a particular segment of the *E. coli* chromosome (between the *gal* and *bio* genes). Viral and bacterial chromosomes break and rejoin at unique sites on each partner, so that viral DNA is spliced into the host DNA, a reaction catalyzed by the viral enzyme integrase (*e*). (Note that the gene order in the provirus is *3, 4, 1, 2,* a cyclic permutation of the viral gene order *1, 2, 3, 4.*) The *E. coli* chromosome is now lysogenic for phage lambda (*f*). After several cell generations radiation or chemically active compounds may induce the provirus to enter the lytic state. When this happens, the lambda repressor, which has so far blocked the expression of most of the viral genes, is inactivated, allowing the synthesis of the viral enzyme excisionase. Together with integrase, excisionase catalyzes the excision of the provirus from the host chromosome, converting it back into the circular form with the original gene order (*g*). The circle of viral DNA replicates, producing multiple copies that are then cleaved by a specific viral enzyme to give rise to the linear form with "sticky" ends (*b*). Each linear DNA segment is then packaged in a virus coat (*a*). When the host cell ruptures, the liberated phages infect healthy cells and the lysogenic cycle begins anew.

his collaborators at Stanford. Lambda DNA is a double helix throughout most of its length, but one end of each polynucleotide chain extends for 12 nucleotides beyond the double helix. These two single-strand chains are complementary to each other and are called "sticky ends." In solution the linear DNA molecules can come to equilibrium with circular molecules formed by the pairing of the two ends. When lambda infects an *E. coli* cell, the open circle formed by the viral DNA is closed by the action of polynucleotide ligase, a bacterial enzyme that seals breaks in one chain of a double helix. This step requires no viral enzymes, and it is not specific to the nucleotide sequences involved. On the other hand, the insertion of viral DNA into the bacterial chromosome requires the recognition and cutting of highly specific nucleotide sequences in both the lambda and the *E. coli* DNA.

Little is known of the biochemistry of insertion, although its genetic control has been intensively explored. At the time I proposed the circular-molecule-intermediate model for the insertion of phage-lambda DNA in 1962, the only known mechanism for breaking or rejoining two DNA molecules at corresponding points was homologous recombination, which requires that the two molecules have similar or identical base sequences in the recombining region. The chemical steps by which homologous recombination takes place are still largely conjectural, but Alvin J. Clark of the University of California at Berkeley and others have isolated bacterial mutants that are unable to carry out this process. Under conditions where homologous recombination is blocked by such mutations, however, phage lambda can still insert its DNA with the normal frequency. Hence the insertion of viral DNA seems to be accomplished not by the same bacterial enzymes that are responsible for homologous recombination but by viral enzymes that cut and join DNA molecules at highly specific sites.

Direct evidence for the existence of such viral enzymes has been provided by the genetic studies of James F. Zissler of the University of Minnesota Medical School and the biochemical investigations of Howard A. Nash of the National Institute of Mental Health, which have shown that the enzyme product of a specific viral gene (dubbed integrase) is required for the insertion of viral DNA; mutant viruses lacking this enzyme are unable to enter the lysogenic state. Similar studies of the reverse process—the excision of viral DNA from the bacterial chromosome—by Gabriel Guarneros and Harrison Echols of the University of California at Berkeley and Susan Gottesman of the National Cancer Institute—have shown that excision requires in addition to integrase the

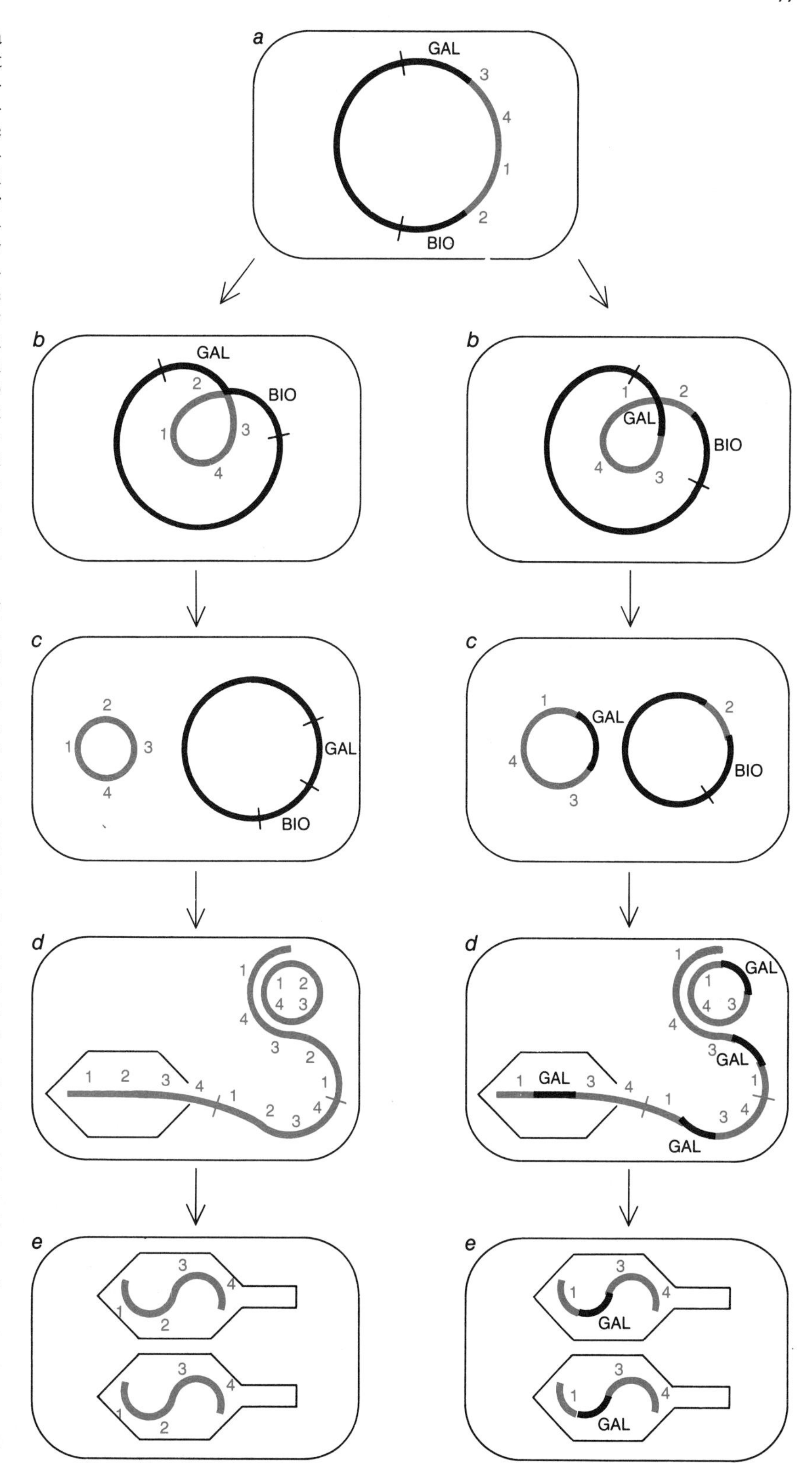

NOVEL VIRUS VARIANTS arise when certain proviruses turn into infectious viruses, killing the bacterium that harbors them and carrying away with them pieces of the DNA of the dead host. The "transduced" bacterial genes linked to the viral chromosome can subsequently replicate at the unrestricted viral pace. Transduction is believed to result from rare errors in the excision of the provirus from the host chromosome. Normal excision (*left panel*) takes place in the vast majority of cells. Viral DNA separates from the bacterial DNA (*b*) to give rise to a circular viral chromosome and a nonlysogenic bacterial chromosome (*c*). The viral chromosome replicates in several stages, ultimately as a rolling circle that generates a long sausage string of DNA in which the entire viral sequence is repeated many times (*d*). Unit DNA lengths are then packaged into infectious virus particles (*e*). In one cell out of 100,000 (*right panel*) abnormal excision generates a circular molecule including some host DNA. Infectious particles are thus formed in which a segment of bacterial DNA has replaced a segment of viral DNA.

product of a second viral gene (called excisionase). The virus thus introduces into the host cell enzymatic machinery for cutting and joining the viral and host DNA at specific sites to bring about the insertion and excision of the provirus. As long as the transcription of the gene coding for excisionase is blocked by the lambda repressor the provirus will remain inserted. When repression is released, the excisionase gene will be expressed and there will be a reciprocal exchange within the lysogenic chromosome, re-creating a circular molecule of lambda DNA and a nonlysogenic bacterial chromosome.

Excision is generally precise: more than 99 percent of the virus particles manufactured by lysogenic cells are identical with the original infecting virus. This fact implies that the DNA breaks at exactly the same point when it comes out of the chromosome as it does when it goes in. About one excision in 100,000, however, is abnormal. Instead of breaking away cleanly the host DNA and the viral DNA break and rejoin to create a circular DNA molecule incorporating some viral DNA and some host DNA. If the size and physical characteristics of the molecule allow it to be recognized by the viral proteins as being suitable for replication and packaging as a virus, it can then give rise to infectious particles in which a segment of host DNA has replaced part of the viral DNA.

M. Laurance Morse of the University of Colorado Medical Center first discovered the existence of these "transducing" virus variants when he found that some of the phage-lambda particles liberated from lysogenic bacteria contained the *gal* genes of the host. When lambda and lambda-*gal* DNA's were hybridized in the test tube, the heteroduplexes showed that nucleotides at the two ends of the molecules were complementary to one another but that in the middle of the duplex there was an unpaired region where the picked-up segment of bacterial DNA (including the *gal* operon) was not complementary to the viral DNA.

The theory that transducing phages are produced by errors in the excision of

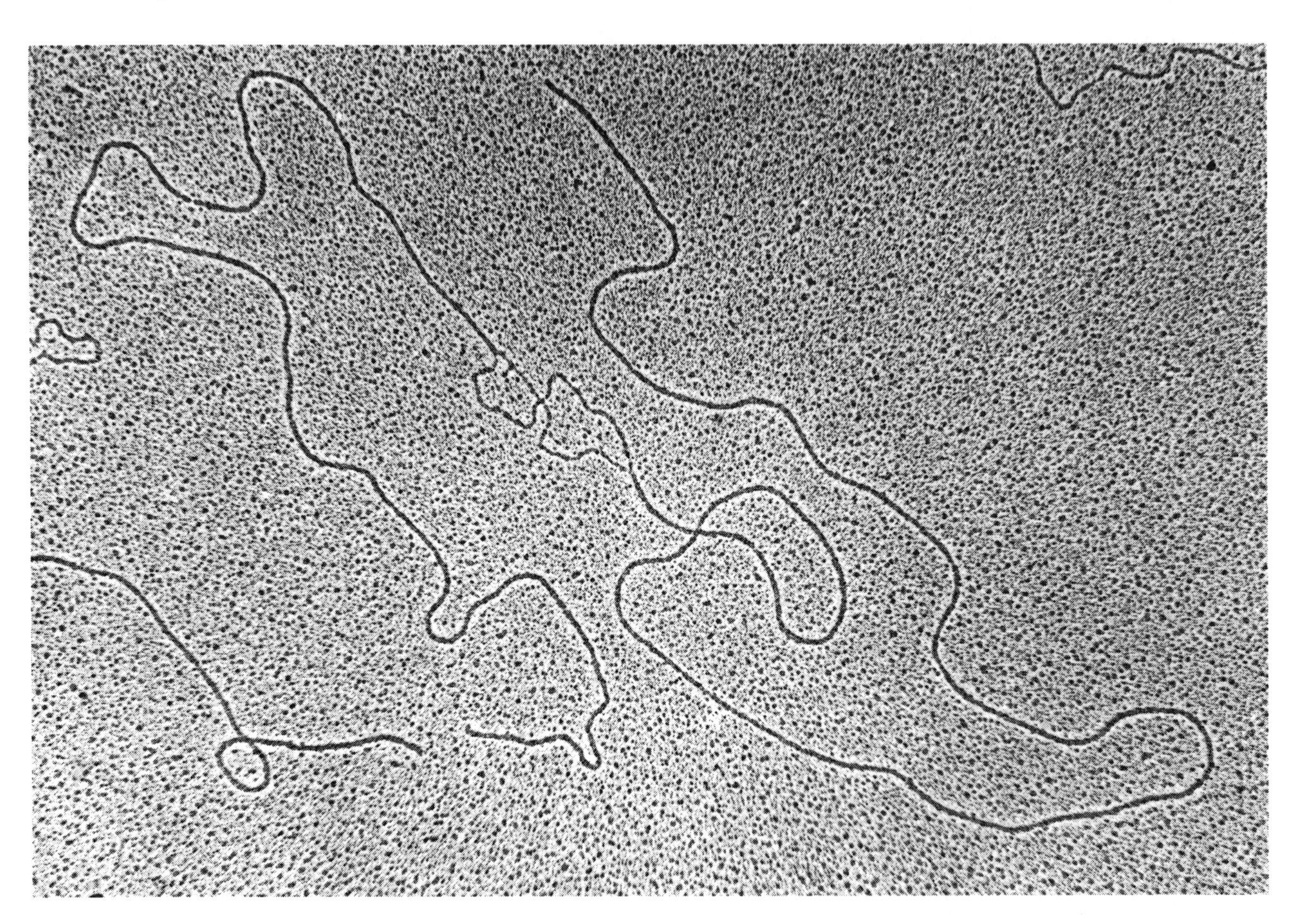

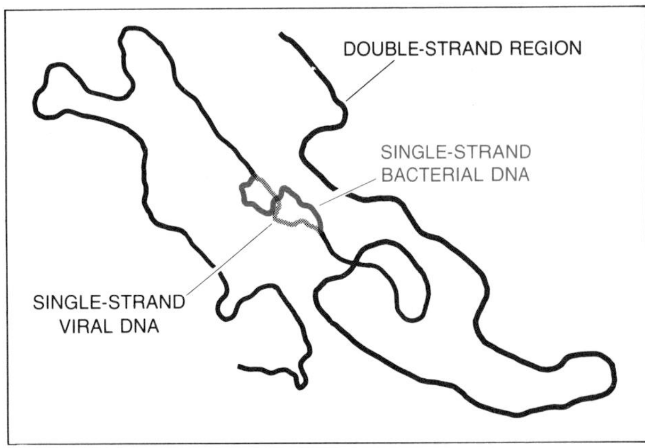

HETERODUPLEX made in the test tube between DNA from a transducing phage designated $\phi 80psu_3$ and DNA from the parental phage $\phi 80$ reveals the location of the inserted bacterial genes in the middle of the viral chromosome. The black lines on the map represent the double-helical regions where the nucleotide sequence on the DNA strands of both partners is complementary. The colored segment is the piece of *E. coli* DNA 3,000 nucleotides long carried by the transducing phage; the gray segment is the piece of viral DNA 2,000 nucleotides long that is present in the normal phage $\phi 80$ but replaced by bacterial DNA in $\phi 80psu_3$. Viral and bacterial DNA sequences on opposing strands are not complementary and cannot pair, forming a substitution bubble. Total length of duplex molecule is about 43,000 nucleotides. Electron micrograph was made by Madeline C. Wu and Norman R. Davidson of California Institute of Technology.

the provirus from the host chromosome was supported by the observation that under normal circumstances lambda can incorporate only genes, such as *gal*, that are within a few thousand nucleotides of its insertion site. This distance is a small fraction (less than 1 percent) of the total length of the host chromosome. Recently K. Shimada and his co-workers at the National Institutes of Health have studied rare bacterial lines in which lambda DNA has inserted itself into a chromosomal site other than the normal one. From such abnormal strains virus variants carrying *gal* are not obtained, but variants carrying genes close to the new attachment site are.

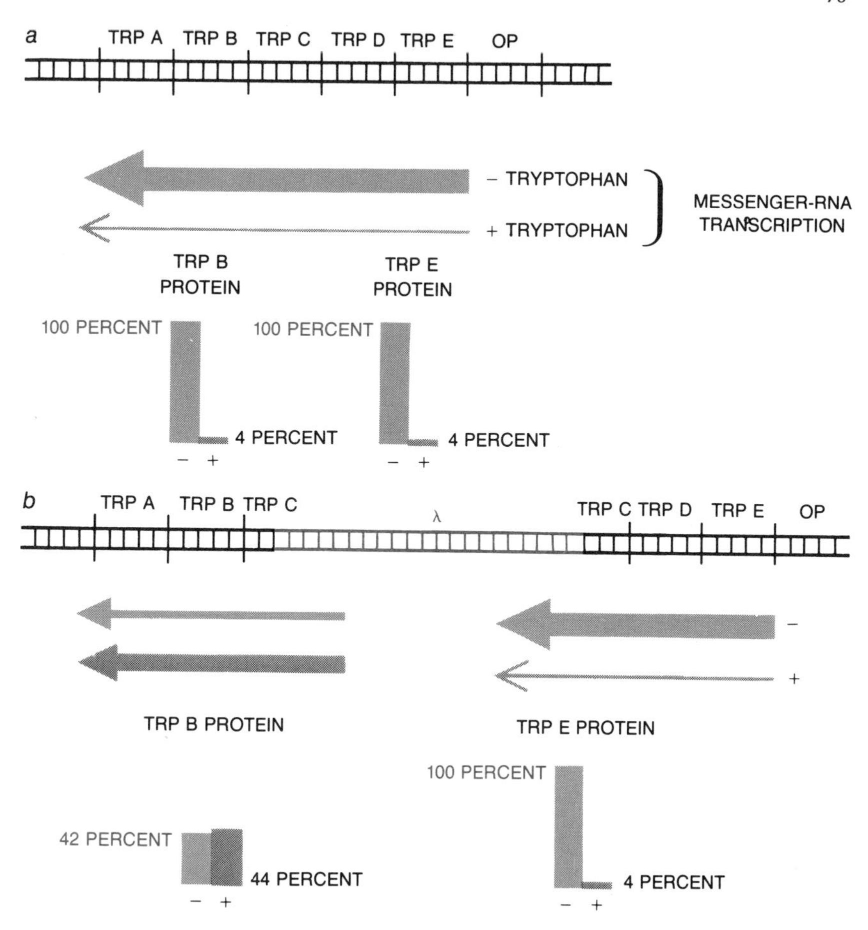

ABNORMAL INSERTION of the phage-lambda chromosome into the *trp* operon (a group of bacterial genes coding for five enzymes in the biosynthetic pathway of the amino acid tryptophan) disrupts the operon's genetic control mechanism. In the normal operon (*a*) transcription of the *trp* genes into messenger RNA is regulated by the tryptophan concentration. When tryptophan levels in the cell are high, transcription is repressed, and when tryptophan levels are low, transcription and synthesis of the *trp* enzymes is high. If phage lambda inserts its DNA into the middle of the *trpC* gene (*b*), the operon as a transcriptional unit will be disrupted. Genes "upstream" from the viral DNA (*trpD* and *trpE*) will continue to be transcribed normally, but the transcript cannot pass through the viral DNA to the genes downstream from the provirus (*trpA* and *trpB*). These genes are expressed to some extent, however, because of transcription arising within the provirus that cannot be repressed by tryptophan. (Protein levels shown are for a mutant of lambda in which the rate of this transcription is abnormally high.)

Why has the virus evolved such a complex and specific mechanism for getting its DNA into and out of chromosomes? The obvious answer is that the ability to do so at appropriate times plays an important role in the virus's survival. Little is known about the selective forces operating on viruses in nature, but one can imagine that it is to the virus's advantage for its DNA to be inserted soon after infection, for the DNA to remain stably inserted while the lysogenic bacterium is growing and for the DNA to be excised while the bacterial genes are repressed. Since insertion and excision have different enzymatic requirements, the virus can control both the direction and the extent of these activities by regulating integrase and excisionase.

The integrase reaction seen in phage lambda and similar viruses is the first case known where two DNA molecules are cut and rejoined at specific sites as part of the normal life cycle of an organism. Enzymatic cleavage and rejoining of DNA molecules in the test tube has become a common pastime of biochemists, but the bacterial restriction enzymes used for this purpose ordinarily function in DNA degradation rather than in genetic recombination. We do not yet know the actual nucleotide sequences recognized and acted on by integrase. The in vivo results require that the viral and bacterial sequences differ from each other, since the genetic requirements for insertion and excision are not the same.

The study of how the phage-lambda DNA inserts itself has provided some useful dividends, among them knowledge of the specific process of the breaking and joining of DNA molecules, which is becoming amenable to biochemical study and opens up new possibilities for the controlled translocation of DNA segments in other organisms. It has also given us the transducing virus variants, which have become workhorses of molecular biologists because they enable one to replicate specific segments of host DNA apart from the rest of the chromosome. In addition understanding the mode of viral DNA insertion has helped to define the ways in which inserted elements and chromosomes can interact functionally. The simplest examples come from instances of abnormal insertion within known genes, such as the insertion of lambda DNA into the *trpC* gene.

The work of Charles Yanofsky of Stanford and others has shown that the *trp* operon consists of five genes, each of which codes for a different enzyme catalyzing a specific step in the biosynthesis of the amino acid tryptophan; the genes are designated *trpA*, *trpB* and so on. This entire stretch of DNA is transcribed into messenger RNA as a unit starting from *trpE* and continuing to *trpA*. The messenger RNA then attaches itself to the ribosomes, the subcellular particles where the enzymes are synthesized. Near the beginning of the transcribed stretch is the specific nucleotide sequence known as an operator. In the presence of high concentrations of tryptophan a repressor protein binds to this sequence and prevents the transcription of the entire DNA segment. All five proteins are hence synthesized together when tryptophan is needed by the cell, but their synthesis is shut off by large amounts of the end product, a feedback control mechanism common to many operons.

How does the insertion of phage-lambda DNA change things? First, since the lambda DNA goes into the middle of the *trpC* gene, the complete protein product of that gene can no longer be formed. A bacterium that carries the abnormally inserted provirus is thus unable to make tryptophan, since one of the enzymes in the biosynthetic pathway is not synthesized. It is possible to recover descendants of this lysogenic bacterium that have lost the provirus. In these bacteria the two halves of the *trpC* gene are rejoined, and active enzyme is again synthesized. None of the DNA of the *trpC* gene has been damaged or permanently lost; it simply cannot code for its normal product when it is cut in two.

Besides disrupting the *trpC* gene, the provirus interrupts the *trp* operon as a functional unit. The transcription of RNA ordinarily proceeds along DNA segments such as the *trp* cluster from a fixed starting point to some stop signal. The precise nature of transcriptional stop signals is not known, but somewhere within the lambda DNA there must be one or more of them. *TrpE* and *trpD* proteins are synthesized normally in these lysogenic bacteria, but the transcription from the *trpE* end never reaches the *trpA* or *trpB* genes.

The provirus can hence constitute a barrier to RNA transcription, although the junction points between provirus DNA and bacterial DNA are not themselves barriers. Whereas bacterial transcription that can be repressed by the product tryptophan does not reach the *trpA* or *trpB* genes, transcription arising from within the provirus does cross the junction between viral DNA and host DNA and produces a low level of tryptophan-independent expression of these genes. The viral transcript that extends across the junction includes only one known gene, the integrase gene of the provirus, which is expressed at a low rate even when the other viral genes are repressed. Thus insertion can not only break up units of transcription but also create new ones.

The transcription of RNA across boundaries in the abnormal lysogenic bacterium illustrates some of the consequences of the viral insertion of DNA both for the regulation of cell function and for evolution. If new regulatory units can be created by insertion, we can be sure that natural selection will then act on them to maximize their selective value to the cell. How extensively DNA of viral origin may have become incorporated into the regulatory systems of existing chromosomes is not known.

There is an old dichotomy between those virologists who view the virus basically as a foreign invader and those who view it more as a cellular component that escapes normal regulatory controls. The argument frequently concerns matters of definition rather than of substance, but it tends to recur at different levels of sophistication as knowledge increases. Lysogenic bacteria have long constituted a prime example for the cellular-component school. At present one can say this much: The conception of the provirus as a normal cellular constituent is at least not a superficial one. The provirus not only behaves like an integral part of the host chromosome; it really is an integral part of the host chromosome. In its manner of attachment there is nothing to distinguish the DNA of the virus from the DNA of the host.

The Isolation of Genes

9

by Donald D. Brown
August 1973

Given a specific RNA product of a gene, it is now possible to find and purify the stretch of DNA that encoded the RNA. The first genes to have been isolated are those that make the RNA's in ribosomes

One of the scientific triumphs of this century has been the discovery that genetic information is encoded along the length of long molecules of deoxyribonucleic acid (DNA). The information is present in discrete segments of the DNA, the genes, each of which determines the structure of a complementary length of ribonucleic acid (RNA). Some of the RNA (messenger RNA) determines the structure of the proteins (primarily enzymes) that constitute or manufacture all the tissues of the organism; the remainder of the RNA (ribosomal RNA and transfer RNA) takes part in the synthesis of protein.

Although a great deal is now known about the nature of DNA and about the functioning of genes, most of this knowledge is based on genetic experiments with microorganisms or on biochemical studies with complex mixtures of DNA known to contain many genes. In the past 10 years a number of investigators have begun to develop methods that make it possible to purify individual genes of known function from the total DNA in an organism.

Techniques for gene purification are now at a stage comparable to those available 25 years ago for fractionating proteins. One difficulty is that, unlike proteins, which are composed of 20 different amino acids, the DNA molecules are long polymers composed of only the four building blocks called nucleotides. The result is that all DNA molecules have very similar chemical and physical properties, even though different arrangements of the four nucleotides can and do provide an almost infinite variety of different polymers. In spite of the chemical simplicity of DNA, techniques have been developed recently for fractionating complex DNA mixtures from animal cells, and so far two DNA components have been purified that contain genes of known function. The genes, which code for the RNA's of the ribosome, have been purified from the DNA of two related amphibians, the African clawed toads *Xenopus laevis* and *Xenopus mulleri.*

Gene purification became possible about 10 years ago with the development of an assay system that could detect the presence of a gene in a complex mixture of related DNA molecules. This assay system, called molecular hybridization, depends on the complementary structure of DNA, which is a long double-strand molecule of which each strand

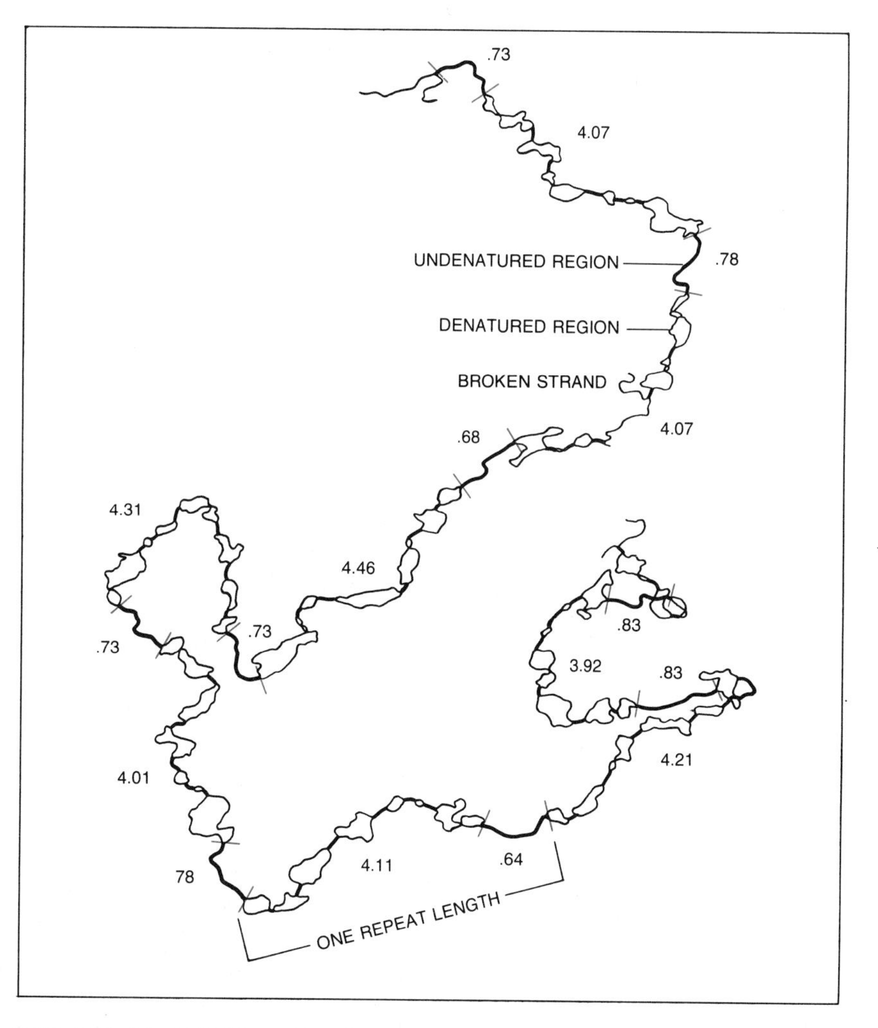

REPEATING PATTERN is evident in a tracing of the molecule on the opposite page. Each of the eight full repeats has one mainly denatured and one undenatured region. Lengths are in microns. Tracings such as this are analyzed to map genes (*see illustrations on page 87*).

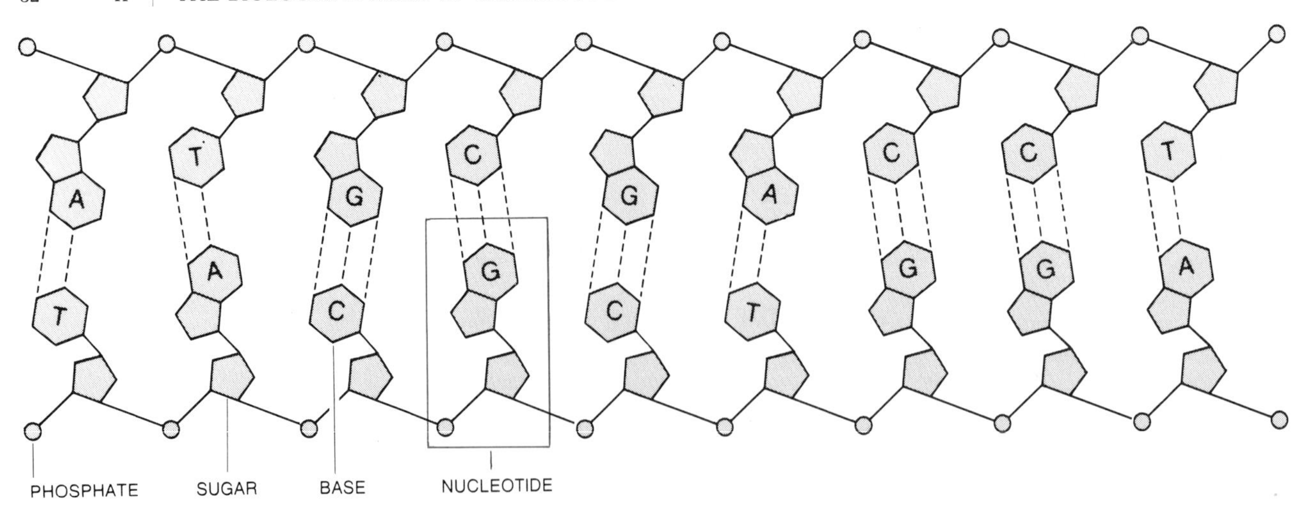

DNA MOLECULE *(gray)* is a chain of nucleotides, each designated by the initial letter of the organic base it contains: adenine *(A)*, guanine *(G)*, cytosine *(C)* and thymine *(T)*. In its usual double-helix form *(left)* DNA consists of two strands of nucleotides connected by hydrogen bonds *(broken lines)* between the bases. The nucleotides in one strand are always opposite the complementary

is a chain of the four nucleotides [*see illustration above*]. Each nucleotide consists of an organic base (adenine, guanine, cytosine or thymine), a sugar (deoxyribose) and a phosphate group. All genes are different combinations of these four nucleotides, which are named for their bases: deoxyadenylic acid (*A*), deoxyguanylic acid (*G*), deoxycytidylic acid (*C*) and thymidylic acid (*T*). The sequence of the nucleotides on one strand of the DNA molecule precisely determines the nucleotide sequence on the other strand: *A* is always opposite *T* and *G* is always opposite *C*. (The complementarity has a well-documented physical basis in hydrogen bonding between the nucleotide pairs, since the bonds either cannot be formed or are formed weakly between other combinations of the nucleotides.) This specificity is what determines the exact nucleotide arrangement within the gene product RNA, a polymer that is chemically very similar to DNA. There are three differences: most RNA molecules have only one strand; RNA has three of the same bases as DNA (*A*, *G* and *C*) but the base thymine (*T*) is replaced by uracil (*U*); the sugar molecule (ribose) in each RNA nucleotide is slightly different from the deoxyribose of DNA. In the living cell one strand of the gene DNA serves as a template along which the RNA nucleotides are lined up according to the complementarity rules.

In the assay system for gene purification the investigator takes advantage of complementarity to determine just how much of a sample of DNA is the gene (or genes) for a specific RNA product [*see illustration below*]. First the DNA is denatured: its two strands are separated by treatment with heat or alkali. Then the DNA is fixed on a piece of filter paper, to which a solution of the RNA, which has been labeled with radioactive atoms, is added. The DNA and the labeled RNA are allowed to interact in the proper salt solution at the correct temperature long enough for the RNA molecules to find any complementary strands of DNA and pair with them, forming stable hybrid molecules in which one strand is DNA and one is RNA. The RNA that does not hybridize is washed away.

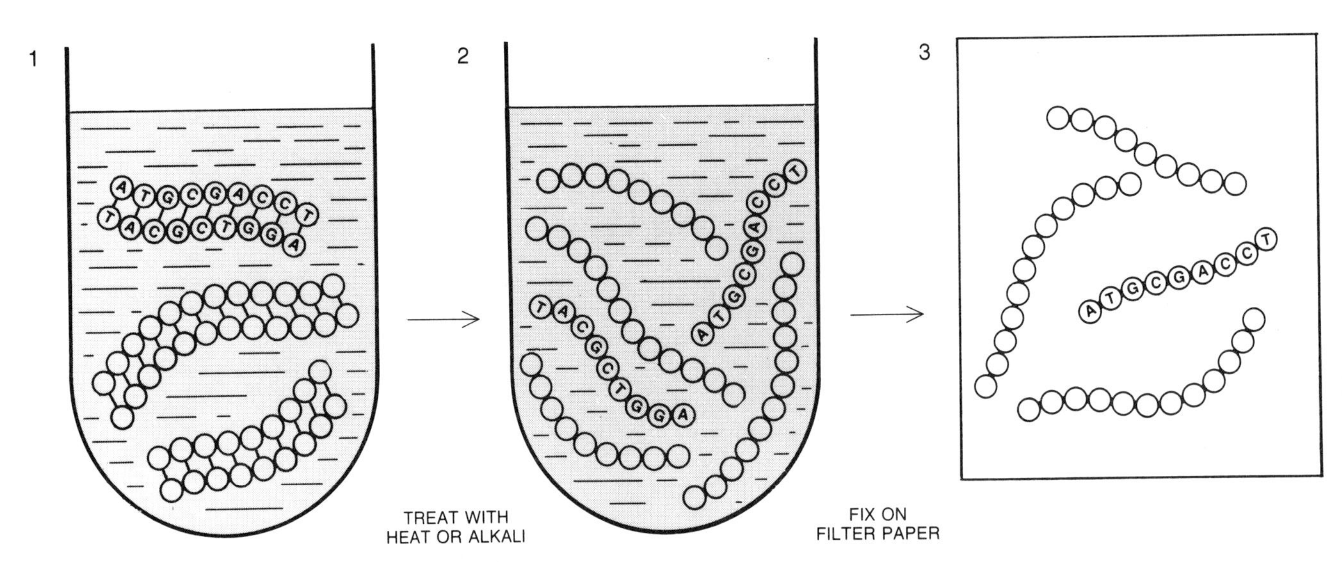

HYBRIDIZATION ASSAY pairs a gene with its product on the basis of complementarity. Double-strand DNA *(1)* is denatured by treatment with heat or alkali and the separated strands *(2)* are fixed on filter paper *(3)*. The filter paper is incubated in a solution of RNA molecules labeled by the incorporation of radioactive uracil *(4)*. The RNA molecules diffuse over the paper and come in con-

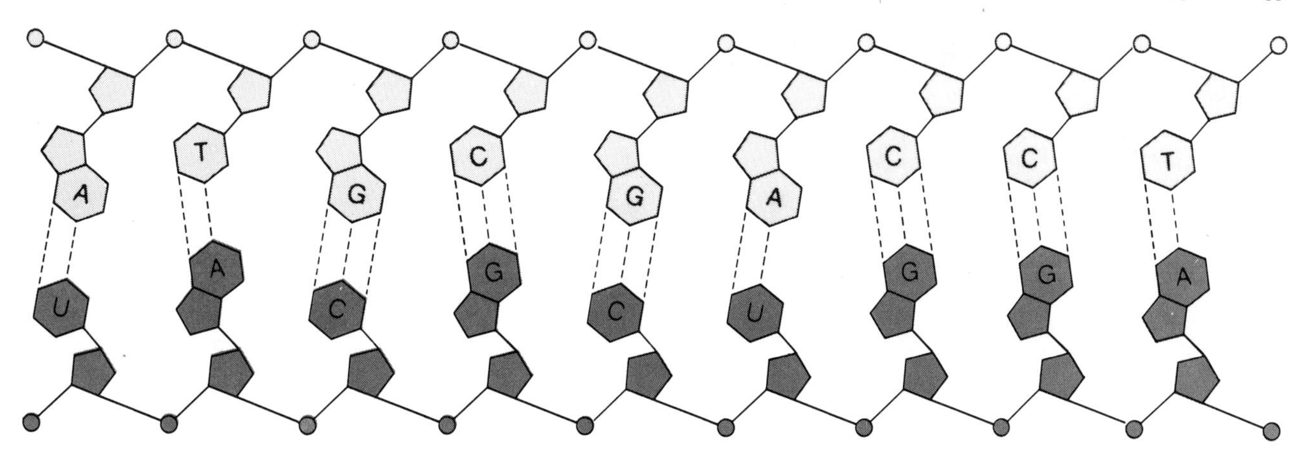

nucleotides in the other strands, with *A* opposite *T* and *G* opposite *C*. This complementarity determines the sequence of the RNA (*color*) that is transcribed from the DNA (*right*). In RNA the sugar is slightly different and uracil (*U*) replaces thymine. Complementarity is the basis of a number of techniques used in gene isolation, including the assay that locates the gene for a specific RNA product.

The next step is to measure the radioactivity on the filter paper. The amount of radioactivity shows how much of the RNA has bound to the DNA and therefore how much of the DNA has a nucleotide sequence that matches the RNA's sequence. That DNA is assumed to be the gene for the RNA. This assay system makes it possible to count the number of copies of a gene that exist in the complete DNA complement of a cell. In the course of gene isolation the assay tells one how close each successive DNA fraction is to containing copies of just a single gene.

Clearly the essential requirement for isolating a gene with such an assay is the availability of the RNA product of that gene in a pure form. About 80 percent of the RNA in a cell is in the ribosomes, particles composed of RNA and protein that serve as the site of protein synthesis. In addition to being abundant, ribosomal RNA is easily identified as such—much more easily, for example, than a messenger RNA that codes for a single protein can be identified. Each ribosome includes three kinds of RNA molecule of differing size, which are easily purified. In *Xenopus* they are called the 28S, 18S and 5S molecules. (S stands for Svedberg units, a measure of the speed with which a molecule sediments when subjected to a centrifugal force.) The three molecules respectively have a molecular weight of 1.5, .7 and .04 million daltons and contain 4,700, 2,200 and 120 nucleotides.

Another characteristic feature of ribosomal DNA that made its isolation possible is its unusual base composition compared with the composition of the rest of *Xenopus* DNA. About 40 percent of the DNA of most animals consists of guanine and cytosine; it is said to be "40 percent *G* + *C*" (and therefore 60 percent *A* + *T*). Some years ago, however, the ribosomal RNA's were found to be very rich in guanine and cytosine; their *G* + *C* content ranges from 45 to 65 percent, depending on the animal. In the case of *Xenopus laevis* the 18S, 28S and 5S RNA's are respectively 53, 63 and 57 percent *G* + *C*, whereas the bulk DNA is only 40 percent *G* + *C*. Since the nucleotide composition of an RNA must be reflected in its gene, a high *G* + *C* content

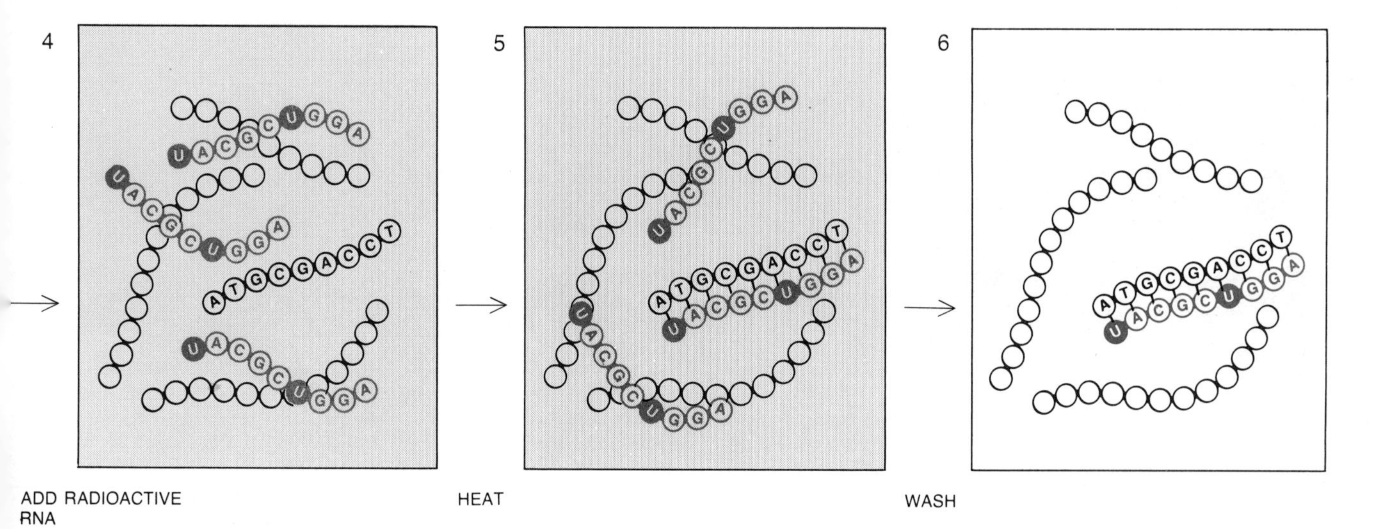

tact with any strands of DNA that are complementary to them, that is, with their gene. The radioactive RNA binds to this DNA, forming DNA-RNA hybrids (*5*). The remaining, unbound RNA is washed away, leaving only RNA that is paired with its gene (*6*). The amount of that RNA, determined by measuring the radioactivity, indicates how much of the DNA is the gene for the RNA.

EIGHT GENES for ribosomal RNA are included in this molecule of DNA from the nucleus of an oöcyte of the African clawed toad *Xenopus mulleri.* DNA containing the genes, isolated by methods described in the article, was treated with alkali and thus partially denatured: the two strands of the DNA double helix were separated. Denaturing occurs first in parts of the molecule that have a relatively low "$G + C$ content," that is, a low proportion of two of the four nucleotides that combine to form DNA. The molecules are fixed, complexed with protein to make them thicker, placed on an electron-microscope grid and shadowed with platinum. The micrograph, in which the molecule is enlarged 25,000 diameters, was made by Pieter Wensink in the author's laboratory. It shows two kinds of regions: those in which the DNA is primarily denatured (*open loops*) and those in which its two strands remain unseparated (*thicker single lines*). Most of each denatured region is gene DNA; the remainder of those regions, together with the unseparated segments, is "spacer" DNA. Spacers alternate with genes and have a generally higher $G + C$ content (*see illustration on page 81*).

5S DNA from *X. laevis* is enlarged 50,000 diameters in an electron micrograph made by Wensink by the same technique as the one on the opposite page: the DNA was partially denatured so that the relatively high *A* + *T* regions separated while those with the highest *G* + *C* content remained unseparated. Again the denaturation pattern is regular along the entire length of the two molecules, as shown in the tracing of a part of one molecule (*right*). The high-*G* + *C* gene for 5S RNA takes up about a third of the length of each undenatured region; the remaining six-sevenths of the repeat (all of the looped region and two-thirds of the undenatured region) is spacer.

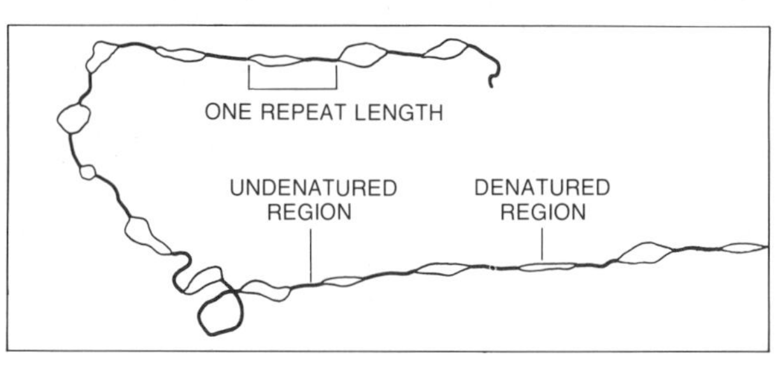

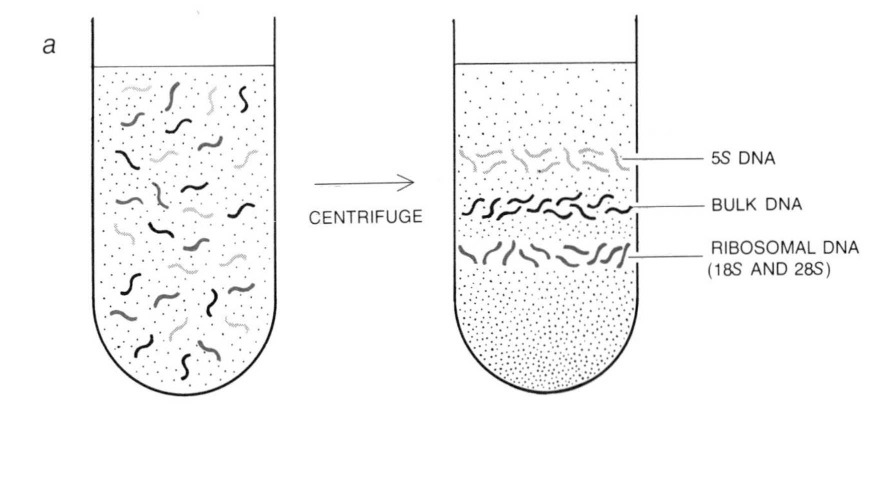

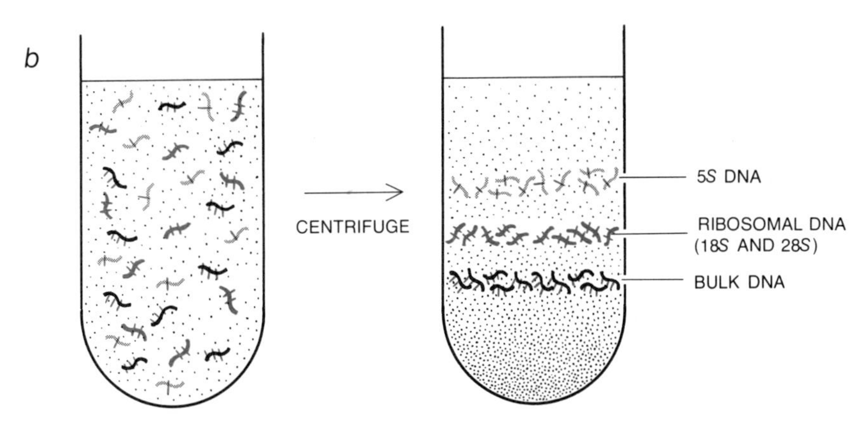

DENSITY-GRADIENT CENTRIFUGATION separates different DNA molecules. In the basic method (*a*) the DNA is spun in a solution of cesium chloride. A density gradient is established by the salt molecules in the solution and each DNA molecule moves to the level at which its density matches that of the solution. The "ribosomal" DNA, containing the genes for 18*S* and 28*S* RNA, forms a band at a denser level than the bulk of the cell's DNA. The 5*S* DNA, on the other hand, is less dense than bulk DNA. It is best separated by a modified process (*b*): centrifugation in cesium sulfate after being complexed with heavy-metal ions (*colored hatching*), silver in the case of *X. laevis* and mercury for *X. mulleri*. The 5*S* DNA forms fewer complexes and is therefore lighter than the cell's other DNA.

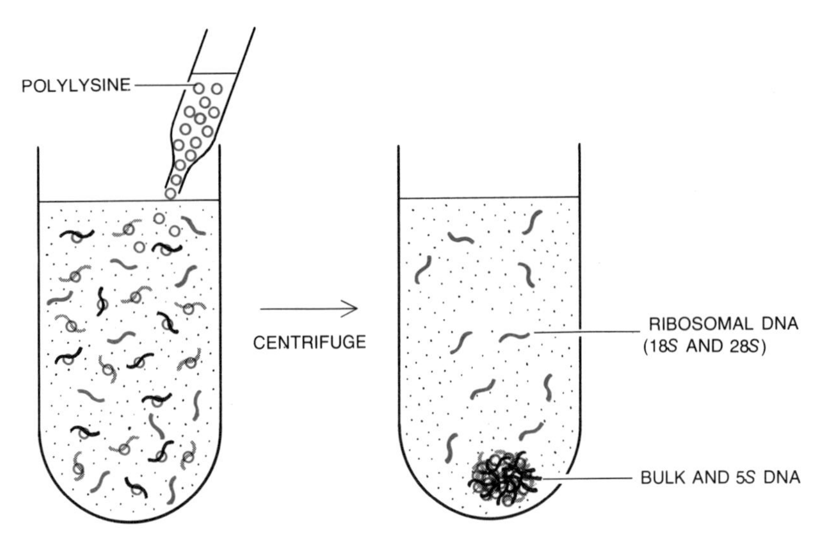

POLYLYSINE METHOD gently separates large ribosomal (18*S* and 28*S*) DNA from other DNA. Polylysine, a synthetic polypeptide, reacts preferentially with DNA that has a high $A + T$ content, forming aggregates in a sodium chloride solution. Centrifugation precipitates the aggregates (bulk DNA and 5*S* DNA), leaving the ribosomal DNA in solution.

is one chemical characteristic that distinguishes the genes for ribosomal RNA from the bulk of the DNA.

That distinction, confirmed by Hugh Wallace and Max Birnstiel of the University of Edinburgh in 1966, is the basis of density-gradient centrifugation, a primary technique for the fractionation of DNA. DNA molecules with different $G + C$ contents have different densities: the greater the $G + C$ content, the higher the density of the DNA. The DNA to be fractionated is mixed with a solution of a cesium salt and spun in a centrifuge at a speed of from 30,000 to 40,000 revolutions per minute. The salt begins to sediment in response to the strong centrifugal force, forming a concentration gradient in the centrifuge tube. Each molecule of DNA moves to a level in the salt gradient at which the salt concentration exactly matches that molecule's density. In this way a mixture of DNA's with different base compositions separates into discrete bands of DNA's at different levels in the salt gradient.

What Wallace and Birnstiel did was to separate *Xenopus laevis* DNA in a centrifuge and test each of the bands by hybridization with radioactive ribosomal RNA. They found that the genes for 18S and 28S ribosomal RNA are denser than the bulk of the cell's DNA. In our laboratory in the Carnegie Institution of Washington we subsequently found that the DNA molecules containing the genes for 5S RNA are lighter than the main-band DNA. (That was surprising, since the 5S RNA gene has a high $G + C$ content. Because the genes themselves are very small we predicted, and later were able to demonstrate, that they are interspersed among stretches of DNA with a very different nucleotide composition, so that the molecule as a whole is low in $G + C$ content.)

The experiments established that two of the three kinds of RNA (18S and 28S) are encoded by genes that band together in cesium chloride and must therefore be on the same molecules (known as ribosomal DNA), whereas the genes for 5S RNA, having a different density, are on other DNA molecules. By measuring the amount of each type of RNA that could be hybridized to *Xenopus* DNA we could estimate the number of each kind of gene in a single set of chromosomes. There are about 450 copies each of the 18S and 28S genes and about 24,000 of the 5S genes in each set of *Xenopus* chromosomes. Since normal body cells have two sets of chromosomes, each cell must contain twice as many of the genes.

Three characteristics of the ribosomal genes made them good candidates for isolation: the availability of homogeneous RNA's that could be radioactively labeled for the assay system during gene purification, the physical differences between these genes and the rest of the DNA, and the presence of the genes in multiple copies. The first small amount of *Xenopus laevis* ribosomal DNA, the DNA containing the genes for 18S and 28S RNA, was isolated in 1967 by Birnstiel and his colleagues. They did it by centrifuging *Xenopus* DNA repeatedly in a cesium chloride gradient, each time collecting only the densest fractions.

Since 1967 the centrifugation method has been improved and there have been several other advances in technique. For one thing, a number of methods other than centrifugation have been devised for purifying DNA that is high in $G + C$ content. One convenient method we sometimes use is based on an observation by Marc Leng and Gary Felsenfeld of the National Institutes of Health that polylysine, the synthetic polymer of the amino acid *L*-lysine, tends to bind to DNA molecules that are high in $A + T$; it precipitates such molecules selectively. Carefully applied, the polylysine technique can precipitate 99 percent of total *Xenopus* DNA, leaving in solution the ribosomal DNA (the DNA molecules that include the genes for 18S and 28S RNA). Another important development was the discovery that the amphibian oöcyte, the precursor of the egg cell, goes through a period of "gene amplification" in which a great deal of ribosomal DNA is synthesized. The ribosomal DNA in such an oöcyte accounts for 75 percent of the DNA in its nucleus, which amounts to a thousandfold amplification and simplifies the task of purification.

These advances in technique made possible the detailed examination of the structure of ribosomal DNA and the genes contained in it. Three laboratories have done most of the work: Birnstiel's at Edinburgh, O. L. Miller's at the Oak Ridge National Laboratory and our group's in the Carnegie Institution. The combined results of several kinds of experiment, including electron microscopy as well as chemical analysis, add up to a rather detailed picture of the DNA that governs the synthesis of the 18S and 28S RNA molecules [*see bottom illustration at right*].

This ribosomal DNA molecule consists of a series of repeating units, each of which includes three major sequences: a gene for the 18S RNA, a gene for the 28S RNA and a "spacer" sequence that is not transcribed into RNA. Chemical evidence indicated that the genes for the

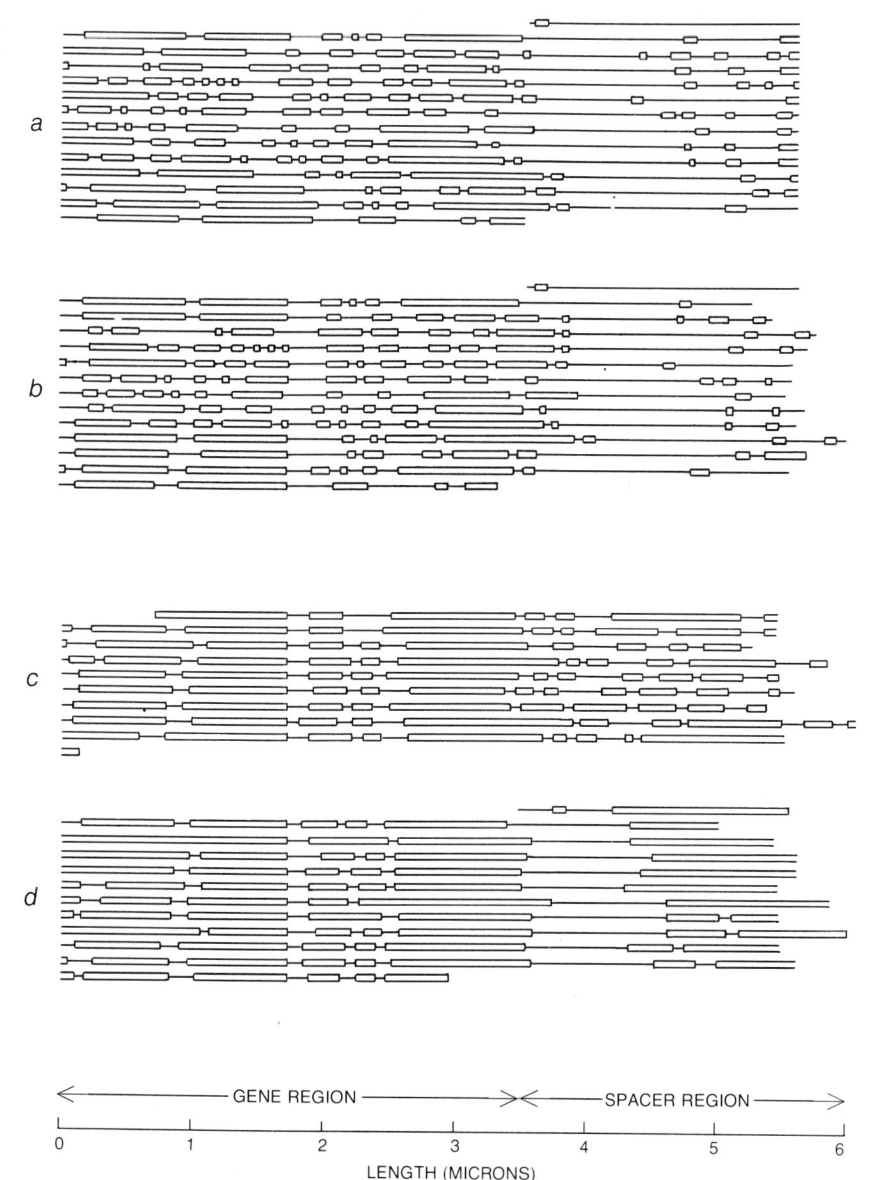

DENATURATION DIAGRAM is made from data in micrographs such as the one on page 84. An average repeat length is determined through computer analysis of measurements from the micrograph and is used to divide the molecule into successive repeats, with the denatured stretches of DNA diagrammed as double lines and the undenatured portions shown as single lines (*a*). It is clear already that there is a largely denatured region (*left*) and a largely unseparated region (*right*). The repeats are adjusted by eye for better alignment (*b*). Two such adjusted diagrams, for *X. laevis* (*c*) and *X. mulleri* (*d*) DNA at an advanced stage of denaturation, are compared. The gene regions are very similar, but the spacer regions are not. In *X. laevis* the spacer is mostly denatured; in *X. mulleri* about half of it is resistant, showing as a single line here and in the micrograph and tracing on pages 81 and 84.

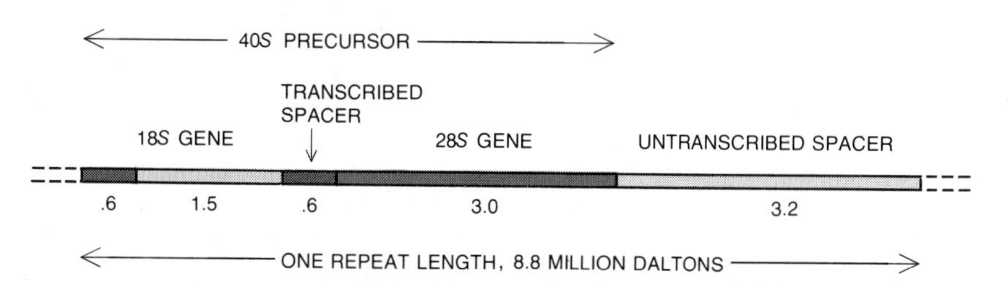

RIBOSOMAL DNA of *Xenopus* is mapped in this diagram of one repeat length based on chemical and electron-micrograph data. The gene region codes for a 40S RNA molecule that is the precursor of the 18S and 28S RNA's, which are cleaved from it in the living cell by enzymes. Two small spacers in the gene region are transcribed into precursor but eliminated in the cleaving process. Then there is the main spacer region, which is not transcribed.

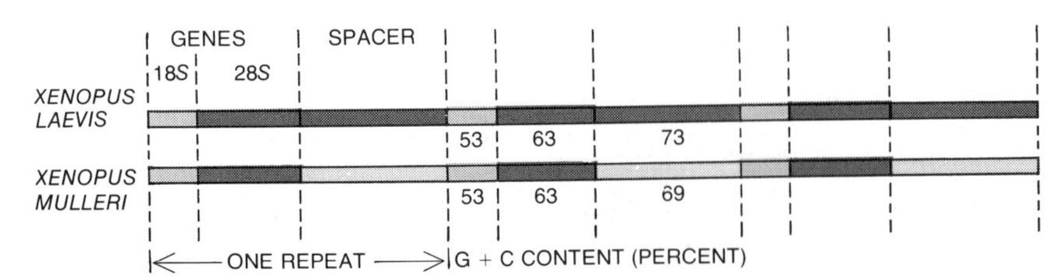

GENE AND SPACER REGIONS of ribosomal DNA from *X. laevis* and *X. mulleri* are compared. The schematic diagram shows three repeats. The gene regions have the same nucleotide compositions as well as the same lengths in the two species; the untranscribed spacers, on the other hand, have the same lengths but very different nucleotide compositions. The transcribed spacers (not diagrammed here) are also different in the two species.

two RNA molecules must be either adjacent or very close together. (Electron-microscope evidence and other experimental results later showed that they are in fact separated by a second very small spacer sequence.) This is consistent with an earlier finding that in the living cell the two large ribosomal RNA's are produced in two steps. First a 40S RNA precursor molecule is transcribed from about 60 percent of the length of one DNA repeat. Then the precursor is cleaved by enzymes; 80 percent of the precursor's length goes to make one 18S and one 28S molecule and the remaining 20 percent is discarded by cellular metabolism. The fraction of the 40S gene sequence that is occupied by the 18S gene and by the 28S gene can be deduced, of course, from the size of the RNA molecules they encode.

The spacer sequence was discovered independently in two laboratories by different methods. At Oak Ridge, Miller isolated functioning genes from living oöcyte nuclei and examined them in the electron microscope. In his micrographs he could see, in a regular repeating pattern along the length of the DNA, arrays of attached RNA molecules: the precursor molecules in the process of being synthesized. Between each of these RNA regions there were stretches of DNA with no RNA attached, and Miller gave the name "spacer" to these apparently inactive stretches [see "The Visualization of Genes in Action," by O. L. Miller, Jr.; March 1973; Offprint 1267]. We had detected the spacer region too, first by chemical means. We found that some of the isolated ribosomal DNA would not form hybrids with either the 18S or the 28S RNA. By doing the hybridizing assay with pieces of DNA of various sizes we could determine that these noncomplementary regions were arranged so as to separate the repetitive gene regions. Further experiment showed that the inactive spacer DNA has an unusual sequence of nucleotides, which not only is different from that of the genes but also is not present anywhere else in the *Xenopus* DNA. The present evidence supports the view that no RNA is synthesized from this spacer DNA, and its function remains a mystery.

We went on to confirm the structure of ribosomal DNA by examining large molecules in the electron microscope. The method, developed by Ross Inman of the University of Wisconsin, depends again on the special nucleotide content of ribosomal DNA. The DNA to be studied is partially denatured by the application of heat or alkali and then fixed with formaldehyde, so that the complementary bases in the open loops where the strands have separated cannot snap back together. Inman had shown that stretches of DNA with a higher $A + T$ content denature first under certain conditions while stretches that are richer in $G + C$ still remain double-stranded. We had chemical evidence that much of the spacer region had an even higher $G + C$ content than the gene region. Pieter Wensink, a graduate student in my laboratory, made micrographs of partially denatured ribosomal DNA [*see illustration on page 84*]. Study of the micrographs disclosed a regular repeating pattern in which denatured loops alternated with solid double-strand DNA. Analysis of the pattern by several methods confirmed the earlier chemical data in detail.

Ribosomal DNA, then, consists of two alternating regions: the gene region, containing genes for the essential 18S and 28S RNA, and the spacer region, which has no known function. Now, if a gene defines an essential structure, it is usually found in at least closely related species and is often found throughout the animal kingdom. The genes for such generally essential components evolve more slowly than those for more specialized functions do, and the genes for ribosomal RNA are highly conserved in evolution. We have tested the nucleotide arrangement of *Xenopus* ribosomal RNA with the ribosomal genes in the cells of more than 50 organisms, including mammals, invertebrates, higher plants, protozoa, fungi and yeasts. In all cases the ribosomal genes have at least some nucleotide sequences in common with *Xenopus* ribosomal DNA.

Unlike the gene regions of ribosomal DNA, the spacers evolve rapidly. We have compared the ribosomal DNA's of *X. laevis* and *X. mulleri*, species so closely related as to produce viable hybrid progeny, some of which reach sexual maturity. The two species' ribosomal RNA's cannot be distinguished from each other by physical or chemical tests, which is to say that there has been no detectable change in the genes since the two species diverged. And yet their spacers are very different. We estimate that at least 10 percent of the bases within the spacers have changed during the period since the divergence of the *Xenopus* species. The evolutionary conservation of the gene regions is really extraordinary in contrast to this rapid divergence of the spacer regions. The genes that code for ribosomal RNA in higher plants are closer in sequence to the genes for that RNA in *Xenopus* than spacer sequences of *X. laevis* are to those of *X. mulleri!*

Since most of the spacer region is not conserved in evolution, we assume that the exact arrangement of bases within at least most of the region had no selective importance for the species and could therefore change rapidly. If this is true, however, we are faced with an extraordinary dilemma. Remember that there are about 450 spacer regions in each set of chromosomes. Our chemical studies show that the multiple spacers in the ribosomal DNA of *X. laevis*, although markedly different from those of *X. mulleri*, are nonetheless very similar to one another—perhaps identical. The same is true of the spacers in *X. mulleri*. How can 450 DNA regions evolve independently but almost identically? We call such simultaneous, parallel evolution of many sequences in the gene complement of a single species "horizontal" evolution; it contrasts sharply with the usual evolution of a single gene in two or more diverging species. Population genetics and evolutionary theory provide no guide for understanding horizontal evolution.

The 5S RNA component is small enough, and available in large enough amounts, so that the exact arrangement of its nucleotides has been determined. This feat was accomplished first by George Brownlee and Frederick

Sanger of the University of Cambridge in 1965 for bacterial 5S RNA; their technology, involving the repeated, selective cleaving of the molecule with different enzymes, has since been applied to the small ribosomal component from a number of animals, including *Xenopus*. The molecule contains 120 nucleotides, and like the 18S and 28S RNA it is highly conserved in evolution: the 5S molecules from all mammals studied to date, from marsupials to man, have exactly the same 120 nucleotides. In the amphibian *Xenopus* different tissues synthesize different kinds of 5S RNA. Body cells synthesize one kind, eight of whose 120 nucleotides are different from those in the mammalian molecule; the oöcytes synthesize a mixture of about four kinds of 5S RNA, of which the predominant form has seven bases that are different from those of the body-cell molecule. How many of the 24,000 gene sequences for 5S RNA code for each version of the molecule is not known.

We have purified 5S DNA from both *X. laevis* and *X. mulleri* cells. Once again the unusual base composition of this DNA is what makes its isolation possible. In this case density-gradient centrifugation is carried out after the DNA has been allowed to react with heavy metal ions: silver for *X. laevis* and mercury for *X. mulleri*. The metal ions react with certain base pairs in the DNA, forming complexes that are denser than the original DNA. Because of its nucleotide content the 5S DNA forms fewer such complexes than the other DNA in the cell, and so it is lighter and can be isolated by centrifugation.

When we analyzed the structure of

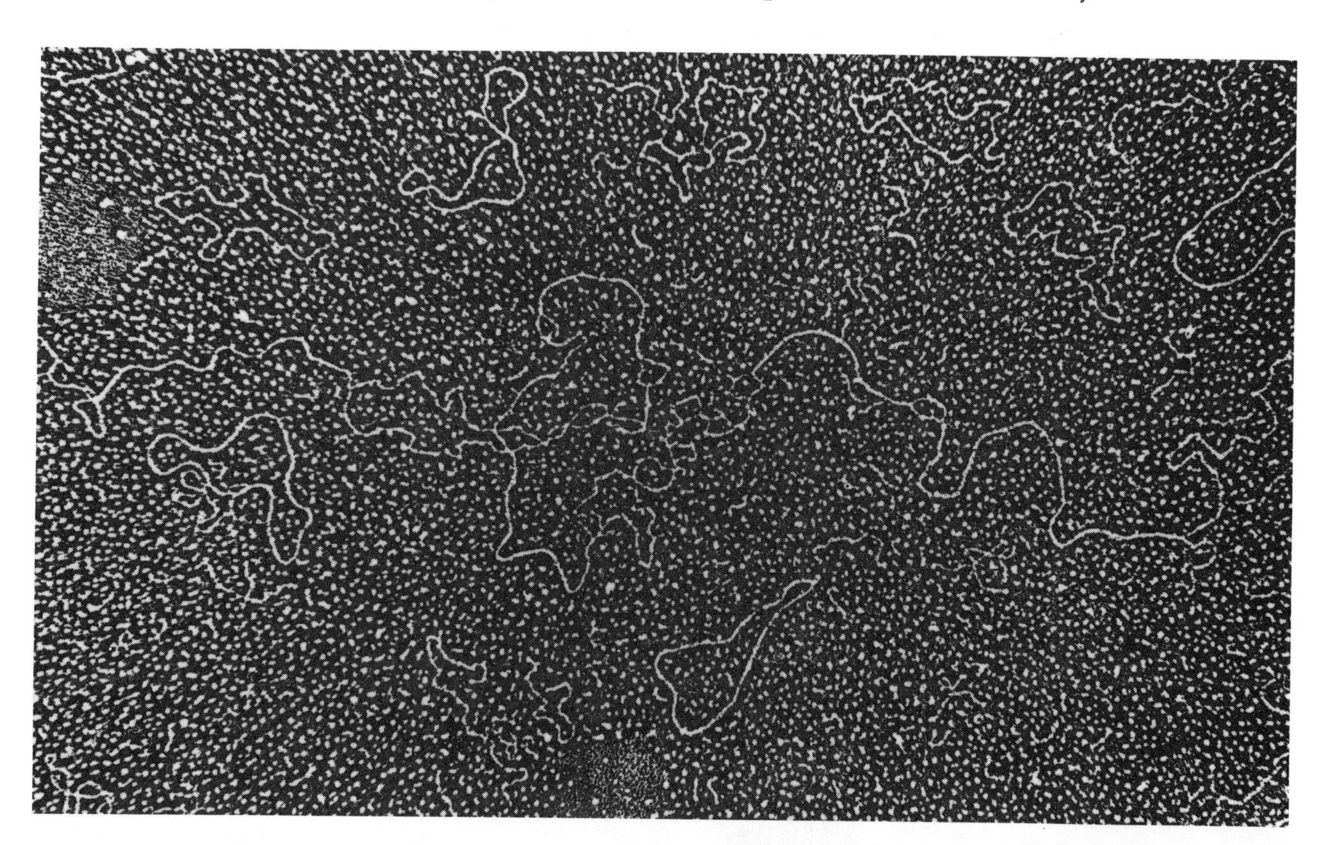

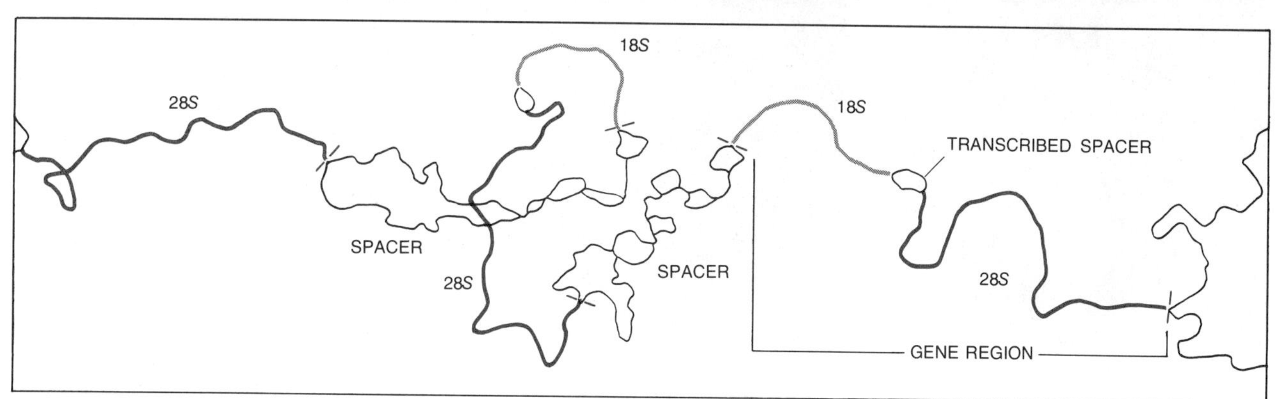

HETERODUPLEX MOLECULE consisting of one strand of ribosomal DNA from *X. laevis* and one from *X. mulleri* was formed and pictured in an electron micrograph by Arleen Forsheit and Norman Davidson of the California Institute of Technology (*top*). The molecule has a structure much like the one formed by denaturing, on page 84, but the structure has a very different origin. DNA from the two species was mixed and completely denatured. Then the single strands of DNA were allowed to pair up. Some strands paired with their complementary strands from the same species, forming complete double-strand molecules. Other strands, such as the ones in the micrograph, came in contact with strands from the other species. Where the nucleotide sequences matched, the two strands became bound; where they did not match, the strands remained separated, forming looped regions, as shown in the tracing (*bottom*). The bound regions and the mismatched, looped regions are the gene and the spacer regions respectively.

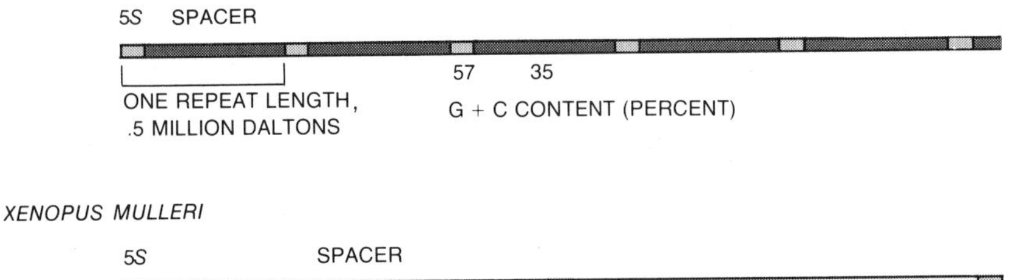

SPACER REGIONS are very different in *X. laevis* and *X. mulleri*, in length as well as in nucleotide composition. The gene regions are the same length in both and appear to synthesize the same 5S RNA. The gene accounts for about a seventh of the total repeat length in *X. laevis* but only for about an eighteenth of the much longer repeat length in *X. mulleri*.

5S DNA from *X. laevis* by many of the techniques described above for ribosomal DNA, it was apparent that other nucleotide sequences besides those of the genes for 5S RNA were present. When the DNA is hybridized with radioactive 5S RNA, only about a seventh of one of the DNA strands binds the RNA, indicating that the remaining six-sevenths of the DNA consists of spacer sequences. When Wensink did the electron-microscope experiment on purified 5S DNA, this structure was confirmed. The micrographs show clearly an arrangement in which loops of strand-separated DNA alternate with natural double-strand regions along each molecule [*see illustration on page 85*]. One repeat length (the combined length of a loop and a double-strand stretch of DNA) is about seven times the predicted length of a gene for one 5S RNA molecule. Similar experiments with purified 5S DNA from *X. mulleri*, carried out with Kazunori Sugimoto, have shown that only about an eighteenth of the DNA molecule encodes the 5S RNA. Since the RNA molecule is the same size in both animals, the difference in repeat length must be due to a much longer spacer [*see top illustration*].

The 5S-DNA spacers are not only different in length in the two species but, like the spacers for the ribosomal genes, also very different in nucleotide sequence. Again, this interspecies difference contrasts with remarkable intraspecies similarity: the many thousands of spacer sequences in the 5S DNA of a single species are not quite identical (as they are in the ribosomal genes) but they are almost the same. The interspecies difference records very rapid evolution, implying that any function the spacers have must surely not depend on precise nucleotide sequence. The rapid evolution of the spacers contrasts, of course, with the 5S RNA genes themselves, which are highly conserved in evolution: we cannot distinguish *mulleri* 5S RNA from *laevis* 5S RNA, and so we assume that their genes are very similar and perhaps identical. Thus the repeating sequences of 5S DNA composed of spacer and gene regions behave as a family of closely related but not identical DNA regions.

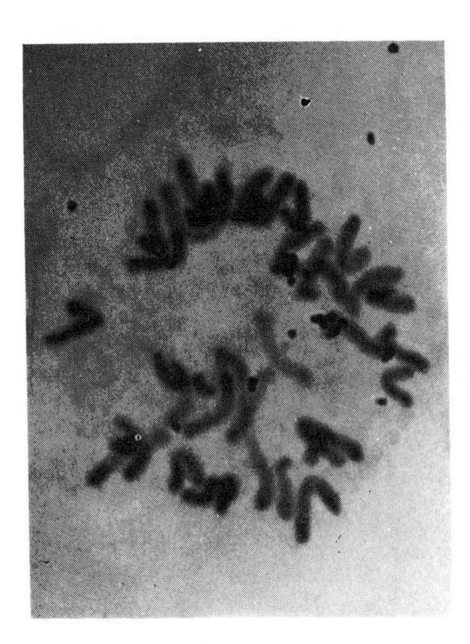

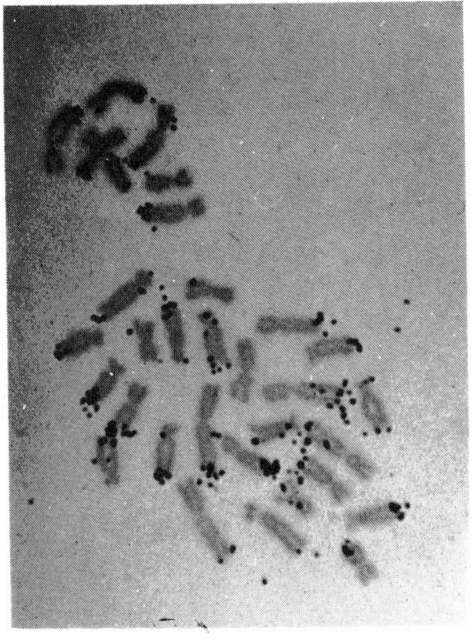

CHROMOSOMAL SITES of large ribosomal DNA and 5S DNA in *X. laevis* are demonstrated in these autoradiographs made by Mary Lou Pardue of the Massachusetts Institute of Technology. Radioactive RNA's are allowed to hybridize to chromosomes fixed on slides and the unbound RNA is washed off. A photographic emulsion is applied and the bound RNA is revealed by the silver grains it develops. The autoradiographs locate the ribosomal DNA at a single site on one chromosome, two copies of which are present in the diploid cell (*left*). 5S DNA, on the other hand, is at the ends of most of the chromosomes (*right*).

The chromosomal location of the 18S and 28S DNA was established first by genetic experiments, which showed that all this DNA was clustered at a single site on one of the 18 chromosomes of *Xenopus*. This has been confirmed by an elegant new mapping technique developed independently by Joseph Gall and Mary Lou Pardue at Yale University and by Kenneth Jones and his colleagues at the University of Edinburgh. In this method a complete set of discrete *Xenopus* chromosomes is fixed on a slide and treated with mild alkali, which denatures the DNA. A drop of radioactive RNA is placed over the slide and hybridization is allowed to occur. Then the slides are washed free of unbound RNA and dipped into a photographic emulsion; wherever there is radioactive RNA, silver grains are developed. When this experiment is done with a radioactive ribosomal RNA, the corresponding genes hybridize to the RNA and thus bind the radioactivity, and in this way the genes are easily located. Large ribosomal DNA turns out to be localized, as predicted, on only one chromosome per set: at the secondary constriction known as the nucleolar organizer, the site where the nucleolus is known to form in the actively synthesizing cell. On the other hand, most if not all of the *Xenopus* chromosomes contain 5S DNA, and it is always at the ends of the chromosomes [*see illustration left*].

In the future, with the development of more sophisticated methods for fractionating DNA, we should be able to purify many genes with a variety of other functions, in particular those that code for specific proteins. With these purified genes it will be possible to analyze their structure, chromosome location and evolution and the control mechanisms that regulate their action in the living cell.

RNA-Directed DNA Synthesis

by Howard M. Temin
January 1972

The discovery that in certain cancer-causing animal viruses genetic information flows "in reverse"—from RNA to DNA—has important implications for studies of cancer in humans

A major goal of present-day biology is to learn how information is coded in molecular structures and how it is transmitted from molecule to molecule in biological systems. Discovery of the rules governing this transmission is an integral part of understanding how embryonic cells differentiate into the hundreds of distinct types of cell observed in plants and animals and how normal healthy cells become cancerous.

It has now been known for nearly 20 years that the genetic information in all living cells is encoded in molecules of deoxyribonucleic acid (DNA) consisting of two long strands of DNA wound in a double helix. The genetic information for each organism is written in a four-letter alphabet, the "letters" being the four different chemical units called bases. In the normal cell short passages of the genetic message (individual genes) are transcribed from DNA into the closely related single-strand molecule ribonucleic acid (RNA). A length of RNA representing a gene is then translated into a particular protein, a molecule constructed with a 20-letter alphabet, the 20 amino acids. When a cell divides, the information contained in each of the two strands of DNA is replicated, thereby equipping the daughter cell with the full genetic blueprint of the parent.

Francis Crick, one of the codiscoverers of the helical structure of DNA, originally proposed that information can be transferred from nucleic acid to nucleic acid and from nucleic acid to protein, but that "once information has passed into protein it cannot get out again," that is, information cannot be transferred from protein to protein or from protein to nucleic acid. These concepts were simplified into what came to be known as the "central dogma" of molecular biology, which held that information is sequentially transferred from DNA to RNA to protein [*see illustration on page 93*]. Although Crick's original formulation contained no proscription against a "reverse" flow of information from RNA to DNA, organisms seemed to have no need for such a flow, and many molecular biologists came to believe that if it were discovered, it would violate the central dogma.

I shall describe experiments that originally hinted at a flow of information from RNA to DNA and that since have provided strong evidence that the "reverse" flow of information not only takes place but also accounts for the puzzling behavior of a sizable group of animal viruses whose genetic information is encoded in RNA rather than in DNA. Many of these viruses also produce cancer in animals. Although they have not yet been linked to cancer in man, their ability to transmit information from RNA to DNA inside the living cell makes it attractive to unify two hypotheses of the cause of human cancer that had previously seemed separate: the genetic hypothesis and the viral hypothesis.

There are two broad classes of viruses: viruses whose genome, or complete set of genes, consists of DNA and viruses whose genome consists of RNA. In the cells that they infect the DNA viruses replicate their DNA into new DNA and transmit information from DNA to RNA and thence into protein. Most RNA viruses, such as the viruses that cause poliomyelitis, the common cold and influenza, replicate RNA directly into new copies of RNA and translate information from RNA into protein; no DNA is directly involved in their replication.

In the past few years it has become apparent that a group of viruses, variously called the RNA tumor viruses, the leukoviruses or the rousviruses (after their discoverer, Peyton Rous), replicate by another mode of information transfer. The rousviruses use information transfer from RNA to DNA in addition to the modes of information transfer (DNA to DNA, DNA to RNA and RNA

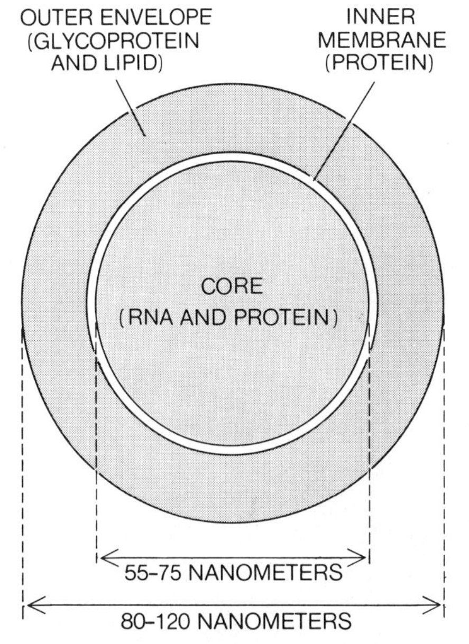

VIRIONS, or individual particles, of an "RNA-DNA virus," an animal-tumor virus that transfers genetic information from RNA (ribonucleic acid) to DNA (deoxyribonucleic acid) in addition to the normal modes of information transfer used by cells and other viruses, are enlarged about 700,000 diameters in the electron micrograph on following page. The particular RNA-DNA virions shown in thin section in the micrograph cause leukemia in mice; they are similar in structure and function to the Rous sarcoma virions discussed in this article. The electron micrograph was made by N. Sarkar of the Institute for Medical Research in Camden, N.J. A diagram of structure of a virion of this type is given above.

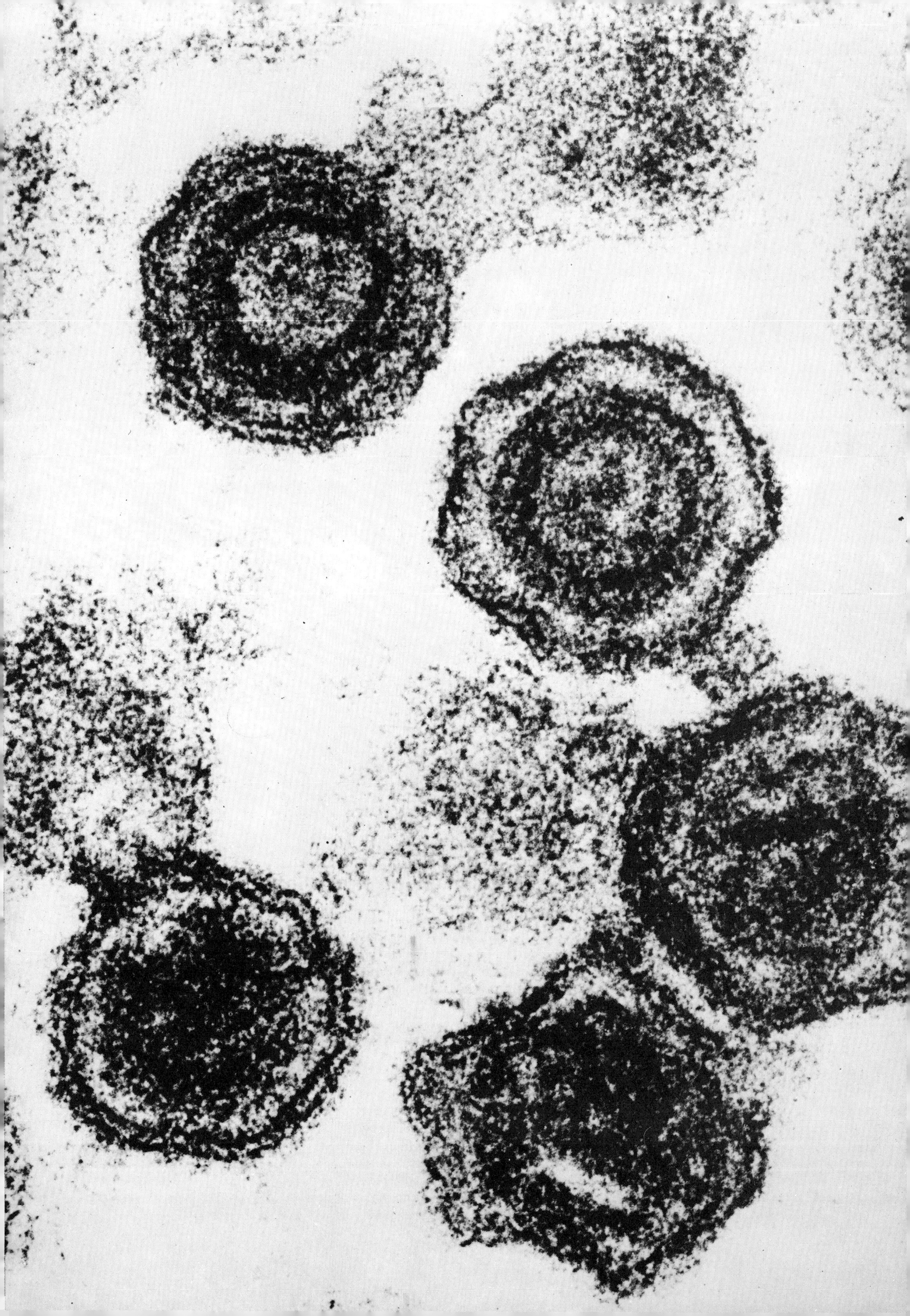

to protein) that are found in cells and in DNA viruses. The rousviruses do not transfer information from RNA to RNA, as other RNA viruses do. The existence of the RNA-to-DNA mode of information transfer in the replication of rousviruses has led some to suggest that there should be three major classes of viruses: DNA viruses, RNA viruses and RNA-DNA viruses [*see top illustration on page 98*].

The prototype RNA tumor virus, the Rous chicken sarcoma virus, was discovered by Rous 61 years ago at the Rockefeller Institute for Medical Research. An RNA tumor virus had actually been found earlier by V. Ellerman and O. Bang of Copenhagen, but their virus was little studied because it caused leukemia in chickens and was harder to work with than Rous's virus. Rous was studying a transplantable tumor of the barred Plymouth Rock hen. Originally he observed that he could transfer the tumor by the transfer of cells. In 1911 he found that the tumor could also be transferred by means of fluid from which the cells had been filtered. Demonstration that a disease can be transmitted by a cell-free filtrate is commonly accepted as evidence that it is caused by a virus. Descendants of the virus originally discovered by Rous are still being worked on in laboratories all over the world. At the time, however, Rous's discovery was met with disbelief, and after 10 years Rous himself stopped working with the tumor. It was not until nearly 30 years later, when Ludwik Gross of the Veterans Administration Hospital in the Bronx discovered that RNA tumor viruses cause leukemia in mice, that the study of rousviruses became popular.

It is now known that viruses in the same group as the virus originally discovered by Rous, or closely related to it, can cause tumors not only in chickens and mice but also in rats, hamsters, monkeys and many other species of animals. Moreover, viruses of the same group have been isolated from nonmammalian species, including snakes. As yet no bona fide human rousvirus has been discovered. It also appears that some members of this group, for example some of the "associated viruses," do not produce cancer.

In the 1950's, with the beginning of the application of cell-culture methods to animal virology, a tissue-culture assay for the Rous sarcoma virus was developed, first by Robert A. Manaker and Vincent Groupé at Rutgers University and subsequently by Harry Rubin and me at the California Institute of Technology. The assay involves adding suspensions of the virus to sparse cultures of cells taken from the body wall of chicken embryos. The Rous sarcoma virus infects some cells and transforms them into tumor cells. The transformed tumor cells differ in morphology and in growth properties from normal cells and therefore create a focus of altered cells. Assays of the same type have been developed for infections that the Rous sarcoma virus causes in cells taken from turkeys, ducks, quail and rats. Similar assays have also been developed for other transforming rousviruses.

The number of foci of transformed cells is proportional to the number of infectious units of the virus added to the cell culture and provides a rapid and reproducible assay for the Rous sarcoma virus. The use of this assay led to the discovery that the Rous sarcoma virus differs from the other viruses that had been studied up to that time in the way it interacts with the cell. The replication of most viruses is incompatible with cell division; in other words, the virus causes the infected cells to die. Chicken cells infected with the Rous sarcoma virus not only survive but also continue to divide and produce new virus particles [*see middle illustration on page 98*]. When the Rous sarcoma virus infects rat cells, there is a slightly different interaction of the cell and the virus. The rat cells are transformed into cancer cells, which divide, but the transformed cells do not produce the Rous sarcoma virus even though the genome (DNA) of the virus can be shown to be present. Production of the Rous sarcoma virus can be induced if the transformed rat cells are fused with normal chicken cells.

In the early 1960's the antibiotic actinomycin D was found to be very useful in unraveling the flow of genetic information in cells infected with RNA viruses. The antibiotic inhibits the synthesis of RNA made on a DNA template but not the synthesis of RNA made on an RNA template. The antibiotic therefore stops all RNA synthesis in cells infected by RNA viruses except for RNA specifically related to the viral genome. With this new tool it became easy to determine which RNA's were specific for the viruses.

When I added actinomycin D to cultures of cells producing Rous sarcoma virus, however, I found that the antibiotic inhibited the production of *all* RNA. One would have expected the replication of RNA on the template of an RNA viral genome to continue without hindrance [*see bottom illustration on page 98*]. This result was the first direct evidence that the molecular biology of the replication of Rous sarcoma virus was different from that of other RNA viruses. Since that observation was made the inhibition of the replication of rousviruses by actinomycin D has been recognized as one of their defining characteristics. The actinomycin D experiments suggested to me that the Rous sarcoma virus might replicate through a DNA intermediate. This hypothesis is called the DNA provirus hypothesis.

Further experiments, carried out by me and by John P. Bader at the National Cancer Institute, demonstrated that if one inhibits the synthesis of DNA in cells immediately after they have been inoculated with Rous sarcoma virus, one can protect the cells from infection. Here the inhibitors were amethopterin, fluorodeoxyuridine and cytosine arabinoside. These experiments appeared to support the idea that infection requires the synthesis of new viral DNA pro-

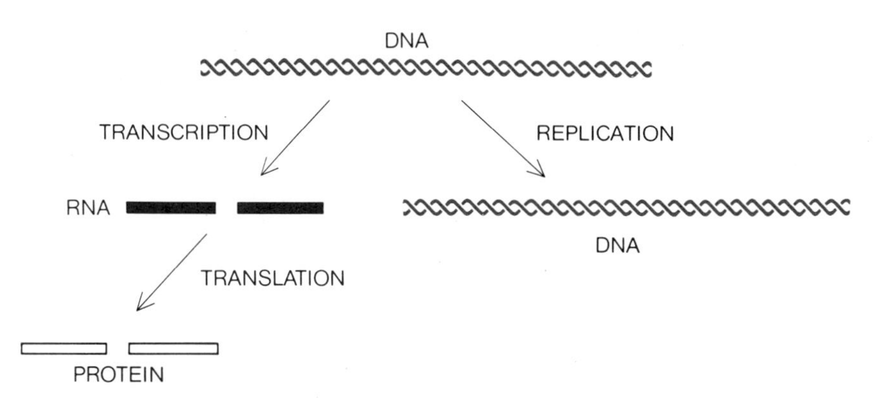

"CENTRAL DOGMA" of molecular biology, originally formulated by Francis Crick, states that within an organism genetic information can be transferred from DNA to DNA or from DNA to RNA to protein, but that it cannot be transferred from protein to protein or from protein to either DNA or RNA. Although a "reverse" flow of genetic information from RNA to DNA was not proscribed in Crick's original formulation, many molecular biologists came to believe that if such a flow were ever discovered, it would violate the central dogma.

duced on an RNA template. This interpretation was not unequivocal, however, because successful production of Rous sarcoma virus requires that the cells divide normally after infection. Therefore the inhibition of DNA synthesis after infection could inhibit production of Rous sarcoma virus not only by blocking possible new viral DNA synthesis but also by preventing normal cell division.

To get around this problem I introduced the idea of infecting cultures of stationary, or nondividing, cells with Rous sarcoma virus. Cells in culture usually require specific factors in blood serum to support their multiplication. If the serum is removed from the medium of the cell cultures, the cells stop dividing. If they are then exposed to Rous sarcoma virus, they become infected but there is no virus production or morphological transformation until serum is added back and the cells divide once again. When such stationary cells are exposed to inhibitors of DNA synthesis, the cells are not killed because they are not making DNA. When the stationary cells are exposed simultaneously to Rous sarcoma virus and to inhibitors of DNA synthesis, the cells are not killed but neither are they infected [*see illustration at right*].

If one now removes the inhibitor of DNA synthesis and adds serum, enabling the cells to divide once more, one finds that the cells remain free of infection. They do not become transformed and they do not produce virus. These experiments supported the hypothesis that after cells are infected by the Rous sarcoma virus new viral DNA is synthesized at a time different from the cell's normal synthesis of DNA. The new viral DNA is evidently synthesized on a template of viral RNA.

A further extension of this approach to understanding the replication of Rous sarcoma virus was carried out by one of my students, David E. Boettiger, and independently by Piero Balduzzi and Herbert R. Morgan at the University of Rochester School of Medicine and Dentistry. It had been found by others that if 5-bromodeoxyuridine, an analogue of the DNA constituent thymidine, is incorporated into DNA, the DNA becomes sensitized so that it can be inactivated by light. Under the same conditions normal thymidine-containing DNA is not affected by light. Boettiger therefore exposed stationary cells to Rous sarcoma virus in the presence of bromodeoxyuridine and then exposed the cells to light. Although the cells were not killed, the treatment prevented their being infected by the virus. When serum was again added to enable the cells to divide, they did not become transformed and did not produce virus [*see illustration on page 96*].

In a related experiment Boettiger showed that the rate of inactivation of the infection by Rous sarcoma virus was dependent on the number of viruses infecting a cell. As he raised the number of viruses infecting each cell, he found that the infection became increasingly resistant to inactivation by light. We interpreted these experiments as showing that each infecting virus makes a new specific DNA, and that the more viruses that infect a cell, the more molecules of new viral DNA that are produced. The experiment seemed to effectively rule out the alternative hypothesis, which was that the infecting virus provokes a new synthesis of some preexisting cellular DNA.

Unfortunately no one has yet been able to unequivocally demonstrate the existence of newly synthesized viral DNA in cells infected with the Rous sarcoma virus. The available techniques are evidently too crude to detect the tiny amounts of new viral DNA expected to be present. Certain results have been reported, however, with transformed cells. One approach has been to bring DNA from infected cells together with labeled viral RNA to see if single strands of the two molecules would coalesce into a double-strand hybrid molecule. Such hybrids are readily created when the base sequences in the DNA are complementary to the base sequences in the RNA, indicating that both carry the same genetic message and hence that each could arise from the transcription of the other.

The hybridization experiments reported thus far have aroused a great deal of controversy. Although some experiments, notably those of Marcel A. Baluda and Debi P. Nayak of the University of California at Los Angeles, have seemed to demonstrate the presence in infected cells of DNA complementary to viral RNA, the results have not been universally accepted. The finding of an intermediate viral DNA is an essential link in the chain of evidence that is still needed to establish firmly the DNA provirus hypothesis.

Meanwhile strong support for the hy-

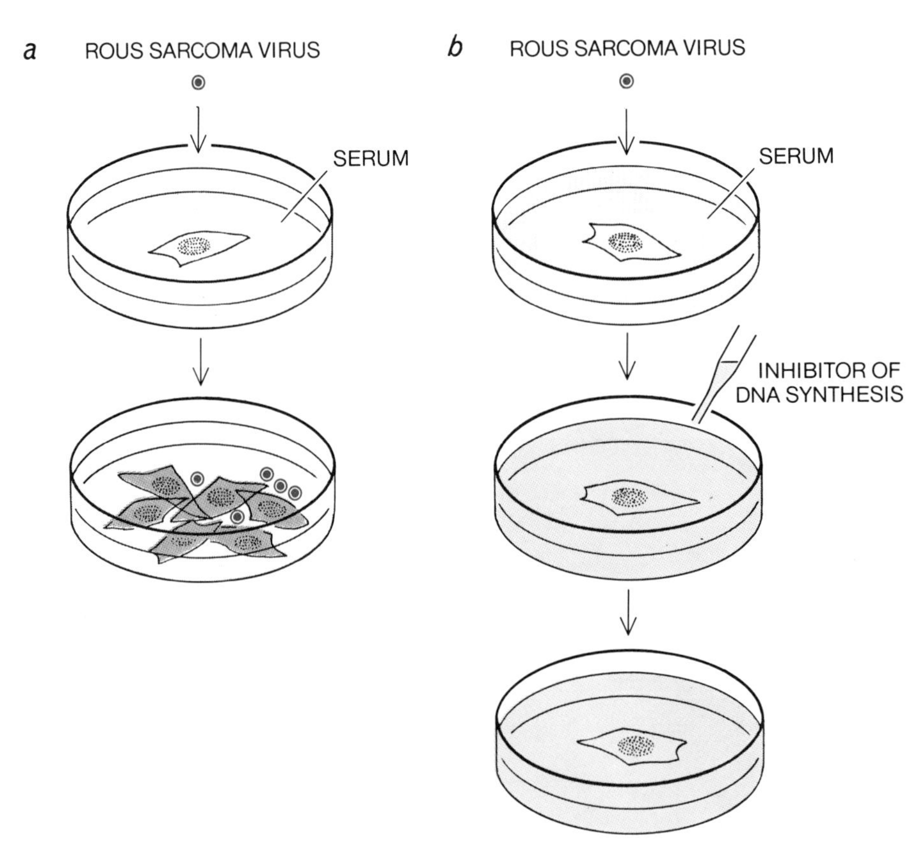

EXPERIMENTS carried out by the author and by John P. Bader at the National Cancer Institute supported the hypothesis that the infection of cells with Rous sarcoma virus requires the synthesis of new viral DNA produced on an RNA template. When the virus is added to cultures of normally dividing cells (*a*), the cells are transformed into cancer cells, which divide and produce new Rous sarcoma virus. By adding a substance that inhibits the synthesis of DNA in the cells immediately after they have been inoculated with Rous sar-

pothesis has come from experiments of a different kind. In 1969 Satoshi Mizutani, who had written his doctoral thesis on bacterial viruses, came to my laboratory for postdoctoral training. We decided to ask the question: What is the origin of the enzyme (a protein) responsible for forming proviral DNA using the viral RNA as template? When Mizutani exposed stationary cells to Rous sarcoma virus in the presence of inhibitors of protein synthesis, he found that the cells still became infected. We interpreted this experiment to mean that the enzyme that synthesizes DNA from the viral RNA template is already in existence before the infection.

Somewhat earlier other workers had fractionated virions—the actual virus particles as distinct from the forms assumed by the virus inside cells—and had found RNA polymerases, enzymes that catalyze the synthesis of RNA from its building blocks: four different ribonucleoside triphosphates. In 1967 Joseph Kates and B. R. McAuslan of Princeton University and William Munyon, E. Paoletti and J. T. Grace, Jr., of the Roswell Park Institute had found RNA polymerases in a poxvirus, a large DNA virus. Other workers had found another RNA polymerase in a reovirus, a double-strand RNA virus. Therefore we decided to look in the virions of Rous sarcoma virus for a DNA polymerase capable of using the viral RNA as a template. After several months of preliminary experiments we succeeded in showing the existence of a DNA polymerase in purified virions of Rous sarcoma virus.

Before discussing this result I should digress briefly to describe the structure of the Rous sarcoma virus [*see illustration, page 92*]. The virion of the Rous sarcoma virus has a diameter of about 100 nanometers, which makes it larger than the particles of the viruses that cause poliomyelitis and smaller than the particles of the viruses that cause smallpox. The virion of the Rous sarcoma virus consists of a lipid-containing envelope (derived by budding from the cell membrane), an inner membrane and a nucleoid, or core, that contains the viral RNA and certain proteins.

In order to demonstrate that the Rous sarcoma virus contains a polymerase capable of producing DNA on an RNA template, we first treated the virion with a detergent to disrupt its lipid-containing envelope. We then added to the disrupted virus the four deoxyribonucleoside triphosphates that are the building blocks of DNA. One of the deoxyribonucleoside triphosphates was radioactively labeled.

When the mixture was incubated at 40 degrees Celsius, it incorporated the radioactive label into an acid-insoluble substance that met the usual tests for DNA. The substance was stable in the presence of alkali and the enzyme ribonuclease, treatments that are known to destroy RNA, whereas it was attacked and fragmented by an enzyme that destroys DNA. When we repeated the experiment with disrupted virions pretreated with ribonuclease, an enzyme that destroys RNA, little or no DNA was produced, indicating that intact viral RNA was needed as the template for the synthesis of DNA [*see top illustration on page 97*].

After we had announced these results at the Tenth International Cancer Congress in Houston in May, 1970, we learned that David Baltimore of the

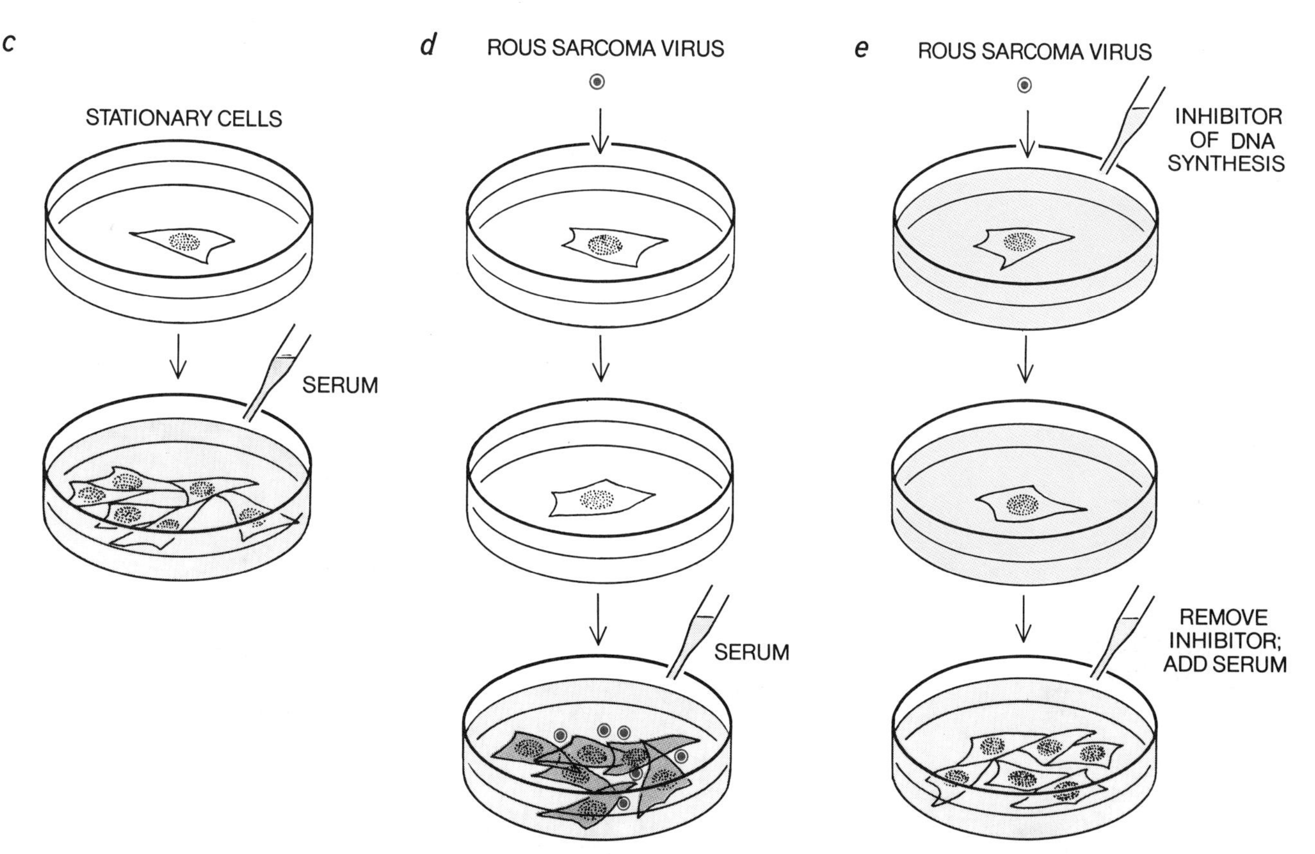

coma virus (*b*), one can protect the cells from infection. In subsequent experiments by the author cultures of stationary, or nondividing, cells were used; when blood serum is added to such cultures (*c*), they divide normally. If such stationary cells are first exposed to Rous sarcoma virus (*d*), however, they become infected but there is no virus production or morphological transformation until serum is added back and the cells divide once again. When the stationary cells are exposed simultaneously to Rous sarcoma virus and to an inhibitor of DNA synthesis (*e*), the cells are not killed but neither are they infected; when the inhibitor of DNA synthesis is removed and serum is added, cells divide normally, are not infected, do not become transformed and do not produce virus.

Massachusetts Institute of Technology had independently made similar observations with the virion of a mouse leukemia virus. The two papers describing these findings were published together in the June 27, 1970, issue of *Nature,* the British scientific weekly. The two publications stimulated an enormous amount of work whose peak is not yet in sight.

In our early papers we called the new viral enzyme RNA-dependent DNA polymerase because the template was RNA and the product was DNA. Subsequently we and others found that the enzyme could also use DNA as a template for DNA synthesis. We therefore decided to change the word "dependent" to "directed," so that we now refer to the enzyme as RNA-directed DNA polymerase. The revised name makes no statement about the origin of the enzyme or its relation to other DNA polymerases. Independently *Nature* began referring to the enzyme as "reverse transcriptase," a name that I do not like because of its ambiguity but that has gained wide currency.

All the later studies confirm the original finding that the virions of RNA tumor viruses contain a DNA polymerase system that is activated by treating the virion with a detergent and that is sensitive to ribonuclease. Moreover, the virion enzyme functions only as a DNA polymerase; it will not act as an RNA polymerase. As I have mentioned, however, other unrelated RNA viruses do contain an RNA polymerase.

If the DNA produced by the RNA-directed DNA polymerase is isolated free of protein, the size of its molecule can be estimated by spinning it at high speed in a sucrose gradient in an ultracentrifuge. The molecule is surprisingly small: less than a tenth as long as one would expect a copy of the complete viral RNA to be. The reason for the small size is still elusive. If the isolated DNA product is centrifuged in a cesium sulfate density gradient, which separates RNA from DNA on the basis of their different densities, one finds that the product has the density of DNA [*see bottom illustration on opposite page*]. Further characterization, for example by treatment with enzymes that specifically attack either single- or double-strand DNA, shows that the product of the DNA polymerase system is a double strand. From such studies one can conclude that the DNA polymerase system of the virion makes short pieces of double-strand DNA.

Many workers have demonstrated that the DNA product of the RNA-directed DNA polymerase system has a base sequence complementary to the viral RNA [*see top illustration on page 99*]. This conclusion is drawn from annealing, or molecular hybridization, experiments. Labeled DNA from the virion polymerase reaction is treated so that the strands of the DNA dissociate. The single-strand DNA is added to unlabeled viral RNA, and the mixture is incubated so that complementary strands can form a hybrid combination. The mixture is then centrifuged in a cesium sulfate density gradient. About half of the product DNA forms a band at a density characteristic of RNA or of hybrid RNA-DNA molecules rather than at a density characteristic of DNA. The test is quite specific and indicates that the DNA polymerase of the virion copies the sequence of the bases of the viral RNA into DNA. This experiment, however, still does not demonstrate that such a copying process takes place in cells infected by Rous sarcoma virus.

The viral DNA polymerase was shown to be present in the core of the virion by the following experiment carried out by George Todaro's group at the National Cancer Institute and by John M. Coffin in my laboratory at the University of Wisconsin. Rousvirus virions were treated with a detergent to disrupt the envelope. Then the disrupted virus was centrifuged in a sucrose density gradient. Most of the viral RNA, about 20 percent of the protein and most of the RNA-directed DNA polymerase activity were found to sediment together in "cores," a term given to structures that are denser than whole virions [*see bottom illustration on page 99*]. Further studies showed that with more extensive disruption of the virion the viral DNA polymerase can be freed from the viral RNA and then purified. The purified enzyme is capable of directing the synthesis of DNA on a variety of templates: synthetic and natural DNA, RNA and RNA-DNA hybrids.

The general conclusion from studies in a number of laboratories is that the

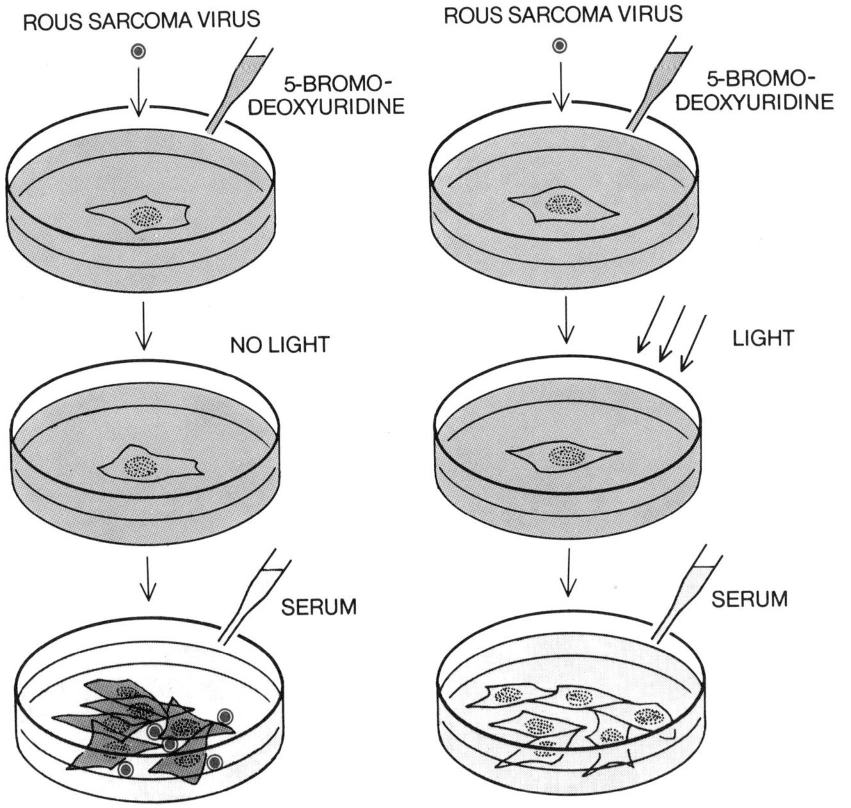

FURTHER EXPERIMENTS, carried out by one of the author's students, David E. Boettiger, and independently by Piero Balduzzi and Herbert R. Morgan at the University of Rochester, involved exposing stationary cells to Rous sarcoma virus in the presence of 5-bromodeoxyuridine, an analogue of the DNA constituent thymidine that, when incorporated into DNA, sensitizes the DNA to inactivation by light. As a control some of the treated cells were first not exposed to light (*left*); after serum was added to these cells to enable them to divide they were transformed into cancer cells and began to produce virus. When another culture of treated cells was exposed to light (*right*), the cells were not killed, but the treatment prevented their infection by the virus. When serum was again added to enable these cells to divide, they did not become transformed and did not produce virus.

rousvirus DNA polymerase closely resembles the other DNA polymerases described above that are present in more familiar biological systems and that catalyze the synthesis of DNA on a DNA template. In other words, it is not a unique property of the rousvirus DNA polymerase to be able to use RNA as a template for DNA synthesis. (This was first proposed several years ago by Sylvia Lee Huang and Liebe F. Cavalieri of the Sloan-Kettering Institute.) What is unique so far is the apparent biological role of RNA-directed DNA synthesis in the replication of rousviruses.

Further work in my laboratory has shown that preparations of purified virions of the Rous sarcoma virus contain other enzymes related to DNA replication. The most unusual of them is an enzyme that is named polynucleotide ligase, which repairs breaks in DNA molecules. It is an attractive hypothesis that the function of the ligase is to join the viral DNA to the chromosomal DNA of the host cell, thus integrating the viral genome with the cell genome. After this integration the genetic information of the virus would be replicated with that of the host and passed from the parent cell to the daughter cell. The Rous sarcoma virus virion also contains many other enzymes whose role is completely unknown. We do not know whether they participate in the life cycle of the virus or whether they are merely accidental contaminants picked up in the formation of the virion.

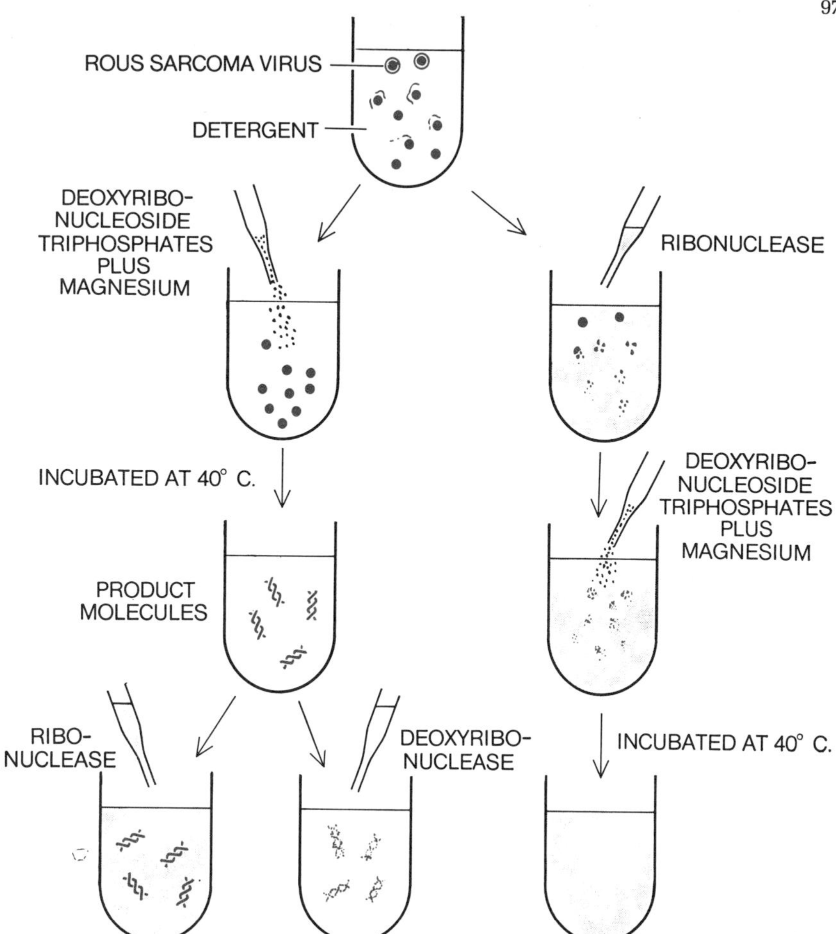

EXISTENCE OF A POLYMERASE capable of producing DNA on an RNA template in RNA tumor viruses was demonstrated by the author and his colleague Satoshi Mizutani (and also independently by David Baltimore of the Massachusetts Institute of Technology). In the experiment conducted by Mizutani and the author purified virions of Rous sarcoma virus were first treated with a detergent to disrupt their lipid-containing envelope. Four deoxyribonucleoside triphosphates, the "building blocks" of DNA, were then added to the disrupted virions. When the mixture was incubated, it incorporated the radioactive label associated with one of the building blocks into an acid-insoluble substance that was stable in the presence of ribonuclease (an enzyme known to destroy RNA), whereas it was fragmented by deoxyribonuclease (an enzyme that destroys DNA). When the experiment was repeated with disrupted virions pretreated with ribonuclease, little or no DNA was produced, indicating that intact viral RNA was needed as template for synthesis of DNA.

After the first discovery of a DNA polymerase in the virions of RNA tumor viruses, a great many other RNA viruses were examined to see if they contain a similar DNA polymerase system. First it was found that all the viruses previously classified in the RNA tumor virus group contain such an enzyme system. This group of RNA viruses includes both the rousviruses that cause tumors and those that do not cause tumors. Even more interesting, it was found that two types of virus that had not been classified in the same group with RNA tumor viruses also contain a DNA polymerase system. One of these viruses is Visna virus, which causes a slowly developing neurological disease in sheep. After the demonstration of a DNA polymerase in virions of the Visna virus, Kenneth Kaname Takemoto and L. B. Stone at the National Institutes of Health showed that the same virus could cause cancerous transformation of mouse cells in culture. Therefore Visna virus can now be considered a transforming rousvirus. The other type of virus that has been found to have a DNA polymerase system is the "foamy," or syncytium-forming, viruses. These viruses, isolated from monkeys and cats, have not been connected with any particular disease but are common contaminants of cell cultures. They have not yet been shown to cause tumors or cancerous transformation.

The DNA polymerase present in RNA tumor viruses may not only explain how these viruses produce stable cancerous transformations in the cells they infect but also account for some viral latency,

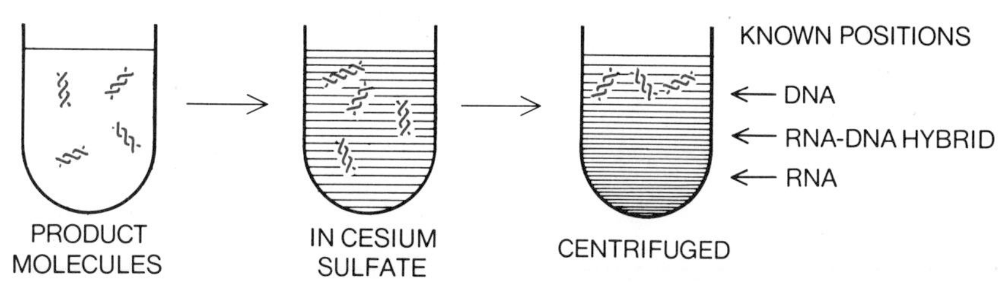

CENTRIFUGATION of the isolated DNA product of the RNA-directed DNA polymerase system in a cesium sulfate density gradient (which separates RNA from DNA on the basis of their different densities) resulted in the finding that the product has the density of DNA. In combination with other findings this result led to the conclusion that the DNA polymerase system of the Rous sarcoma virus virion makes short pieces of double-strand DNA.

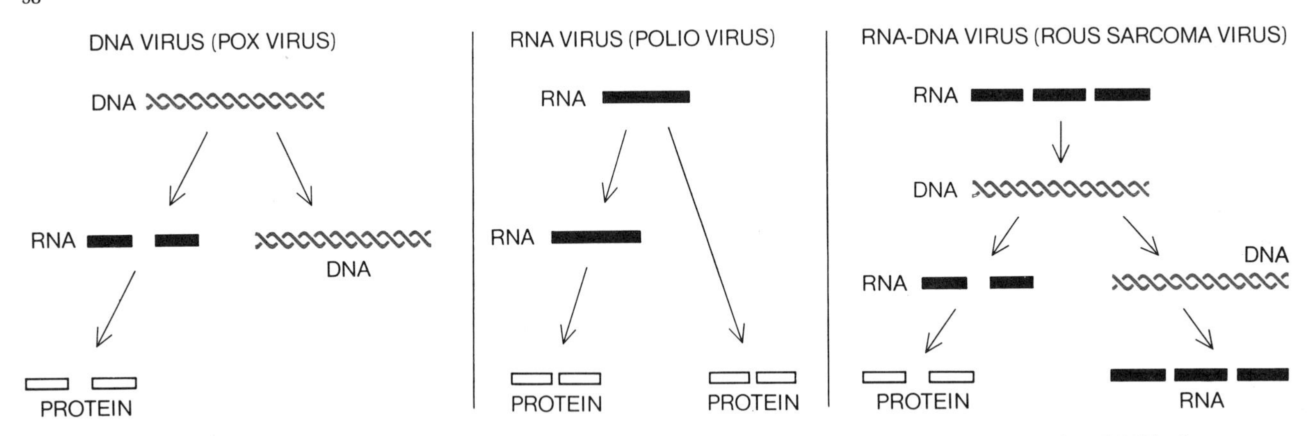

VIRUSES CAN BE GROUPED into three major classes: DNA viruses (*left*), whose genome, or complete set of genes, consists of DNA; RNA viruses (*middle*), whose genome consists of RNA, and RNA-DNA viruses (*right*), the most recently discovered group, whose genome consists alternately of RNA and DNA. A prototype virus in each major class is indicated in parentheses next to the class name. The diagrams illustrate the mode of information transfer that characterizes the replication of viruses in each class.

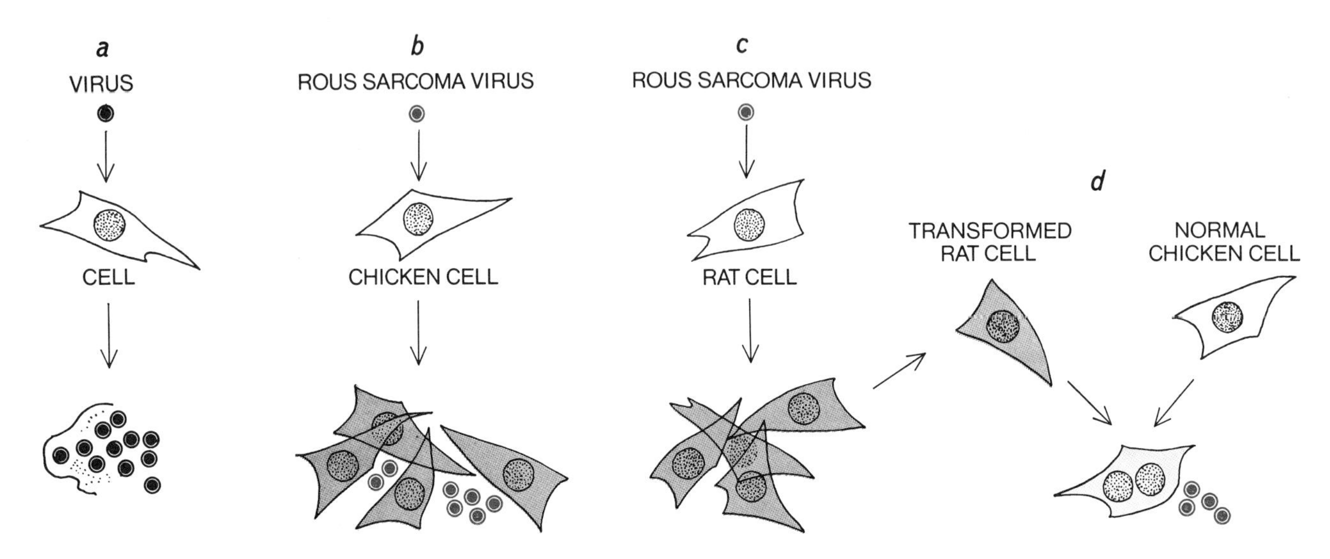

VIRUS-CELL INTERACTION usually leads to the death of the infected cell (*a*), since the replication of most viruses is incompatible with cell division. The Rous sarcoma virus, however, interacts with cells in a different way. Chicken cells infected with the Rous sarcoma virus (*b*) not only survive but also are transformed into cancer cells, which continue to divide and produce new virions. Rat cells infected with the Rous sarcoma virus (*c*) are transformed into cancer cells, which divide but do not produce new virions. By fusing the transformed rat cells with normal chicken cells the production of Rous sarcoma virions can be induced (*d*).

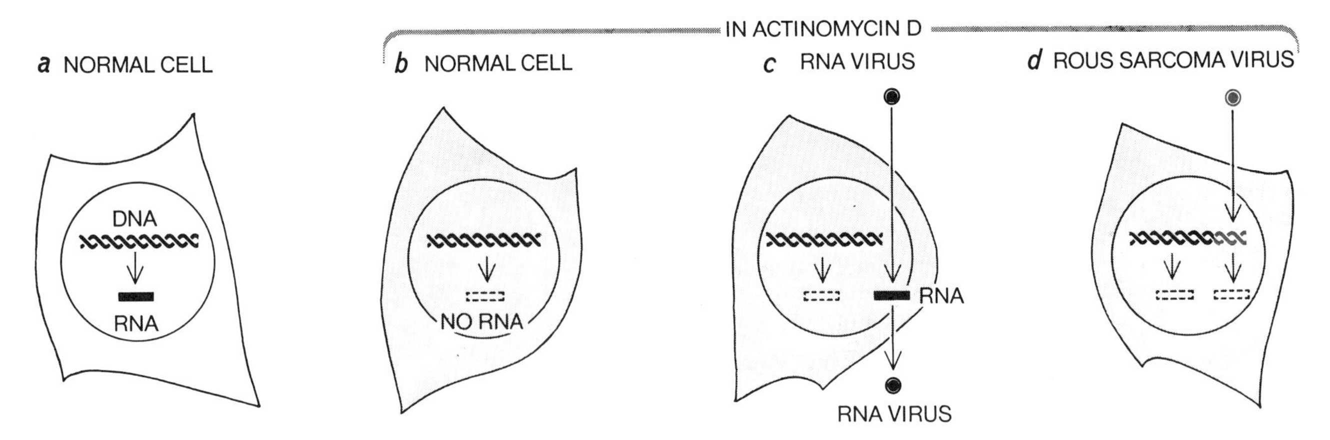

SYNTHESIS OF RNA on a DNA template in normal cells (*a*) is inhibited by the addition of the antibiotic actinomycin D (*b*). Since the antibiotic does not affect the synthesis of RNA made on an RNA template, however, it does not stop RNA synthesis specifically related to the viral genome in cells infected by most RNA viruses (*c*). The finding that actinomycin D inhibited the production of *all* RNA in cells producing Rous sarcoma virus (*d*) was the first direct evidence that the molecular biology of the replication of Rous sarcoma virus was different from that of other RNA viruses. The actinomycin D experiments led the author to propose the DNA provirus hypothesis, which holds that rousviruses such as the Rous sarcoma virus replicate through a DNA intermediate.

the phenomenon in which a virus disappears after infecting an organism only to reappear months or years later. Once an RNA virus has transferred its genetic information to DNA, it would be able to remain latent in a cell and be replicated by the cellular enzyme systems that replicate and repair the cell DNA. After some later activation the virus could appear again as infectious virus particles [*see top illustration on following page*].

About a year ago considerable public excitement was generated by the reported discovery of "RNA-dependent DNA polymerases" in human tumor cells. The general conclusion I would draw now from most of this work has been stated above: All DNA polymerases are capable, under the appropriate conditions, of transcribing information from RNA into DNA. At present we lack generally accepted criteria for determining whether or not such syntheses have any biological role or any relation to rousviruses.

In my laboratory we have taken a slightly different approach to the question of RNA-directed DNA synthesis in cells. We have used detergent activation and ribonuclease sensitivity as criteria in a broad search for DNA polymerase systems in a variety of animal cells. That is, we have looked in cells for a DNA polymerase system similar to viral "cores." Coffin has found such a DNA polymerase system in normal, uninfected rat embryo cells. So far we do not know the full significance of this discovery, but it suggests that ribonuclease-sensitive DNA polymerase systems are present in cells other than tumor cells or virus-infected cells.

For many years I have favored the idea that RNA-directed DNA synthesis may be important in normal cellular processes, particularly those involved in the embryonic differentiation of cells. This idea has been expanded in the form of the protovirus hypothesis [*see bottom illustration on following page*]. The general idea is that in normal cells there are regions of DNA that serve as templates for the synthesis of RNA, and that this RNA serves in turn as a template for the synthesis of DNA that subsequently becomes integrated with the cellular DNA. By this means certain regions of DNA can be amplified. With additional processes that introduce changes in the DNA, the DNA of different cells can be made different. This difference might serve as a means of distinguishing different cells.

What, then, are the general implications of this work for the prevention or treatment of human cancer? We can

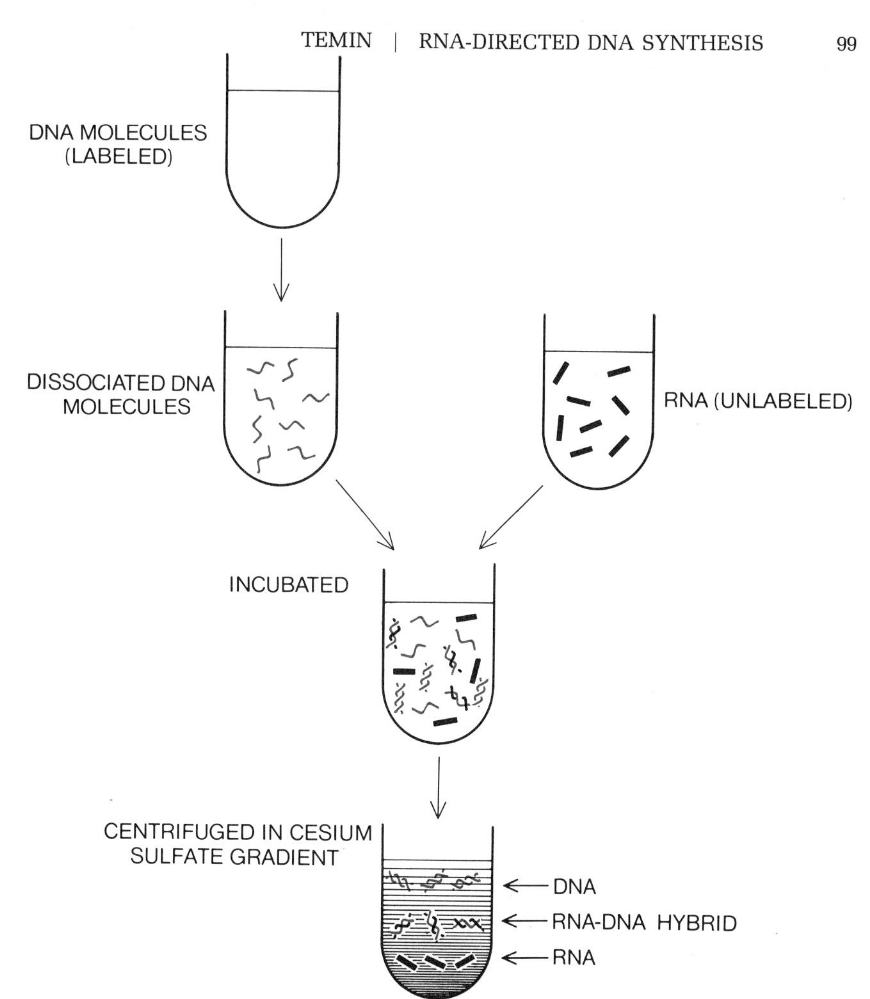

MOLECULAR-HYBRIDIZATION EXPERIMENTS demonstrated that the DNA product of the RNA-directed DNA polymerase system within the virion copies the sequence of bases of the viral RNA into DNA. Labeled DNA from the virion polymerase reaction was first treated so that strands of the DNA dissociated. The single-strand DNA was then added to unlabeled viral RNA, and the mixture was incubated at high temperature so that complementary strands could form a hybrid combination. When the resulting "annealed" mixture was centrifuged in a cesium sulfate density gradient, about half of the product DNA was observed to form a band at a density characteristic of hybrid RNA-DNA molecules.

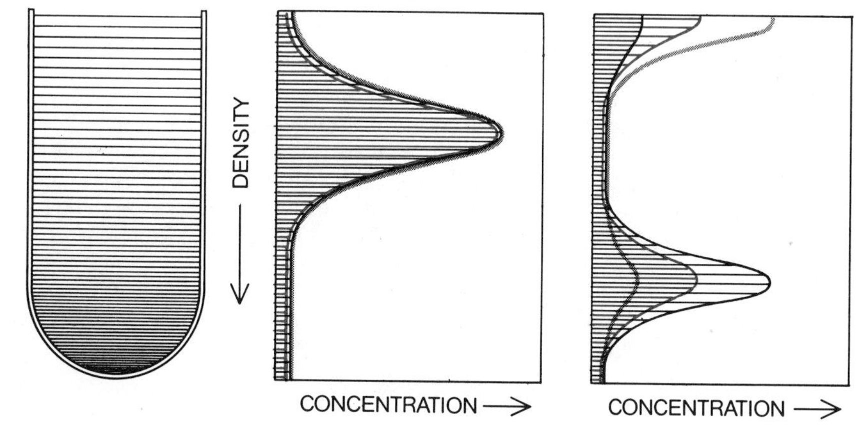

PRESENCE OF VIRAL DNA POLYMERASE in the cores of the Rous sarcoma virus virions was demonstrated by John M. Coffin in the author's laboratory. The curves at center show the density distribution of various radioactively labeled constitutents of the whole virions as determined by centrifugation in a sucrose density gradient (*left*). The curves at right show the density distribution of the same constituents determined by centrifugation after the virions were treated with a detergent to disrupt their envelopes. Most of the viral RNA (*black curves*), about 20 percent of the protein (*gray curves*) and most of the RNA-directed DNA polymerase activity (*colored curves*) of the disrupted virions were found to sediment together at a higher density than the corresponding constituents of the whole virions, indicating that these constituents are concentrated in the cores of the virions.

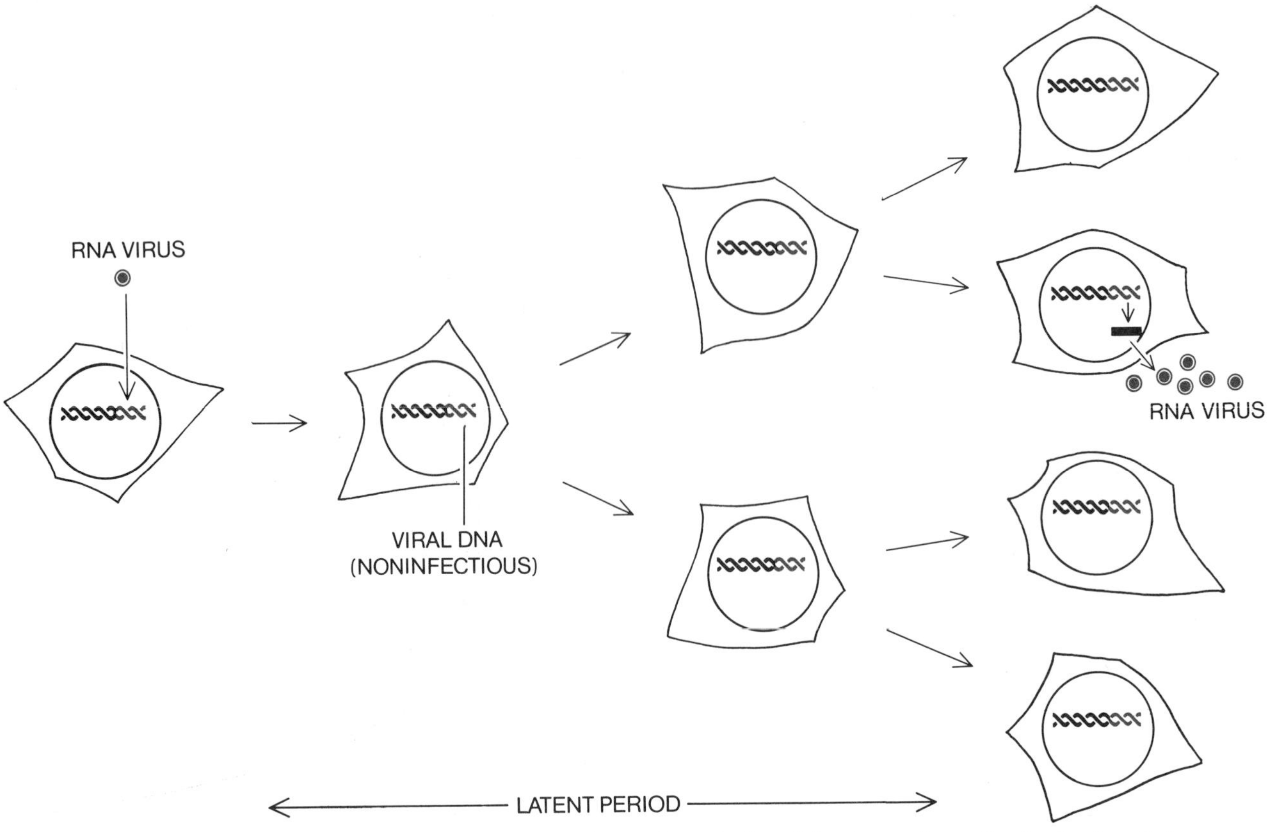

LATENCY OF RNA VIRUSES after infecting an organism may be attributable to the DNA polymerase system present in the cores of such viruses. After transferring its genetic information to DNA in the cell nucleus (*left*), the RNA virus would disappear, remaining latent in the cell by virtue of its replication by the cellular enzyme systems that replicate and repair the cell's DNA (*left center, right center*). Months or years later some form of activation could then cause the infectious RNA virions to appear again (*upper right*).

conclude only that some biological systems utilize a previously undescribed mode of information transfer: from RNA to DNA. It is an interesting coincidence that this new mode of information transfer was first discovered in tumor-causing viruses. We cannot say, however, that RNA-directed DNA synthesis is an exclusive property of such viruses. What the discovery of RNA-directed DNA synthesis does mean is that we now have some simple biochemical tests to determine whether or not newly discovered human viruses are members of the same group as the RNA viruses that produce tumors and cancerous transformations in animal cells, and to look for information related to these viruses in human cancers. We cannot now say that inhibitors of RNA-directed DNA synthesis would have any effect on human cancer. In rousvirus-induced tumors in animals the synthesis of new viral DNA appears to be important only at the initial stage of cancerous transformation, not thereafter.

Probably the most important implication of this discovery for the understanding of cancer in man has been the removal of the dichotomy between viral and genetic theories of the origin of cancer. At a time when genes were thought to consist of DNA alterable only by mutation, and when most of the known cancer-causing animal viruses were of the RNA type, it was hard to imagine common features of genetic and viral theories. Now that we have uncovered evidence that cancer-causing RNA viruses can produce a DNA transcript of the viral RNA, one can readily formulate hypotheses in which elements related to viral RNA are attached to the genome of the cell and transmitted genetically to become activated at some future time and cause "spontaneous" cancer. Experiments designed to test this idea are now in progress in a number of laboratories around the world.

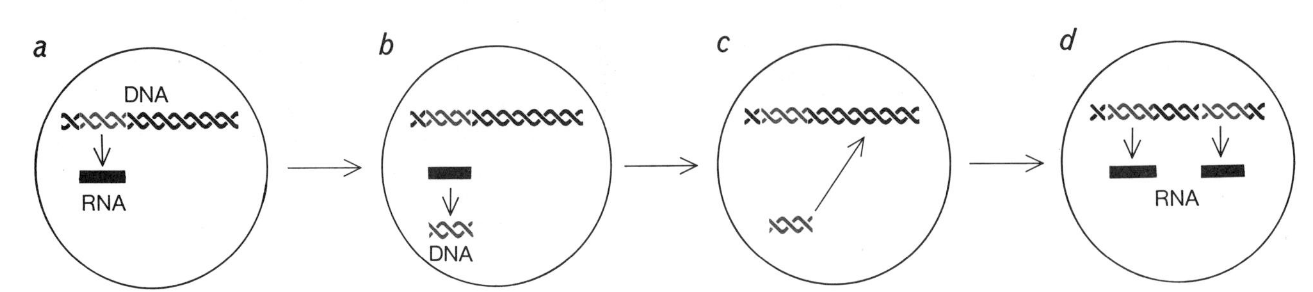

PROTOVIRUS HYPOTHESIS, put forward by the author, embodies the idea that RNA-directed DNA synthesis may be important in normal cellular processes. According to this view, there are regions of DNA in normal cells that serve as templates for the synthesis of RNA (*a*). This RNA serves in turn as a template for the synthesis of DNA (*b*), which later becomes integrated with the cellular DNA (*c*). The amplification of certain regions of DNA resulting from the repetition of the process (*d*) may, in conjunction with additional processes that introduce changes in the DNA, play an important role in the embryonic differentiation of cells.

The Synthesis of DNA

by Arthur Kornberg
October 1968

Test-tube synthesis of the double helix that controls heredity climaxes a half-century of effort by biochemists to re-create biologically active giant molecules outside the living cell

My colleagues and I first undertook to synthesize nucleic acids outside the living cell, with the help of cellular enzymes, in 1954. A year earlier James Watson and Francis Crick had proposed their double-helix model of DNA, the nucleic acid that conveys genetic information from generation to generation in all organisms except certain viruses. We attained our goal within a year, but not until some months ago—14 years later—were we able to report a completely synthetic DNA, made with natural DNA as a template, that has the full biological activity of the native material.

Our starting point was an unusual single-strand form of DNA found in the bacterial virus designated ϕX174. The single strand is in the form of a closed loop. When ϕX174 infects cells of the bacterium *Escherichia coli,* the single-strand loop of DNA serves as the template that directs enzymes in the synthesis of a second loop of DNA. The two loops form a ring-shaped double helix similar to the DNA helixes found in bacterial cells and higher organisms. In our laboratory at the Stanford University School of Medicine we succeeded in reconstructing the synthesis of the single-strand DNA copies of viral DNA and finally in making a completely synthetic double helix. The way now seems open for the synthesis of DNA from other sources: viruses associated with human disease, bacteria, multicellular organisms and ultimately the DNA of vertebrates such as mammals.

An Earlier Beginning

The story of the cell-free synthesis of DNA does not start with the revelation of the structure of DNA in 1953. It begins around 1900 with the biochemical understanding of how the fermentation of fruit juices yields alcohol. Some 40 years earlier Louis Pasteur had convinced his contemporaries that the living yeast cell played an essential role in the fermentation process. Then Eduard Buchner observed in 1897 that a cell-free juice obtained from yeast was just as effective as intact cells for converting sugar to alcohol. This observation opened the era of modern biochemistry.

During the first half of this century biochemists resolved the overall conversion of sucrose to alcohol into a sequence of 14 reactions, each catalyzed by a specific enzyme. When this fermentation proceeds in the absence of air, each molecule of sucrose consumed gives rise to four molecules of adenosine triphosphate (ATP), the universal currency of energy exchange in living cells. The energy represented by the fourfold output of ATP per molecule of sucrose is sufficient to maintain the growth and multiplication of yeast cells. When the fermentation takes place in air, the oxidation of sucrose goes to completion, yielding carbon dioxide and water along with 18 times as much energy as the anaerobic process does. This understanding of how the combustion of sugar provides energy for cell metabolism was succeeded by similar explanations of how enzymes catalyze the oxidation of fatty acids, amino acids and the subunits of nucleic acids for the energy needs of the cell.

By 1950 the enzymatic dismantling of large molecules was well understood. Little thought or effort had yet been invested, however, in exploring how the cell makes large molecules out of small ones. In fact, many biochemists doubted that biosynthetic pathways could be suc-

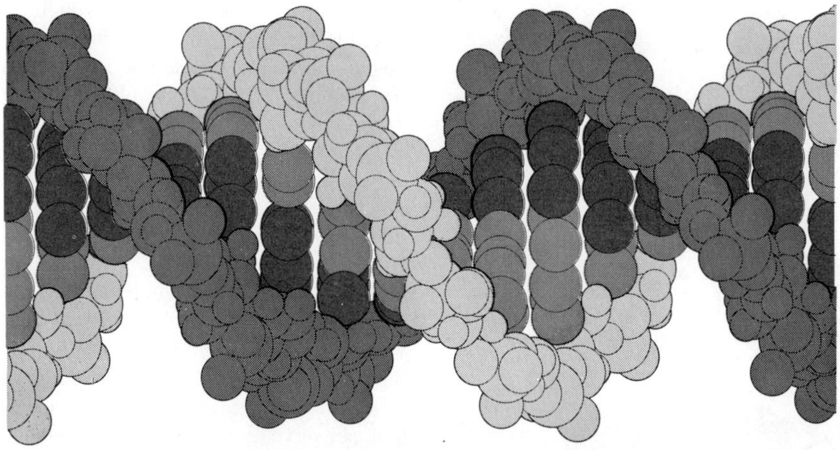

DOUBLE HELIX, the celebrated model of deoxyribonucleic acid (DNA) proposed in 1953 by James D. Watson and F. H. C. Crick, consists of two strands held together by crossties (*color*) that spell out a genetic message, unique for each organism. The Watson-Crick model explained for the first time how each crosstic consists of two subunits, called bases, that form obligatory pairs (*see illustrations on page 102*). Thus each strand of the double helix and its associated sequence of bases is complementary to the other strand and its bases. Consequently each strand can serve as a template for the reconstruction of the other strand.

ADENINE GUANINE

THYMINE CYTOSINE

DEOXYRIBOSE PHOSPHATE

DNA CONSTITUENTS are bases of four kinds, deoxyribose (a sugar) and a simple phosphate. The bases are adenine (*A*) and thymine (*T*), which form one obligatory pair, and guanine (*G*) and cytosine (*C*), which form another. Deoxyribose and phosphate form the backbone of each strand of the DNA molecule. The bases provide the code letters of the genetic message. For purposes of tagging synthetic DNA, thymine can be replaced by 5'-bromouracil, which contains a bromine atom where thymine contains a lighter CH_3 group.

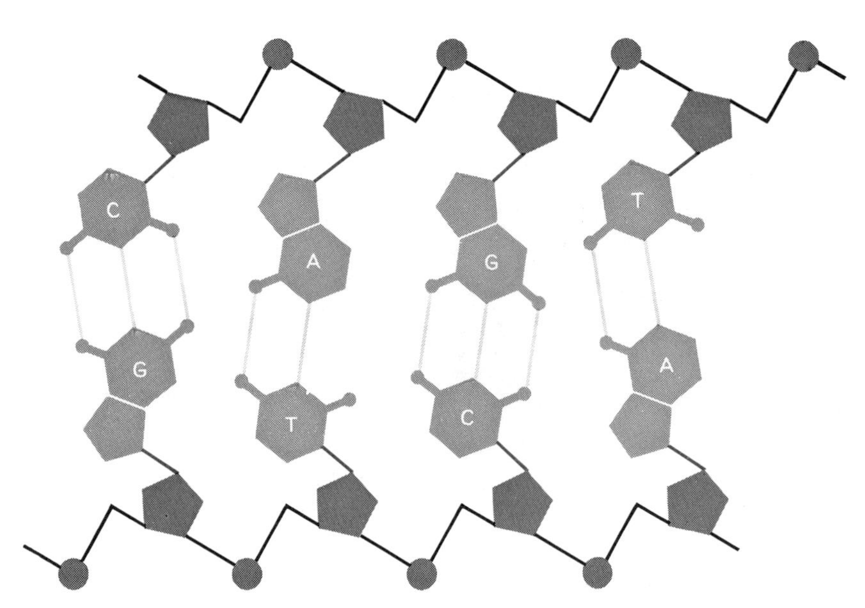

DNA STRUCTURE resembles a ladder in which the side pieces consist of alternating units of deoxyribose and phosphate. The rungs are formed by the bases paired in a special way, A with T and G with C, and held together respectively by two and three hydrogen bonds.

cessfully reconstructed in cell-free systems. Since then nearly two decades of intensive study have been devoted to the cell-free biosynthesis of large molecules. Two things above all have been made clear.

The first is that large molecules can be assembled in cell-free systems with the aid of purified enzymes and coenzymes. The second is that the routes of biosynthesis are different from those of degradation. Some biochemists had speculated that the routes of breakdown were really two-way streets whose flow might somehow be reversed. Now we know that the molecular traffic in cells flows on distinctive and divided highways. All cells have the enzymatic machinery to manufacture most of the subunits of large molecules from simple nutrients such as glucose, ammonia and carbon dioxide. Cells also have the capacity to salvage preformed subunits when they are available. On the basis of what has been learned the prospects are that in this century biochemists will assemble in the test tube complex viruses and major components of the cell. Perhaps the next century will bring the synthesis of a complete cell.

The Nucleotides

My co-workers and I were at Washington University in St. Louis when we made our first attempts to synthesize a nucleic acid in the test tube. By that time the constituents of nucleic acid were well known [*see illustrations at left*]. If one regards DNA as a chain made up of repeating links, the basic link is a structure known as a nucleotide [*see illustration on opposite page*]. It consists of a phosphate group attached to the five-carbon sugar deoxyribose, which is linked in turn to one of four different nitrogen-containing bases. The four bases are adenine (A), thymine (T), guanine (G) and cytosine (C). In the double helix of DNA the phosphate and deoxyribose units alternate to form the two sides of a twisted ladder. The rungs joining the sides consist of two bases: A is invariably linked to T and G is invariably linked to C. This particular pairing arrangement was the key insight of the Watson-Crick model. It means that if the two strands of the helix are separated, uncoupling the paired bases, each half can serve as a template for re-creating the missing half. Thus if the bases projecting from a single strand follow the sequence A, G, G, C, A, T . . . , one immediately knows that the complementary bases on the missing strand are T, C, C, G, T, A. . . . This base-pairing mechanism enables the cell to make accurate copies of the DNA molecule however many times the cell may divide.

When a strand of DNA is taken apart link by link (by treatment with acid or certain enzymes), the phosphate group of the nucleotide may be found attached to carbon No. 3 of the five-carbon deoxyribose sugar. Such a structure is called a 3′-nucleoside monophosphate. We judged, however, that better subunits for purposes of synthesis would be the 5′-nucleoside monophosphates, in which the phosphate linkage is to carbon No. 5 of deoxyribose.

This judgment was based on two lines of evidence. The first had just emerged from an understanding of how the cell itself made nucleotides from glucose, ammonia, carbon dioxide and amino acids. John M. Buchanan of the Massachusetts Institute of Technology had shown that nucleotides containing the bases A and G were naturally synthesized with a 5′ linkage. Our own work had shown the same thing for nucleotides containing T and C. The second line of evidence came from earlier studies my group had conducted at the National Institutes of Health. We had found that certain coenzymes, the simplest molecules formed from two nucleotides, were elaborated from 5′ nucleotide units. For the enzymatic linkage to take place the phosphate of the nucleotide had to be activated by an additional phosphate group [*see illustration on next page*]. Thus it seemed reasonable that activated 5′ nucleotides (nucleoside 5′ triphosphates) might combine with each other, under the proper enzymatic guidance, to form long chains of nucleic acid.

Our initial attempts at nucleic acid synthesis relied principally on two techniques. The first involved the use of radioactive atoms to label the nucleotide so that we could detect the incorporation of even minute amounts of it into nucleic acid. We sought the enzymatic machinery for synthesizing nucleic acids in the juices of the thymus gland, bone marrow and bacterial cells. Unfortunately such extracts also have a potent capacity for degrading nucleic acids. We added our labeled nucleotides to a pool of nucleic acids and hoped that a few synthesized molecules containing a labeled nucleotide would survive by being mixed into the pool. Even if there were net destruction of the pool of nucleic acids, the synthesis of a few molecules trapped in this pool might still be detected. The second technique exploited the fact that the nucleic acid could be precipitated by making the medium strongly acidic, whereas the nucleotide precursors remained behind in solution.

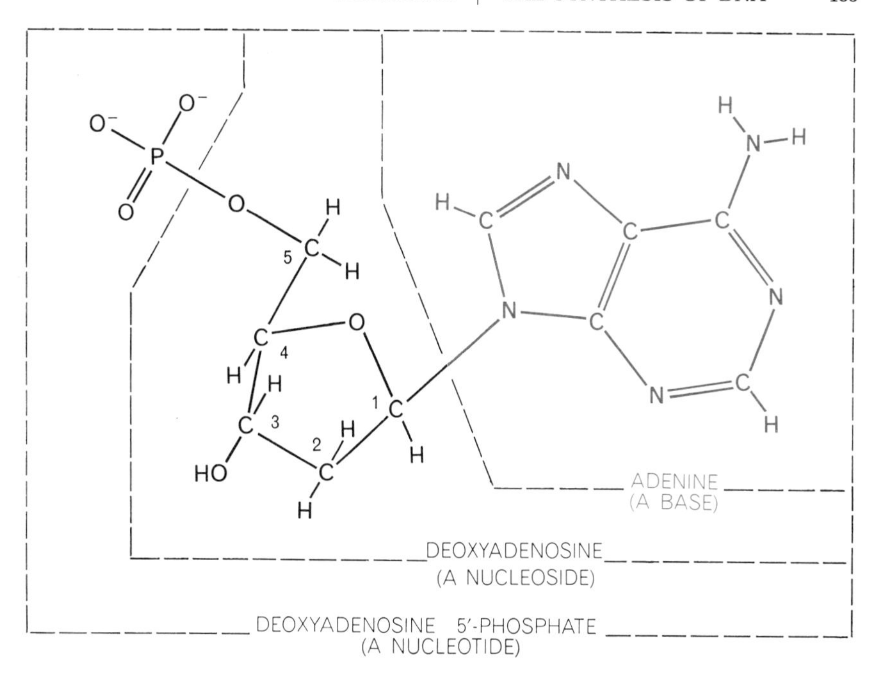

DNA BUILDING BLOCK, the monomer from which DNA polymers are constructed, is termed a nucleotide. There are four nucleotides, one for each of the four bases A, T, G and C. Deoxyadenosine 5′-phosphate, the nucleotide incorporating adenine, is shown here. If the phosphate group is replaced by a hydrogen atom, the structure is called a nucleoside.

Our first experiments with animal-cell extracts were uniformly negative. Therefore we turned to *E. coli*, which has the virtue of reproducing once every 20 minutes. Here we saw a glimmer. In samples to which we had added a quantity of labeled nucleotides whose radioactive atoms disintegrated at the rate of a million per minute we detected about 50 radioactive disintegrations per minute in the nucleic acid fraction that was precipitated by acid. Although the amount of nucleotide incorporated into nucleic acid was minuscule, it was nonetheless significantly above the level of background "noise." Through this tiny crack we tried to drive a wedge. The hammer was enzyme purification, a technique that had matured during the elucidation of alcoholic fermentation.

DNA Polymerase

In these experiments Uriel Littauer, a Fellow of the Weizmann Institute in Israel, and I observed the incorporation of adenylate (a nucleotide) from ATP into ribonucleic acid (RNA), in which the five-carbon sugar in the backbone of the chain is ribose rather than deoxyribose. Actually the first definitive demonstration of synthesis of an RNA-like molecule in a cell-free system had been achieved in the laboratory of Severo Ochoa in 1955. Working at the New York University School of Medicine, he and Marianne Grunberg-Manago were investigating an aspect of energy metabolism and made the unexpected observation that one of the reactants, adenosine diphosphate (ADP), had been polymerized by cell juices into a chain of adenylates resembling RNA.

In our first attempts to achieve DNA synthesis in a cell-free system we used the deoxyribonucleoside called deoxythymidine. To Morris E. Friedkin, who was then at Washington University, we are grateful not only for supplying the radioactively labeled compound but also for the knowledge that the compound was readily incorporated into DNA by bone marrow cells and other animal cells. We were hopeful that extracts of *E. coli* would be able to incoporate deoxythymidine into nucleic acid by converting it first into the 5′ deoxynucleotide and then activating the deoxynucleotide to the triphosphate form. I found this to be the case. In subsequent months Ernest Simms and I were able to prepare separately deoxythymidine 5′-triphosphate and the other deoxynucleoside triphosphates, using enzymes or chemical synthetic routes. (In what follows the various deoxynucleosides in their 5′ triphosphate form will be designated simply by the initial of the base followed by an asterisk. Thus deoxythymidine 5′-triphosphate will be T*.)

In November, 1955, I. Robert Lehman, who is now at Stanford, started on the purification of the enzyme system in *E. coli* extracts that is responsible for converting T* into DNA. We were joined by Maurice J. Bessman some weeks later. Those were eventful days in which the enzyme, now given the name DNA polymerase, was progressively separated from other large molecules. With each step in purification the character of this DNA synthetic reaction became clearer. By June, 1956, when we participated at a conference on the chemical basis of heredity held at Johns Hopkins University, we could report two important facts about DNA synthesis in vitro, although we still lacked the answers to many important questions.

We reported first that preformed DNA had to be present along with DNA polymerase, and that all four of the deoxynucleotides that occur in DNA (A, G, T and C) had to be furnished in the activated triphosphate form. We also reported that DNA from virtually any source—virus, bacterium or animal—could serve with the *E. coli* enzyme. What we still did not know was whether the synthetic DNA was a new molecule or an extension of a preexisting one. There were other questions. Did the synthetic DNA have the same chemical backbone and physical structure as natural DNA? Did it have a chemical composition typical of DNA, in which A equals T and G equals C, and in which, therefore, A plus G equals T plus C? Finally, and crucially: Did the chemical composition of the synthetic DNA reflect the composition of the particular natural DNA used to direct the reaction?

During the next three years these questions and related ones were resolved by the efforts of Julius Adler, Sylvy Kornberg and Steven B. Zimmerman. The synthetic DNA was shown to be a molecule with the chemical structure typical of DNA and the same ratio of A-T pairs to G-C pairs as the particular DNA used to prime, or direct, the reaction [*see illustration on page 106*]. The relative starting amounts of the four deoxynucleoside triphosphates had no influence whatever on the composition of the new DNA. The composition of the synthetic DNA was determined solely by the composition of the DNA that served as a template. An interesting illustration of this last fact justifies a slight digression.

Howard K. Schachman of the University of California at Berkeley spent his sabbatical year of 1957–1958 with us at Washington University examining the physical properties of the synthetic DNA. It had the high viscosity, the comparatively slow rate of sedimentation and other physical properties typical of natural DNA. The new DNA, like the natural one, was therefore a long, fibrous polymer molecule. Moreover, the longer the mixture of active ingredients was allowed to incubate, the greater the viscosity of the product was; this was direct evidence that the synthetic DNA was continuing to grow in length and in amount. However, we were startled to find one day that viscosity developed in a control test tube that lacked one of the essential triphosphates, G*. To be sure, no reaction was observed during the standard incubation period of one or two hours. On prolonging the incubation for several more hours, however, a viscous substance materialized!

Analysis proved this substance to be a DNA that contained only A and T nucleotides. They were arranged in a perfect alternating sequence. The isolated polymer, named dAT, behaved like any other DNA in directing DNA synthesis: it led to the immediate synthesis of more dAT polymer. Would any G* and C* be polymerized if these nucleotides were present in equal or even far greater amounts than A* and T* in a synthesis directed by dAT polymer? We found no detectable incorporation of G or C under conditions that would have measured the inclusion of even one G for every 100,000 A or T nucleotides polymerized. Thus DNA polymerase rarely, if ever, made the mistake of matching G or C with A or T.

The DNA of a chromosome is a linear array of many genes. Each gene, in turn,

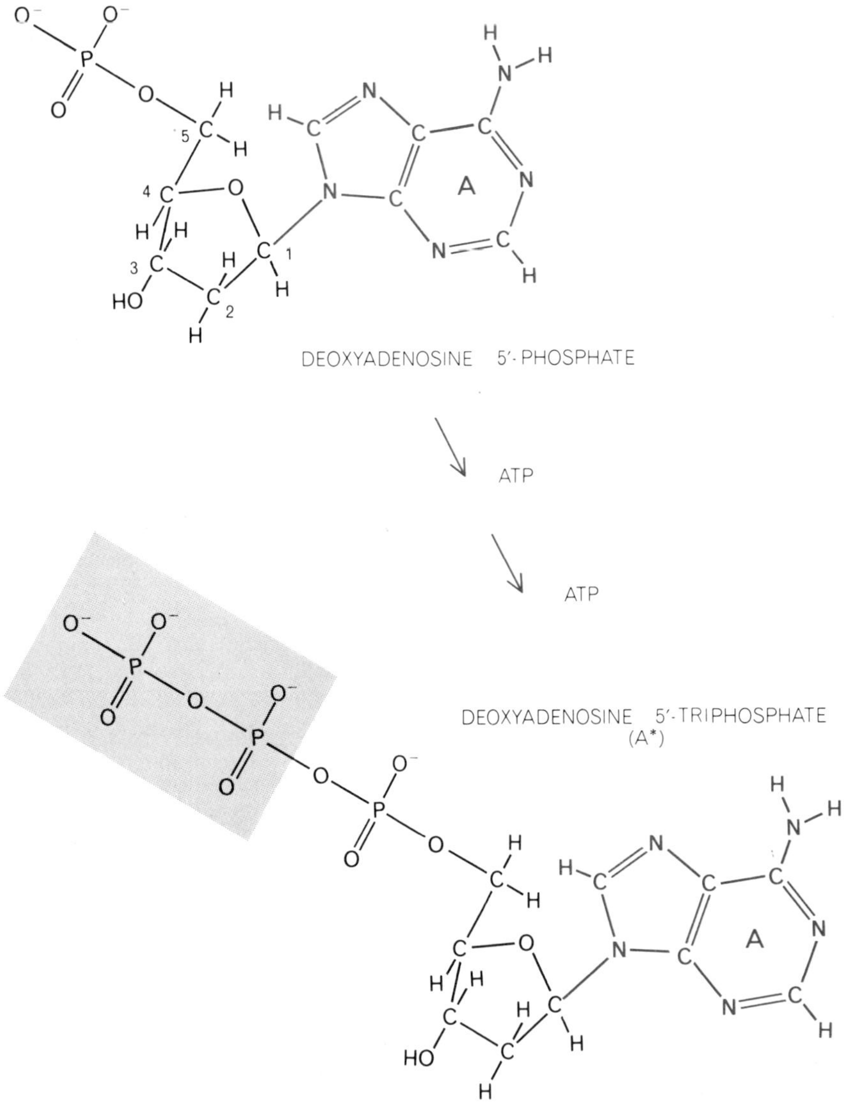

DEOXYADENOSINE 5'-PHOSPHATE

ATP

ATP

DEOXYADENOSINE 5'-TRIPHOSPHATE (A*)

ACTIVATED BUILDING BLOCK is required when synthesizing DNA on a template of natural DNA with the aid of enzymes. The activated form of the nucleotide containing adenine is deoxyadenosine 5'-triphosphate, symbolized in this article by "A*." It is made from deoxyadenosine 5'-monophosphate by two different enzymes in two steps. Each step involves the donation of a terminal phosphate group from adenosine triphosphate (ATP).

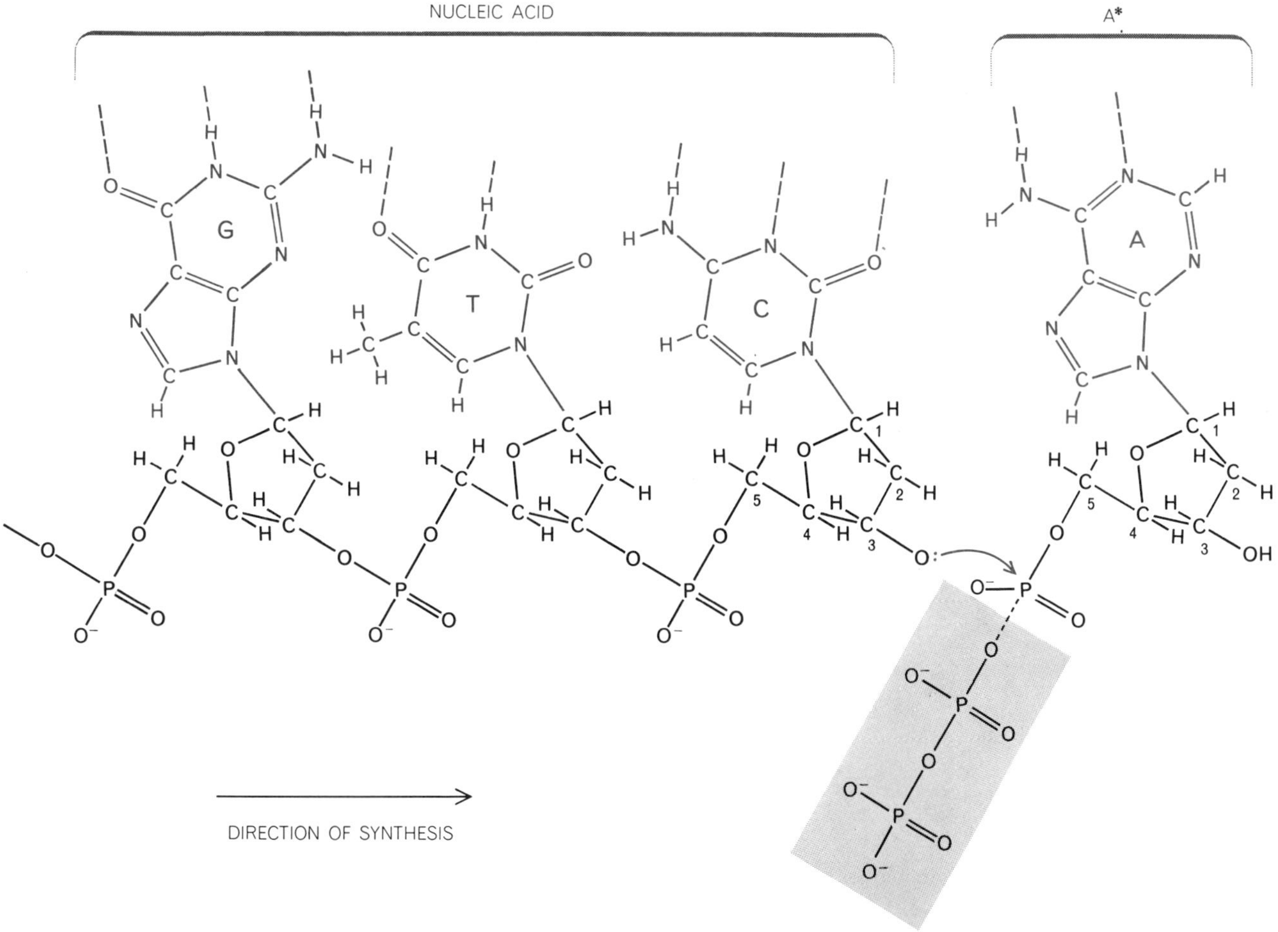

SYNTHESIS OF DNA involves the stepwise addition of activated nucleotides to the growing polymer chain. In this illustration deoxyadenosine 5′-triphosphate (A*) is being coupled through a phosphodiester bond that links the 3′ carbon in the deoxyribose portion of the last nucleotide in the growing chain to the 5′ carbon in the deoxyribose portion of the newest member of the chain.

is a chain of about 1,000 nucleotides in a precisely defined sequence, which when translated into amino acids spells out a particular protein or enzyme. Does DNA polymerase in its test-tube synthesis of DNA accurately copy the sequential arrangement of nucleotides by base-pairing (A = T, G = C) without errors of mismatching, omission, commission or transposition? Unfortunately techniques are not available for determining the precise sequence of nucleotides of even short DNA chains. Because it is impossible to spell out the base sequence of natural DNA or any copy of it, we have resorted to two other techniques to test the fidelity with which DNA polymerase copies the template DNA. One is "nearest neighbor" analysis. The other is the duplication of genes with demonstrable biological activity.

Nearest-Neighbor Analysis

The nearest-neighbor analysis devised by John Josse, A. Dale Kaiser and myself in 1959 determines the relative frequency with which two nucleotides can end up side by side in a molecule of synthetic DNA. There are 16 possible combinations in all. There are four possible nearest-neighbor sequences of A (AA, AG, AT and AC), four for G (GA, GG, GT and GC) and similarly four for T and four for C. How can the frequency of these dinucleotide sequences be determined in a synthetic DNA chain? The procedure is to use a triphosphate labeled with a radioactive phosphorus atom in conducting the synthesis and to treat the synthesized DNA with a specific enzyme that cleaves the DNA and leaves the radioactive phosphorus atom attached to its nearest neighbor. For example, DNA synthesis is carried out with A* labeled in the innermost phosphate group, the group that will be included in the DNA product. This labeled phosphate group now forms a normal linkage (10^{16} times in a typical experiment!) with the nucleotide next to it in the chain—its nearest neighbor [*see illustration on page 107*]. After the synthetic DNA is isolated it is subjected to degradation by an enzyme that cleaves every bond between the 5′ carbon of deoxyribose and the phosphate, leaving the radioactive phosphorus atom attached to the neighboring nucleotide rather than to the one (A) to which it had originally been attached. The nucleotides of the degraded DNA are readily separated by electrophoresis or paper chromatography into the four types of which DNA is composed: A, G, T and C. Radioactive assay establishes the radioactive phosphorus content in each of these nucleotides and at once indicates the frequency with which A is next to A, to G, to T and to C.

The entire experiment is repeated, this time with the radioactive label in G* instead of A*. The second experiment yields the frequency of GA, GG, GT and GC dinucleotides. Two more experiments with radioactive T* and C* complete the analysis and establish the 16 possible nearest-neighbor frequencies.

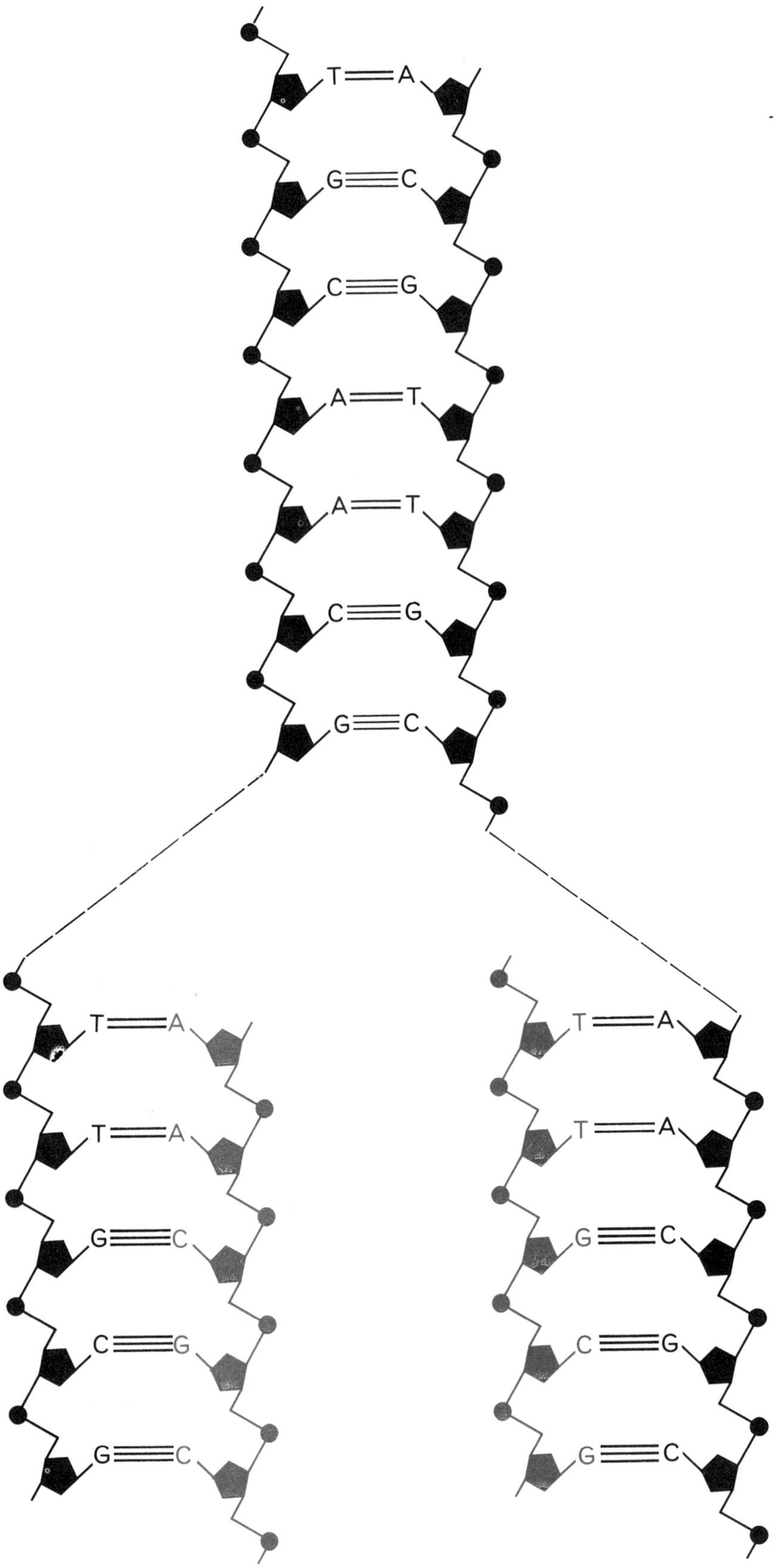

SYNTHESIS OF DUPLEX CHAIN OF DNA yields two hybrid molecules, consisting of a parental strand and a daughter strand, that are identical with each other and with the original duplex molecule. During the replicating process the parental duplex (*black*) separates into two strands, each of which then serves as the template for assembly of a daughter strand (*color*). The pairing of A with T and G with C guarantees faithful reproduction.

Many such experiments were performed with DNA templates obtained from viruses, bacteria, plants and animals. The DNA of each species guided the synthesis of DNA with what proved to be a distinctive assortment of nearest-neighbor frequencies. What is more, when a synthetic DNA was used as a template for a new round of replication, it gave rise to DNA with a nearest-neighbor frequency distribution identical with itself. Among the other insights obtained from these analyses was the recognition of a basic fact about the structure of the double helix. In replication the direction of the DNA chain being synthesized was found to run opposite to that of its template. By inference we can conclude that the chains of the double helix in natural DNA, as surmised by Watson and Crick, must also run in opposite directions.

Even with considerable care the accuracy of nearest-neighbor frequency analysis cannot be better than about 98 percent. Consequently we were still left with major uncertainties as to the precision of copying chains that contain 1,000 nucleotides or more, corresponding to the length of genes. An important question thus remained unanswered: Does DNA that is synthesized on a genetically or biologically active template duplicate the activity of that template?

One way to recognize the biological activity of bacterial DNA is to see if it can carry out "transformation," a process in which DNA from one species of bacteria alters the genetic endowment of a second species. For example, DNA from a strain of *Bacillus subtilis* resistant to streptomycin can be assimilated by a strain susceptible to the antibiotic, whereupon the recipient bacterium and all its descendants carry the trait of resistance to streptomycin. In other words, DNA molecules carrying the genes for a particular characteristic can be identified by their capacity for assimilation into the chromosome of a cell that previously lacked that trait. Yet when DNA was synthesized on a template of DNA that had transforming ability, the synthetic product invariably lacked that ability.

Part of the difficulty in synthesizing biologically active DNA lay in the persistence of trace quantities of nuclease enzymes in our DNA polymerase preparations. Nucleases are enzymes that degrade DNA. The introduction by a nuclease of one break in a long chain of DNA is enough to destroy its genetic activity. Further purification of DNA polymerase was indicated. Efforts over several years by Charles C. Richardson, Thomas Jovin, Paul T. Englund and LeRoy L. Bertsch resulted in a new procedure that was both simple and efficient. Finally, in April, 1967, with the assistance of the personnel and large-scale equipment of the New England Enzyme Center (sponsored by the National Institutes of Health at the Tufts University School of Medicine), we processed 100 kilograms of *E. coli* bacterial paste and obtained about half a gram of pure enzyme, free of the nuclease that puts random breaks in a DNA chain.

Unfortunately even this highly purified DNA polymerase has proved incapable of producing a biologically active DNA from a template of bacterial DNA. The difficulty, we believe, is that the DNA we extract from a bacterium such as *B. subtilis* provides the enzyme with a poor template. A proper template would be the natural chromosome, which is a double-strand loop about one millimeter in circumference. During its isolation from the bacterium the chromosome is broken, probably at random, into 100 or more fragments. The manner in which DNA polymerase and its related enzymes go about the replication of a DNA molecule as large and complex as the *B. subtilis* chromosome is the subject of current study in many laboratories.

The Virus ϕX174

It occurred to us in 1964 that the problem of synthesizing biologically active DNA might be solved by dealing with a simpler form of DNA that also has genetic activity. This is represented in viruses, such as ϕX174, whose DNA core is a single-strand loop. This "chromosome" not only is simpler in structure but also it is so small (about two microns in circumference) that it is fairly easy to extract without breakage. We also knew from the work of Robert L. Sinsheimer at the California Institute of Technology that when the DNA of ϕX174 invades *E. coli,* the first stage of infection involves the "subversion" of one of the host's enzymes to convert the single-strand loop into a double-strand helical loop. Sinsheimer called this first-stage product a "replicative form." Could the host enzyme that copies the viral DNA be the same DNA polymerase we had isolated from *E. coli?*

In undertaking the problem of copying a closed-loop DNA we could foresee some serious obstacles. Would it be possible for DNA polymerase to orient itself and start replication on a DNA template if the template had no ends? Shashanka Mitra and later Peter Reichard succeeded in finding conditions under which the enzyme, as judged by electron microscope pictures, appeared to copy the single-strand loop. We then wondered if in spite of appearances in the electron micrographs, the DNA of ϕX174 was really just a simple loop. Perhaps, as had been suggested by other workers, it was really more like a necklace with a clasp, the clasp consisting of substances unrelated to the nucleotides we were supplying. Finally, we were aware from Sinsheimer's work that the DNA of ϕX174 had to be a completely closed loop in order to be infectious. We knew that our polymerase could only catalyze the synthesis of linear DNA molecules. How could we synthesize a genuinely closed loop? We were still missing either the clasplike component to insert into our product or, if the clasp was a mistaken hypothesis, a new kind of enzyme to close the loop.

Fortunately the missing factor was provided for us by work carried on independently in five different laboratories. The discovery in 1966 of a polynucleotide-joining enzyme was made almost simultaneously by Martin F. Gellert and his co-workers at the National Institutes of Health, by Richardson and Bernard Weiss at the Harvard Medical

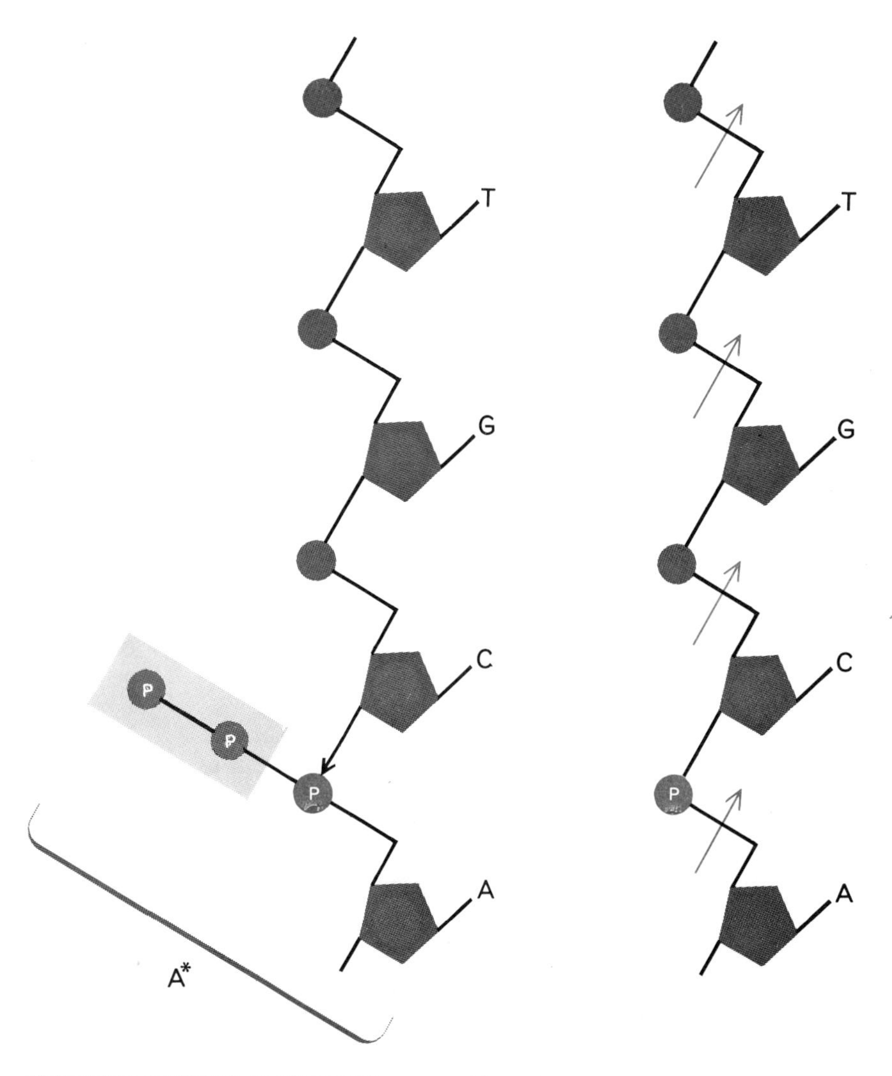

"NEAREST NEIGHBOR" ANALYSIS can reveal how often any of the four bases is located next to any other base in a single strand of synthetic DNA. Thus one can learn how often A is next to A, T, G or C, and so on. A radioactive phosphorus atom (*color*) is placed in the innermost position of one of the activated nucleotides, for example A*. The finished DNA molecule is then treated with an enzyme (*right*) that cleaves the chain between every phosphate and the 5′ carbon of the adjacent deoxyribose. Thus the phosphate is separated from the nucleotide on which it entered the chain and ends up attached to the nearest neighbor instead, C in the above example. The four kinds of nucleotide are separated by paper chromatography and the radioactivity associated with each is measured. The experiment is repeated with radioactive phosphorus linked to the other activated nucleotides.

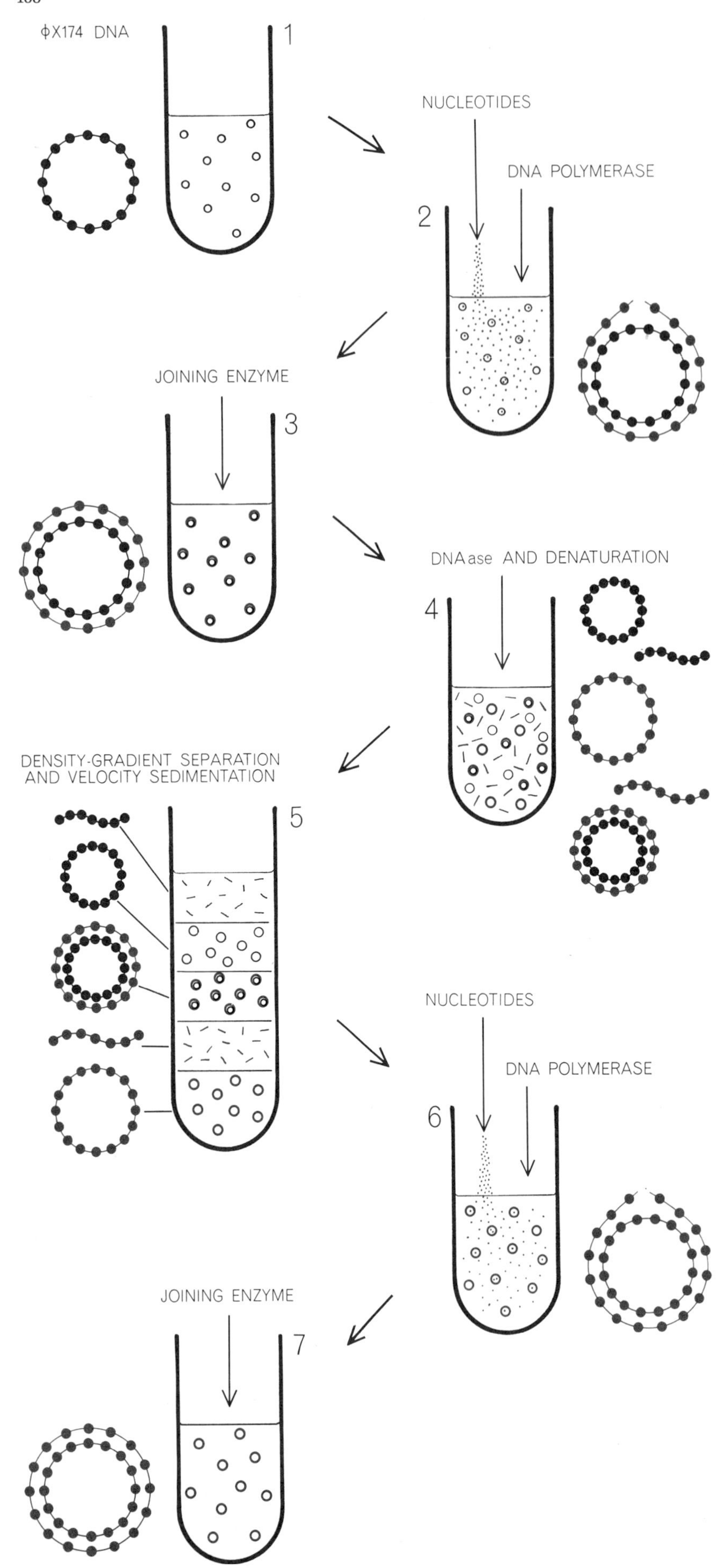

School, by Jerard Hurwitz and his colleagues at the Albert Einstein College of Medicine in New York, by Lehman and Baldomero M. Olivera at Stanford and by Nicholas R. Cozzarelli in my own group. It was the Lehman-Olivera preparation that we now employed in our experiments.

The polynucleotide-joining enzyme has the ability to repair "nicks" in the DNA strand. The nicks occur where there is a break in the sugar-phosphate backbone of one strand of the DNA molecule. The enzyme can repair a break only if all the nucleotides are intact and if what is missing is the covalent bond in the DNA backbone between a sugar and the neighboring phosphate. Provided with the joining enzyme, we were now in a position to find out whether it could work in conjunction with DNA polymerase to synthesize a completely circular and biologically active virus DNA.

By using the DNA of ϕX174 as a template we gained an important advantage over experiments based on transforming ability. Even if we were successful in synthesizing a DNA with transforming activity, this would still be of relatively limited significance. We could then say only that a restricted section of the DNA—a section as small as a part of a gene—had been assimilated by the recipient cell to replace a comparable section of its chromosome, substituting a proper sequence for a defective or incorrect one. However, Sinsheimer had demonstrated with the DNA of ϕX174 that a change in even one of its 5,500 nucleotides is sufficient to make the virus noninfective. Therefore the demonstration of infectivity in a completely synthetic

SYNTHESIS OF ϕX174 DNA was accomplished by the following steps. Circular single-strand ϕX174 DNA, tagged with tritium, served as a template (*1*). Activated nucleotides containing A, G, C and 5′-bromouracil instead of T were added to the template, together with DNA polymerase. One of the activated nucleotides was tagged with radioactive phosphorus. The DNA synthesized on the template was complete but not yet joined in a loop (*2*). The loop was closed by the joining enzyme (*3*). Enough nuclease was now added to cut one strand in about half of all the duplex loops (*4*). This left a mixture of complete duplex loops, template loops, synthetic loops, linear template strands and linear synthetic strands. Since the synthetic strands contained 5′-bromouracil, they were heavier than the template strands and could be separated by centrifugation (*5*). The synthetic loops were then isolated and used as templates for making wholly synthetic duplex loops (*6 and 7*).

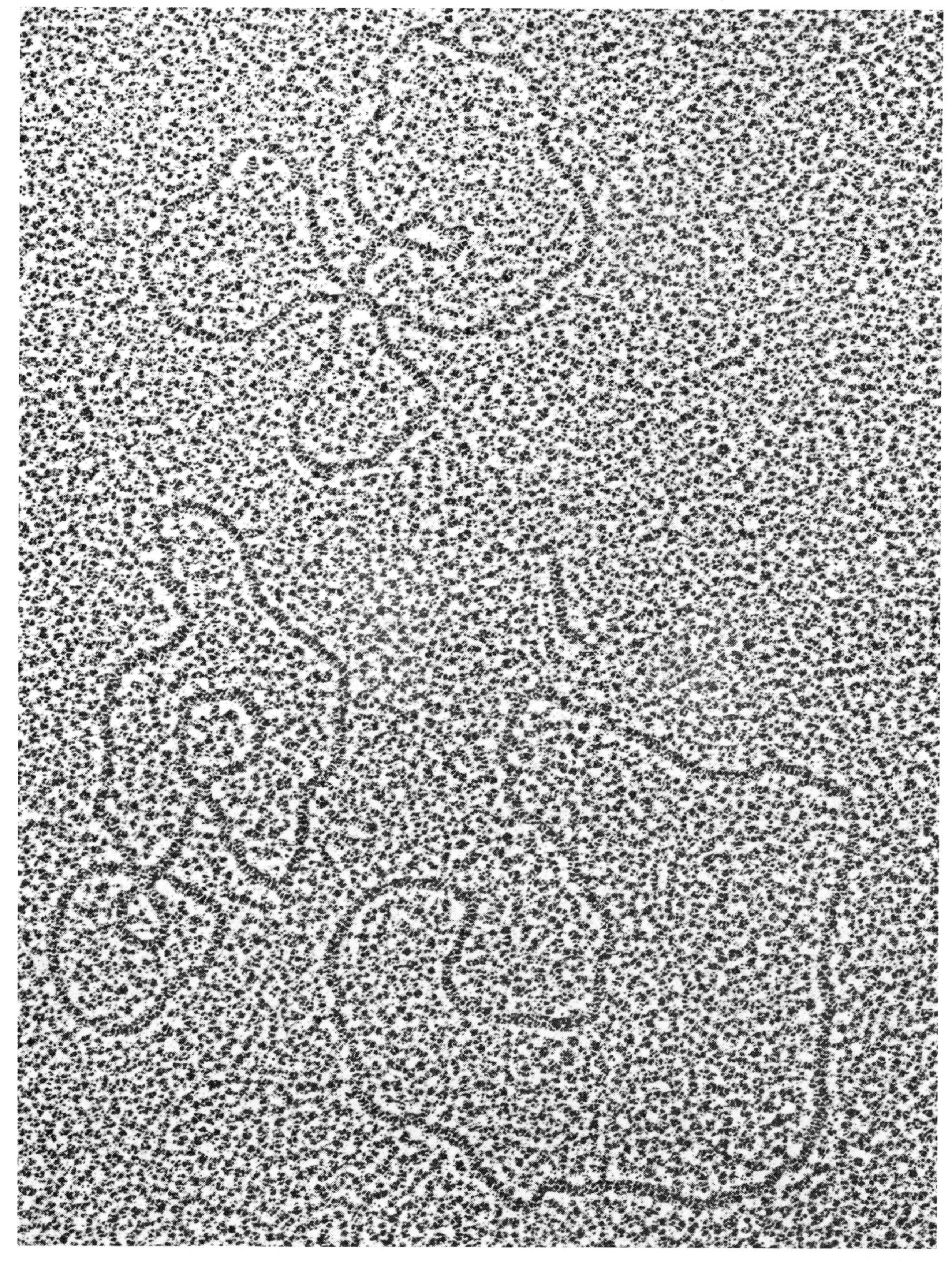

THREE CLOSED LOOPS OF DNA, each a complete double helix, are shown in this electron micrograph made in the author's laboratory at the Stanford University School of Medicine. One strand of each loop is the natural single-strand DNA of the bacterial virus ϕX174, which served as a template for the test-tube synthesis, carried out by enzymes, of a synthetic complementary strand. The hybrid molecules are biologically active. The enlargement is about 200,000 diameters. Each loop contains some 5,500 pairs of bases. If enlarged to the scale of the model on the opposite page, each loop of DNA would form a circle roughly 150 feet in circumference.

virus DNA would conclusively prove that we had carried out virtually error-free synthesis of this large number of nucleotides, comprising the five or six genes that carry out the virus's biological function.

In less than a year the test-tube synthesis of ϕX174 DNA was achieved. The steps can be summarized as follows. Template DNA was obtained from ϕX174 and labeled with tritium, the radioactive isotope of hydrogen. Tritium would thereafter provide a continuing label identifying the template. To the template were added DNA polymerase, purified joining enzyme and a cofactor (diphosphopyridine nucleotide), together with A*, T*, G* and C*. One of the nucleoside triphosphates was labeled with radioactive phosphorus. The radioactive phosphorus would thus provide a label for synthetic material analogous to the tritium label for the template. The interaction of the reagents then proceeded until the number of nucleotide units polymerized was exactly equal to the number of nucleotides in the template DNA. This equality was readily determined by comparing the radioactivity from the tritium in the template with the radioactivity from the phosphorus in the nucleotides provided for synthesis.

Such comparison showed that the experiments had progressed to an extent adequate for the formation of complementary loops of synthetic DNA. Complementary loops were designated (−) to distinguish them from the template loop (+). We had to demonstrate that the synthetic (−) loops were really loops. Had the polymerase made a full turn around the template and had the two ends of the chain been united by the joining enzyme? Several physical measurements, including electron microscopy, assured us that our product was a closed loop coiled tightly around the virus-DNA template and that it was identical in size and other details with the replicative form of DNA that appears in the infected cells. We could now exclude the possibility that some clasp material different from the nucleotide-containing compounds we had employed was involved in closing the virus-DNA loop.

The critical questions remaining were whether the synthetic (−) loops had biological activity—that is, infectivity—and whether the synthetic loops could in turn act as templates for the formation of a completely synthetic "duplex" DNA analogous to the replicative forms that were produced naturally inside infected cells. In order to answer the first of these questions we had to isolate the synthetic DNA strands from the partially synthetic duplexes. For reasons that will be apparent below, we substituted bromouracil, a synthetic but biologically active analogue of thymine, for thymine [*see top illustration on page 102*]. We then introduced just enough nuclease to produce a single nick in one strand of about half the population of molecules. The duplex loops that had been nicked would release a single linear strand of DNA; these single strands could be separated from their circular companions and from unnicked duplex loops by heating. Thus we were left with a mixture that contained (+) template loops, (−) synthetic loops, (+) template linear forms, (−) synthetic linear forms—all in about equal quantities—and full duplex loops.

It was at this point that the substitution of bromouracil for thymine became useful. Because bromouracil contains a bromine atom in place of the methyl group of thymine, it is heavier than the thymine. Therefore a molecule containing bromouracil can be separated from one containing thymine by high-speed centrifugation in a heavy salt solution (the density-gradient technique perfected by Jerome R. Vinograd of Cal Tech). In this system the denser a substance is, the lower in the centrifuge tube it will settle. Thus from top to bottom of the centrifuge tube we obtained fractions containing the light single strands of thymine-containing (+) template DNA, the duplex hybrids of intermediate weight and finally the single-strand synthetic (−) DNA "weighted down" with bromouracil. The reliability of this fractionation was confirmed by three separate peaks of radioactivity corresponding to each of the fractions. We were further reassured by observations that the mean density of each fraction corresponded almost exactly to the mean density of standard samples of virus DNA containing bromouracil or thymine.

Still another physical technique involving density-gradient sedimentation was employed to separate the synthetic linear forms from the synthetic circular forms. The circular forms could then be used in tests of infectivity, by methods previously developed by Sinsheimer to demonstrate the infectivity of circular ϕX174 DNA. We tested our (−) loops by incubating them with *E. coli* cells whose walls had been removed by the action of the enzyme lysozyme. Infectivity is assayed by the ability of the virus to lyse, or dissolve, these cells when they are "plated" on a nutrient medium. Our synthetic loops showed almost exactly the same patterns of infectivity as their natural counterparts had. Their biological activity was now demonstrated.

One further set of experiments remained in which the (−) synthetic loops were employed as the template to determine if we could produce completely synthetic duplex circular forms analogous to the replicative forms found in cells infected with natural ϕX174 virus. Because the synthetic (−) loops were labeled with radioactive phosphorus, this time we added tritium to one of the nucleotide-containing subunits (C*). The remaining procedures were essentially the same as the ones described above, and we did produce fully synthetic duplex loops of ϕX174. The (+) loops were then separated and were found to be identical in all respects with the (+) loops of natural ϕX174 virus. Their infectivity could also be demonstrated. Sinsheimer had previously shown that, under these assay conditions, a change in a single nucleotide of the virus gave rise to a mutant of markedly decreased infectivity. Therefore the correspondence between the infectivity of our synthetic forms and their natural counterparts attested to the precision of the enzymatic operation.

Future Directions

The total synthesis of infective virus DNA by DNA polymerase with the four deoxynucleoside triphosphates not only demonstrates the capacity of this enzyme to copy a small chromosome (of five or six genes) without error but also shows that this chromosome, at least, is as simple and straightforward as a linear sequence of the standard four deoxynucleotide units. It is a long step to the human chromosome, some 10,000 times larger, yet we are encouraged to extrapolate our current conceptions of nucleotide composition and nucleotide linkage from the tiny ϕX174 chromosome to larger ones.

What are the major directions this research will take? I see at least three immediate and productive paths. One is the exploration of the physical and chemical nature of DNA polymerase in order to understand exactly how it performs its error-free replication of DNA. Without this knowledge of the structure of the enzyme and how it operates under defined conditions in the test tube, our understanding of the intracellular behavior of the enzyme will be incomplete.

A second direction is to clarify the

control of DNA replication in the cell and in the animal. Why is DNA synthesis arrested in a mature liver cell and what sets it in motion 24 hours after part of the liver is removed surgically? What determines the slow rate of DNA replication in adult cells compared with the rate in embryonic or cancer cells? The time is ripe for exploration of the factors that govern the initiation and rate of DNA synthesis in the intact cell and animal. Finally, there are now prospects of applying our knowledge of DNA structure and synthesis directly to human welfare. This is the realm of genetic engineering, and it is our collective responsibility to see that we exploit our great opportunities to improve the quality of human life.

An obvious area for investigation would be the synthesis of the polyoma virus, a virus known to induce a variety of malignant tumors in several species of rodents. Polyoma virus in its infective form is made up of duplex circular DNA and presumably replicates in this form on entering the cell. On the basis of our experience it would appear quite feasible to synthesize polyoma virus DNA. If this synthesis is accomplished, there would seem to be many opportunities for modifying the virus DNA and thus determining where in the chromosome its tumor-producing capacity lies. With this knowledge it might prove possible to modify the virus in order to control its tumor-producing potential.

Our speculations can extend even to large DNA molecules. For example, if a failure in the production of insulin were to be traced to a genetic deficit, then administration of the appropriate synthetic DNA might conceivably provide a cure for diabetes. Of course, a system for delivering the corrective DNA to the cells must be devised. Even this does not seem inconceivable. The extremely interesting work of Stanfield Rogers at the Oak Ridge National Laboratory suggests a possibility. Rogers has shown that the Shope papilloma virus, which is not pathogenic in man, is capable of inducing production of the enzyme arginase in rabbits at the same time that it induces tumors. Rogers found that in the blood of laboratory investigators working with the virus there is a significant reduction of the amino acid arginine, which is destroyed by arginase. This is apparently an expression of enhanced arginase activity. Might it not be possible, then, to use similar nonpathogenic viruses to carry into man pieces of DNA capable of replacing or repairing defective genes?

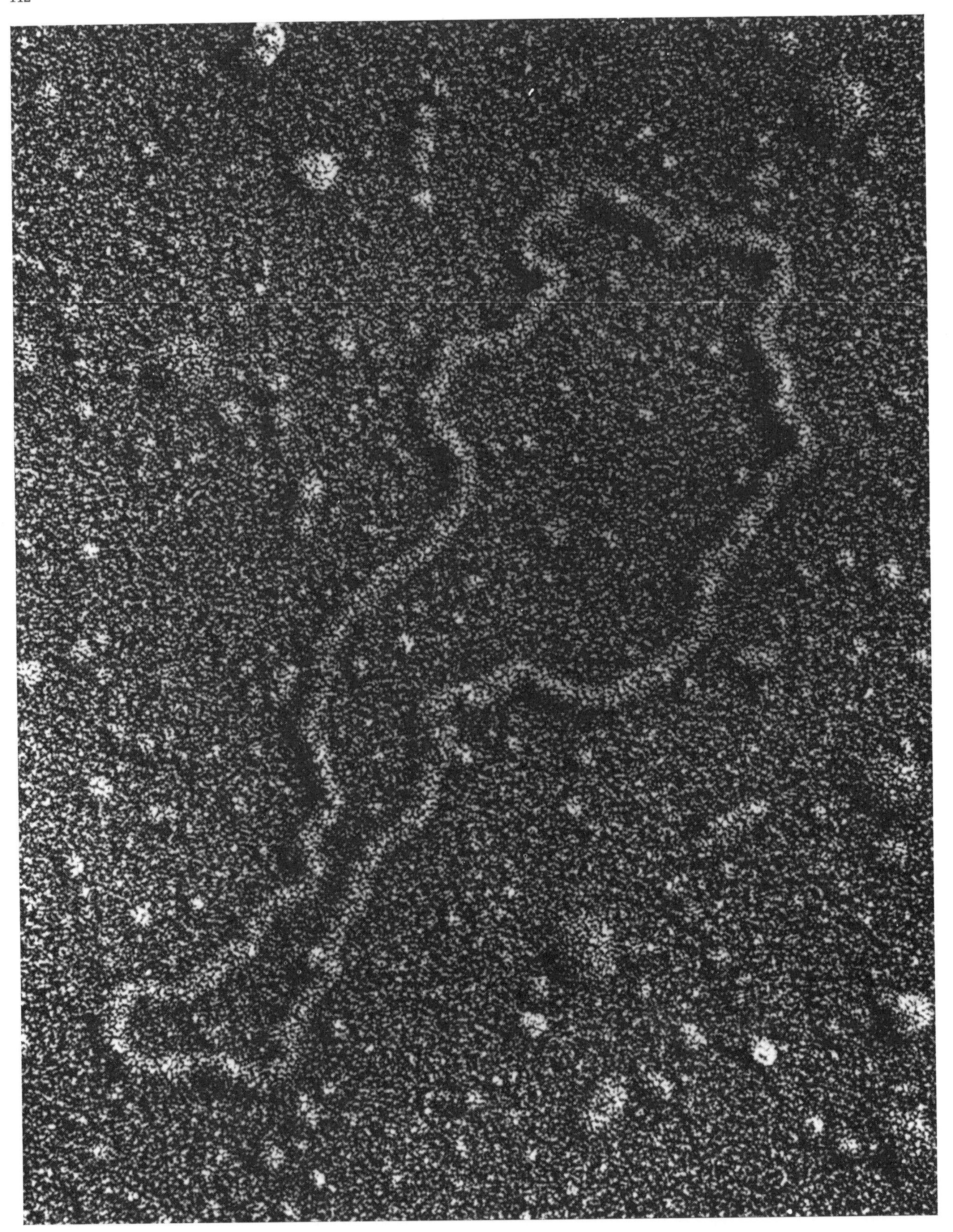

PLASMID *pSC*101 is shadowed with platinum-palladium and enlarged 230,000 diameters in an electron micrograph made by the author. A plasmid is a molecule of DNA that exists apart from the chromosome in a bacterium and replicates on its own, often carrying the genes for some supplementary activity such as resistance to antibiotics. This plasmid, a small one made by shearing a larger plasmid native to the bacterium *Escherichia coli,* is a circular, or closed-loop, molecule of DNA about three micrometers in circumference that carries the genetic information for replicating itself in *E. coli* and for conferring resistance to the antibiotic tetracycline. It was the "vehicle" for the first gene-manipulation experiments by the author and his colleagues. Foreign DNA was spliced to it and the plasmid was introduced into *E. coli,* where it replicated and expressed both its own and the foreign DNA's genetic information.

The Manipulation of Genes

12

by Stanley N. Cohen
July 1975

Techniques for cleaving DNA and splicing it into a carrier molecule make it possible to transfer genetic information from one organism to an unrelated one. There the DNA replicates and expresses itself

Mythology is full of hybrid creatures such as the Sphinx, the Minotaur and the Chimera, but the real world is not; it is populated by organisms that have been shaped not by the union of characteristics derived from very dissimilar organisms but by evolution within species that retain their basic identity generation after generation. This is because there are natural barriers that normally prevent the exchange of genetic information between unrelated organisms. The barriers are still poorly understood, but they are of fundamental biological importance.

The basic unit of biological relatedness is the species, and in organisms that reproduce sexually species are defined by the ability of their members to breed with one another. Species are determined and defined by the genes they carry, so that in organisms that reproduce asexually the concept of species depends on nature's ability to prevent the biologically significant exchange of genetic material—the nucleic acid DNA—between unrelated groups.

The persistence of genetic uniqueness is perhaps most remarkable in simple organisms such as bacteria. Even when they occupy the same habitat most bacterial species do not exchange genetic information. Even rather similar species of bacteria do not ordinarily exchange the genes on their chromosomes, the structures that carry most of their genetic information. There are exceptions, however. There are bits of DNA, called plasmids, that exist apart from the chromosomes in some bacteria. Sometimes a plasmid can pick up a short segment of DNA from the chromosome of its own cell and transfer it to the cell of a related bacterial species, and sometimes the plasmid and the segment of chromosomal DNA can become integrated into the chromosome of the recipient cell. This transfer of genes between species by extrachromosomal elements has surely played some role in bacterial evolution, but apparently it has not been widespread in nature. Otherwise the characteristics of the common bacterial species would not have remained so largely intact over the huge number of bacterial generations that have existed during the era of modern bacteriology.

In 1973 Annie C. Y. Chang and I at the Stanford University School of Medicine and Herbert W. Boyer and Robert B. Helling at the University of California School of Medicine at San Francisco reported the construction in a test tube of biologically functional DNA molecules that combined genetic information from two different sources. We made the molecules by splicing together segments of two different plasmids found in the colon bacillus *Escherichia coli* and then inserting the composite DNA into *E. coli* cells, where it replicated itself and expressed the genetic information of both parent plasmids. Soon afterward we introduced plasmid genes from an unrelated bacterial species, *Staphylococcus aureus*, into *E. coli*, where they too expressed the biological properties they had displayed in their original host; then, applying the same procedures with John F. Morrow of Stanford and Howard M. Goodman in San Francisco, we were able to insert into *E. coli* some genes from an animal: the toad *Xenopus laevis*.

We called our composite molecules DNA chimeras because they were conceptually similar to the mythological Chimera (a creature with the head of a lion, the body of a goat and the tail of a serpent) and were the molecular counterparts of hybrid plant chimeras produced by agricultural grafting. The procedure we described has since been used and extended by workers in several laboratories. It has been called plasmid engineering, because it utilizes plasmids to introduce the foreign genes, and molecular cloning, because it provides a way to propagate a clone, or line of genetically alike organisms, all containing identical composite DNA molecules. Because of the method's potential for creating a wide variety of novel genetic combinations in microorganisms it is also known as genetic engineering and genetic manipulation. The procedure actually consists of several distinct biochemical and biological manipulations that were made possible by a series of independent discoveries made in rapid succession in the late 1960's and early 1970's. There are four essential elements: a method of breaking and joining DNA molecules derived from different sources; a suitable gene carrier that can replicate both itself and a foreign DNA segment linked to it; a means of introducing the composite DNA molecule, or chimera, into a functional bacterial cell, and a method of selecting from a large population of cells a clone of recipient cells that has acquired the molecular chimera.

In 1967 DNA ligases—enzymes that can repair breaks in DNA and under certain conditions can join together the loose ends of DNA strands—were discovered almost simultaneously in five laboratories. A DNA strand is a chain of nucleotides, each consisting of a deoxyribose sugar ring, a phosphate group and one of four organic bases: adenine, thymine, guanine and cytosine. The sugars and phosphates form the backbone of the strand, from which the bases project. The individual nucleotide building blocks are connected by phosphodiester bonds between the carbon atom at position No. 3 on one sugar and the carbon atom at position No. 5 on the adjacent sugar. Double-strand DNA, the form found in most organisms, consists of two

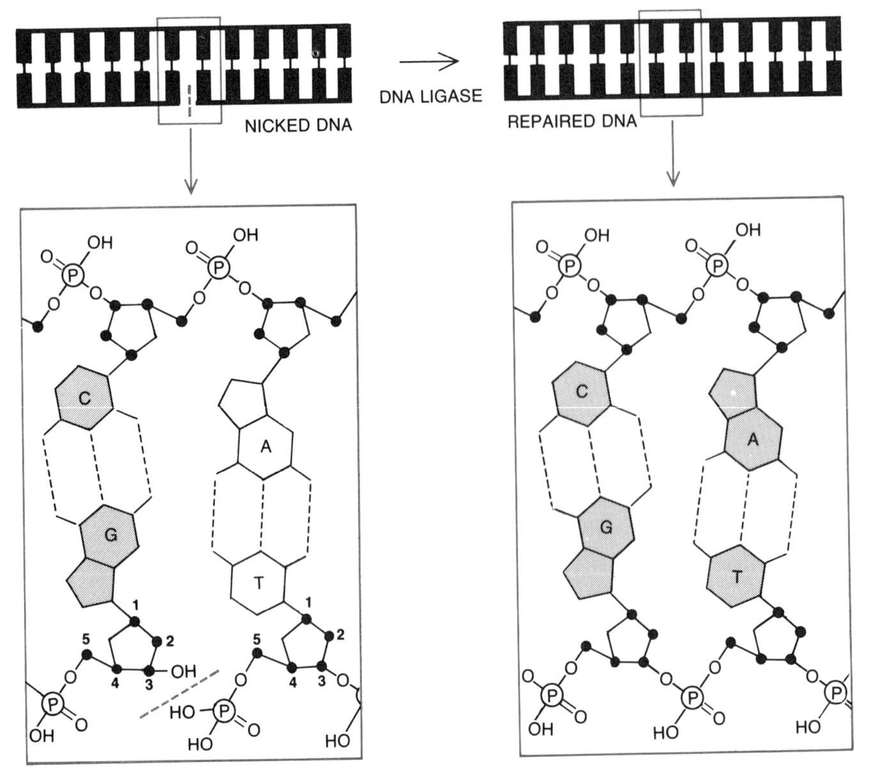

DNA LIGASE is an enzyme that repairs "nicks," or breaks in one strand of a double-strand molecule of DNA (*top*). A strand of DNA is a chain of nucleotides (*bottom*), each consisting of a deoxyribose sugar and a phosphate group and one of four organic bases: adenine (*A*), thymine (*T*), guanine (*G*) and cytosine (*C*). The sugars and phosphates constitute the backbone of the strand, and paired bases, linked by hydrogen bonds (*broken black lines*), connect two strands. The ligase catalyzes synthesis of a bond at the site of the break (*broken colored line*) between the phosphate of one nucleotide and the sugar of the next nucleotide.

chains of nucleotides linked by hydrogen bonds between their projecting bases. The bases are complementary: adenine (*A*) is always opposite thymine (*T*), and guanine (*G*) is always opposite cytosine (*C*). The function of the ligase is to repair "nicks," or breaks in single DNA strands, by synthesizing a phosphodiester bond between adjoining nucleotides [*see illustration above*].

In 1970 a group working in the laboratory of H. Gobind Khorana, who was then at the University of Wisconsin, found that the ligase produced by the bacterial virus T4 could sometimes catalyze the end-to-end linkage of completely separated double-strand DNA segments. The reaction required that the ends of two segments be able to find each other; such positioning of two DNA molecules was a matter of chance, and so the reaction was inefficient. It was clear that efficient joining of DNA molecules required a mechanism for holding the two DNA ends together so that the ligase could act.

An ingenious way of accomplishing this was developed and tested independently in two laboratories at Stanford: by Peter Lobban and A. Dale Kaiser and by David Jackson, Robert Symons and Paul Berg. Earlier work by others had shown that the ends of the DNA molecules of certain bacterial viruses can be joined by base-pairing between complementary sequences of nucleotides that are naturally present on single-strand segments projecting from the ends of those molecules: *A*'s pair with *T*'s, *G*'s pair with *C*'s and the molecules are held together by hydrogen bonds that form between the pairs. The principle of linking DNA molecules by means of the single-strand projections had been exploited in Khorana's laboratory for joining short synthetic sequences of nucleotides into longer segments of DNA.

The Stanford groups knew too that an enzyme, terminal transferase, would catalyze the stepwise addition, specifically at what are called the 3′ ends of single strands of DNA, of a series of identical nucleotides. If the enzyme worked also with double-strand DNA, then a block of identical nucleotides could be added to one population of DNA molecules and a block of the complementary nucleotides could be added to another population from another source. Molecules of the two populations could then be annealed by hydrogen bonding and sealed together by DNA ligase. The method was potentially capable of joining any two species of DNA. While Lobban and Kaiser tested the terminal-transferase procedure with the DNA of the bacterial virus P22, Jackson, Symons and Berg applied the procedure to link the DNA of the animal virus SV40 to bacterial-virus DNA.

The SV40 and bacterial-virus DNA molecules Berg's group worked with are closed loops, and the loops had first to be cleaved to provide linear molecules with free ends for further processing and linkage [*see illustration on opposite page*]. (As it happened, the particular enzyme chosen to cleave the loops was the *Eco RI* endonuclease, which was later to be used in a different procedure for making the first biologically functional gene combinations. At the time, however, the enzyme's special property of producing complementary single-strand ends all by itself had not yet been discovered.)

The cleaved linear molecules were treated with an enzyme, produced by the bacterial virus lambda, called an exonuclease because it operates by cutting off nucleotides at the end of a DNA molecule. The lambda exonuclease chewed back the 5′ ends of DNA molecules and thus left projecting single-strand ends that had 3′ termini to which the blocks of complementary nucleotides could be added. The next step was to add, with the help of terminal transferase, a block of *A*'s at the 3′ end of one of the two DNA species to be linked and a block of *T*'s at the 3′ ends of the other species. The species were mixed together. Fragments having complementary blocks at their ends could find each other, line up and become annealed by hydrogen bonding, thus forming combined molecules. To fill the gaps at the 5′ ends of the original segments the investigators supplied nucleotides and two more enzymes: exonuclease III and DNA polymerase. Finally the nicks in the molecules were sealed with DNA ligase.

The method of making cohesive termini for joining DNA molecules in the first successful genetic-manipulation experiments was conceptually and operationally different from the terminal-transferase procedure. It was also much simpler. It depended on the ability of one of a group of enzymes called restriction endonucleases to make complementary-ended fragments during the cleavage of DNA at a site within the molecule, instead of requiring the addition of new blocks of complementary nucleotides to DNA termini.

Viruses grown on certain strains of *E. coli* were known to be restricted in their ability to grow subsequently on other strains. Investigations had shown that this restriction was due to bacterial enzymes that recognize specific sites on a "foreign" viral DNA and cleave that DNA. (To protect its own DNA the bacterial cell makes a modification enzyme that adds methyl groups to nucleotides constituting the recognition sites for the restriction endonuclease, making them resistant to cleavage.) Restriction endonucleases (and modification methylases) are widespread in microorganisms; genes for making them were found on viral chromosomes and extrachromosomal plasmid DNA as well as on many bacterial chromosomes. During the early 1970's the nucleotide sequences at the cleavage sites recognized by several re-

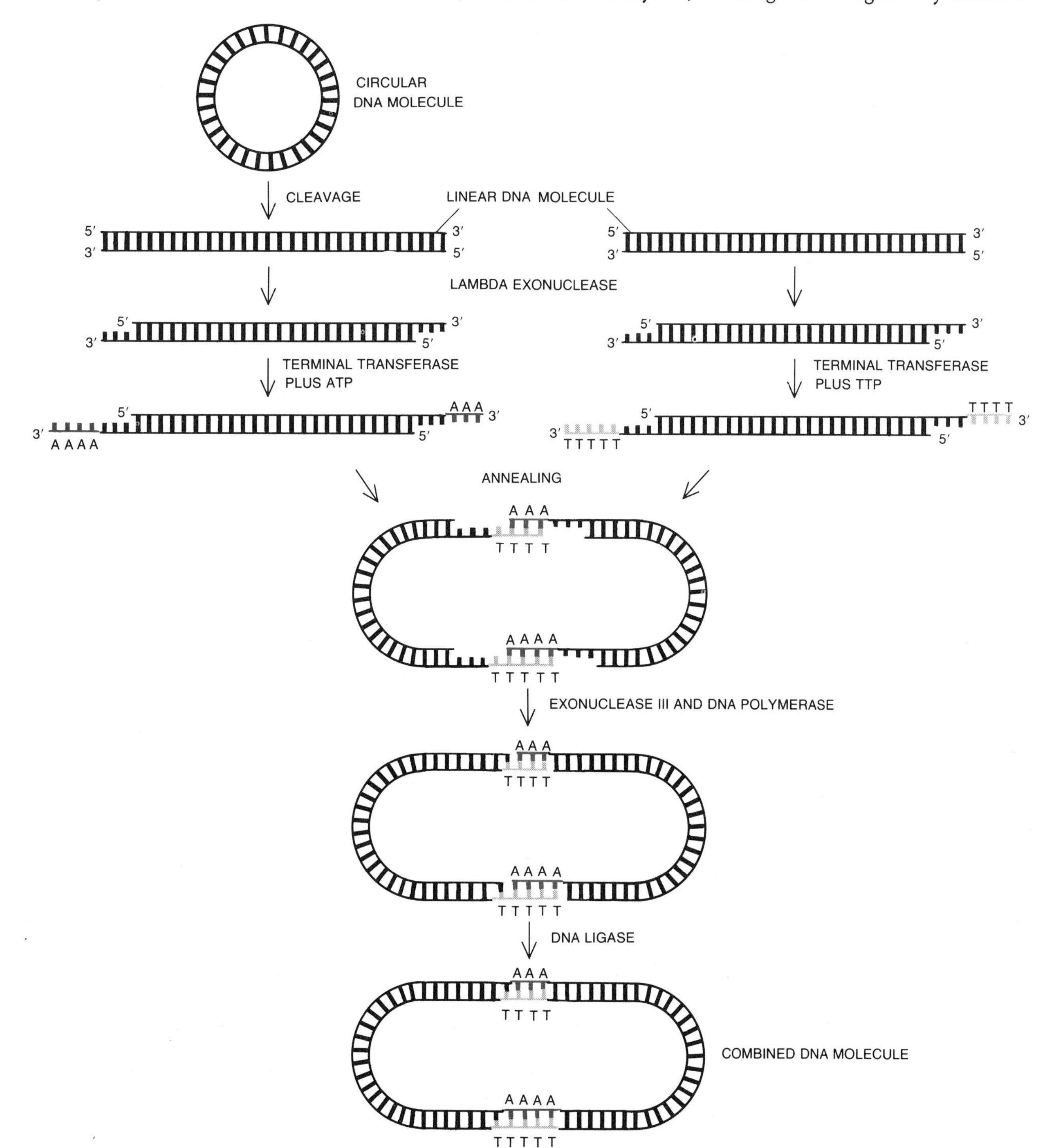

TERMINAL-TRANSFERASE procedure for joining DNA molecules involves a number of steps, each dependent on a different enzyme. If one of the molecules to be joined is a closed loop, it must first be cleaved. The linear molecules are treated with lambda exonuclease, an enzyme that cuts nucleotides off the 5′ end of DNA strands (the end with a phosphate group on the No. 5 carbon). Then specific nucleotides are added to the 3′ end (the end with an OH group on the No. 3 carbon) by the action of the enzyme terminal transferase. One DNA species is supplied with adenosine triphosphate (ATP), the other with thymidine triphosphate (TTP), so that *A* nucleotides are added to one species and complementary *T* nucleotides to the other. When the two species are mixed, the complementary bases pair up, annealing the molecules. Nucleotides and the enzymes DNA polymerase and exonuclease III are added to fill gaps and DNA ligase is added to seal the DNA backbones. The result is a double molecule composed of two separate DNA segments.

striction endonucleases were identified. In every instance, it developed, the cleavage was at or near an axis of rotational symmetry: a palindrome where the nucleotide base sequences read the same on both strands in the 5′-to-3′ direction [*see illustration below*].

In some instances the breaks in the DNA strands made by restriction enzymes were opposite each other. One particular endonuclease, however, the *Eco RI* enzyme isolated by Robert N. Yoshimori in Boyer's laboratory in San Francisco, had a property that was of special interest. Unlike the other nucleases known at the time, this enzyme introduced breaks in the two DNA strands that were separated by several nucleotides. Because of the symmetrical, palindromic arrangement of the nucleotides in the region of cleavage this separation of the cleavage points on the two strands yielded DNA termini with projecting complementary nucleotide sequences: "sticky" mortise-and-tenon termini. The *Eco RI* enzyme thus produced in one step DNA molecules that were functionally equivalent to the cohesive-end molecules produced by the complicated terminal-transferase procedure.

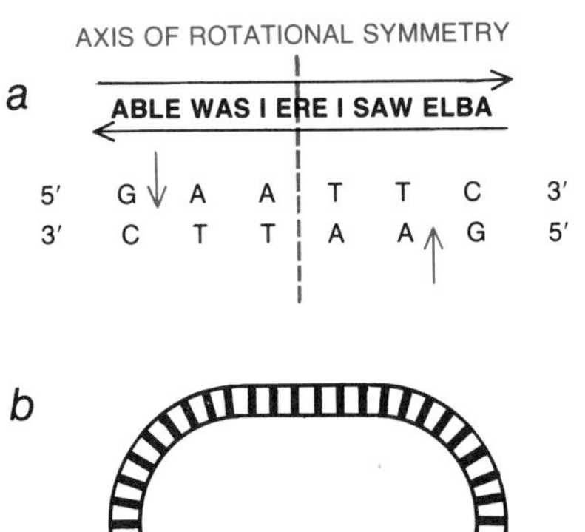

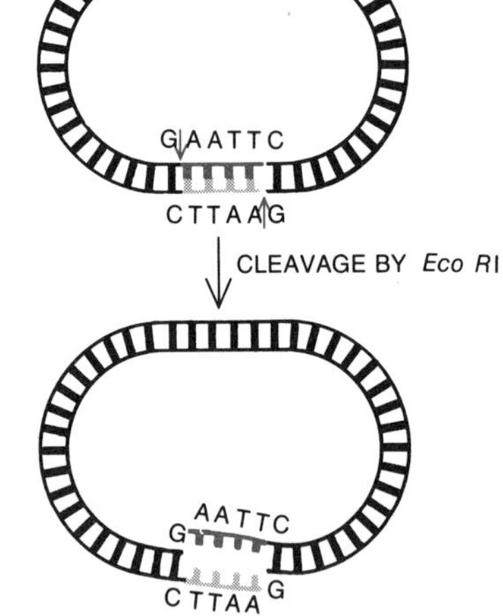

RESTRICTION ENDONUCLEASES cleave DNA at sites where complementary nucleotides are arranged in rotational symmetry: a palindrome, comparable to a word palindrome (*a*). The endonuclease *Eco RI* has the additional property of cleaving complementary strands of DNA at sites (*colored arrows*) four nucleotides apart. Such cleavage (*b*) yields DNA fragments with complementary, overlapping single-strand ends. As a result the end of any DNA fragment produced by *Eco RI* cleavage can anneal with any other fragment produced by the enzyme.

The experiments that led to the discovery of the capabilities of *Eco RI* were reported independently and simultaneously in November, 1972, by Janet Mertz and Ronald W. Davis of Stanford and by another Stanford investigator, Vittorio Sgaramella. Sgaramella found that molecules of the bacterial virus P22 could be cleaved with *Eco RI* and would then link up end to end to form DNA segments equal in length to two or more viral-DNA molecules. Mertz and Davis observed that closed-loop SV40-DNA molecules cleaved by *Eco RI* would reform themselves into circular molecules by hydrogen bonding and could be sealed with DNA ligase; the reconstituted molecules were infectious in animal cells growing in tissue culture. Boyer and his colleagues analyzed the nucleotide sequences at the DNA termini produced by *Eco RI*, and their evidence confirmed the complementary nature of the termini, which accounted for their cohesive activity.

In late 1972, then, several methods were available by which one could join double-strand molecules of DNA. That was a major step in the development of a system for manipulating genes. More was necessary, however. Most segments of DNA do not have an inherent capacity for self-replication; in order to reproduce themselves in a biological system they need to be integrated into DNA molecules that can replicate in the particular system. Even a DNA segment that can replicate in its original host was not likely to have the specific genetic signals required for replication in a different environment. If foreign DNA was to be propagated in bacteria, as had long been proposed in speculative scenarios of genetic engineering, a suitable vehicle, or carrier, was required. A composite DNA molecule consisting of the vehicle and the desired foreign DNA would have to be introduced into a population of functional host bacteria. Finally, it would be necessary to select, or identify, those cells in the bacterial population that took up the DNA chimeras. In 1972 it still seemed possible that the genetic information on totally foreign DNA molecules might produce an aberrant situation that would prevent the propagation of hybrid molecules in a new host.

Molecular biologists had focused for many years on viruses and their relations with bacteria, and so it was natural that bacterial viruses were thought of as the most likely vehicles for genetic manipulation. For some time there had been speculation and discussion about using viruses, such as lambda, that occasionally acquire bits of the *E. coli* chromosome by natural recombination mechanisms for cloning DNA from foreign sources. It was not a virus, however, but a plasmid that first served as a vehicle for introducing foreign genes into a bacterium and that provided a mechanism for the replication and selection of the foreign DNA.

A ubiquitous group of plasmids that confer on their host bacteria the ability to resist a number of antibiotics had been studied intensively for more than a decade. Antibiotic-resistant *E. coli* isolated in many parts of the world, for example, were found to contain plasmids, designated *R* factors (for "resistance"), carrying the genetic information for products that in one way or another could interfere with the action of specific antibiotics [see "Infectious Drug Resistance," by Tsutomu Watanabe; SCIENTIFIC AMERICAN, December, 1967]. Double-strand circular molecules of *R*-factor DNA had been separated from bacterial chromosomal DNA by centrifugation in density gradients and had been characterized by biochemical and physical techniques [see "The Molecule of Infectious Drug Resistance," by Royston C. Clowes, beginning on page 59].

In 1970 Morton Mandel and A. Higa of the University of Hawaii School of Medicine had discovered that treatment of *E. coli* with calcium salts enabled the bacteria to take up viral DNA. At Stanford, Chang and I, with Leslie Hsu, found that if we made the cell membranes of *E. coli* permeable by treating them with calcium chloride, purified *R*-factor DNA could be introduced into them [*see illustration on opposite page*]. The *R*-factor DNA is taken up in this transformation process by only about one bacterial cell in a million, but those few cells can be selected because they live and multiply in the presence of the antibiotics to which the *R* factor confers resistance, whereas other cells die. Each transformed cell gives rise to a clone that contains exact replicas of the parent plasmid DNA molecules, and so we reasoned that plasmids might serve as vehicles for propagating new genetic information in a line of *E. coli* cells.

In an effort to explore the genetic and molecular properties of various regions of the *R*-factor DNA we had begun to take plasmids apart by shearing their DNA mechanically and then transforming *E. coli* with the resulting fragments. Soon afterward we began to cleave the plasmids with the *Eco RI* enzyme, which had been shown to produce multiple site-specific breaks in several viruses. It might therefore be counted on to cleave all molecules of a bacterial plasmid in the same way, so that any particular species of DNA would yield a specific set of cleavage fragments, and do so reproducibly. The fragments could then be separated and identified according to the different rates at which they would migrate through a gel under the influence of an electric current.

When the DNA termini produced by *Eco RI* endonuclease were found to be cohesive, Chang and I, in collaboration with Boyer and Helling in San Francisco, proceeded to search for a plasmid that the enzyme would cleave without affecting the plasmid's ability to replicate or to confer antibiotic resistance. We hoped that if such a plasmid could be found, we could insert a segment of foreign DNA at the *Eco RI* cleavage site, and that it might be possible to propagate the foreign DNA in *E. coli.*

In our collection at Stanford there was a small plasmid, *pSC*101, that had been isolated following the mechanical shearing of a large plasmid bearing genes for multiple antibiotic resistance. It was less than a twelfth as long as the parent plasmid, but it did retain the genetic information for its replication in *E. coli* and for conferring resistance to one antibiotic, tetracycline. When we subjected *pSC*101 DNA to cleavage by *Eco RI* and analyzed the products by gel electrophoresis, we found that the enzyme had cut the plasmid molecule in only one place, producing a single linear fragment. We were able to join the ends of that fragment again by hydrogen bonding and reseal them with DNA ligase, and when we introduced the reconstituted circular DNA molecules into *E. coli* by transformation, they were biologically functional plasmids: they replicated and conferred tetracycline resistance.

The next step was to see if a fragment of foreign DNA could be inserted at the cleavage site without interfering with replication or expression of tetracycline resistance and thus destroying the plasmid's ability to serve as a cloning vehicle. We mixed the DNA of another *E. coli* plasmid, which carried resistance to the antibiotic kanamycin, with the *pSC*101 DNA. We subjected the mixed DNA to cleavage by *Eco RI* and then to ligation, transformed *E. coli* with the resulting DNA and found that some of the transformed bacteria were indeed resist-

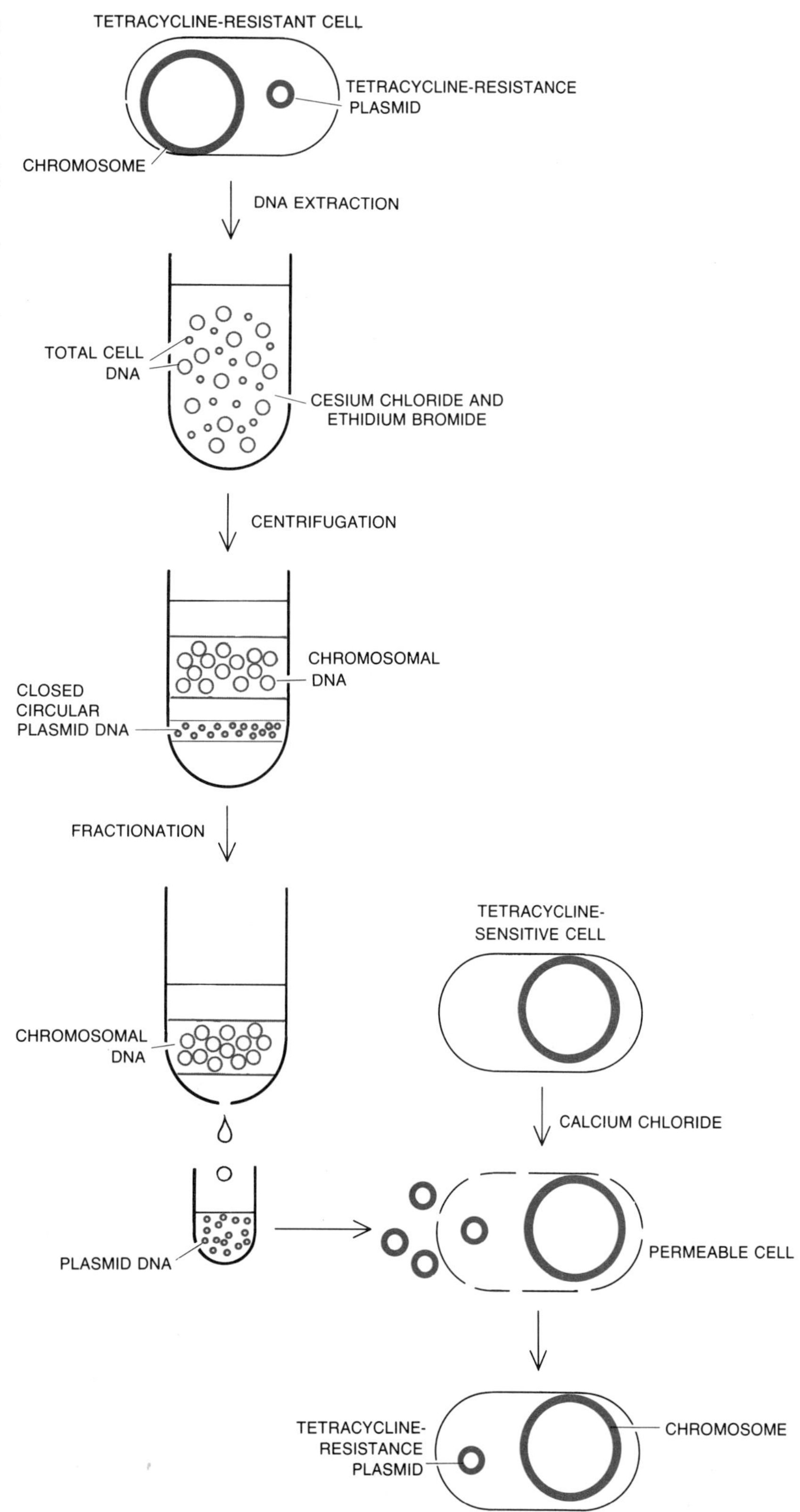

PLASMID DNA can be introduced into a bacterial cell by the procedure called transformation. Plasmids carrying genes for resistance to the antibiotic tetracycline (*top left*) are separated from bacterial chromosomal DNA. Because differential binding of ethidium bromide by the two DNA species makes the circular plasmid DNA denser than the chromosomal DNA, the plasmids form a distinct band on centrifugation in a cesium chloride gradient and can be separated (*bottom left*). The plasmid DNA is mixed with bacterial cells that are not resistant to tetracycline and that have been made permeable by treatment with a calcium salt. The DNA enters the cells, replicates there and makes the cells resistant to tetracycline.

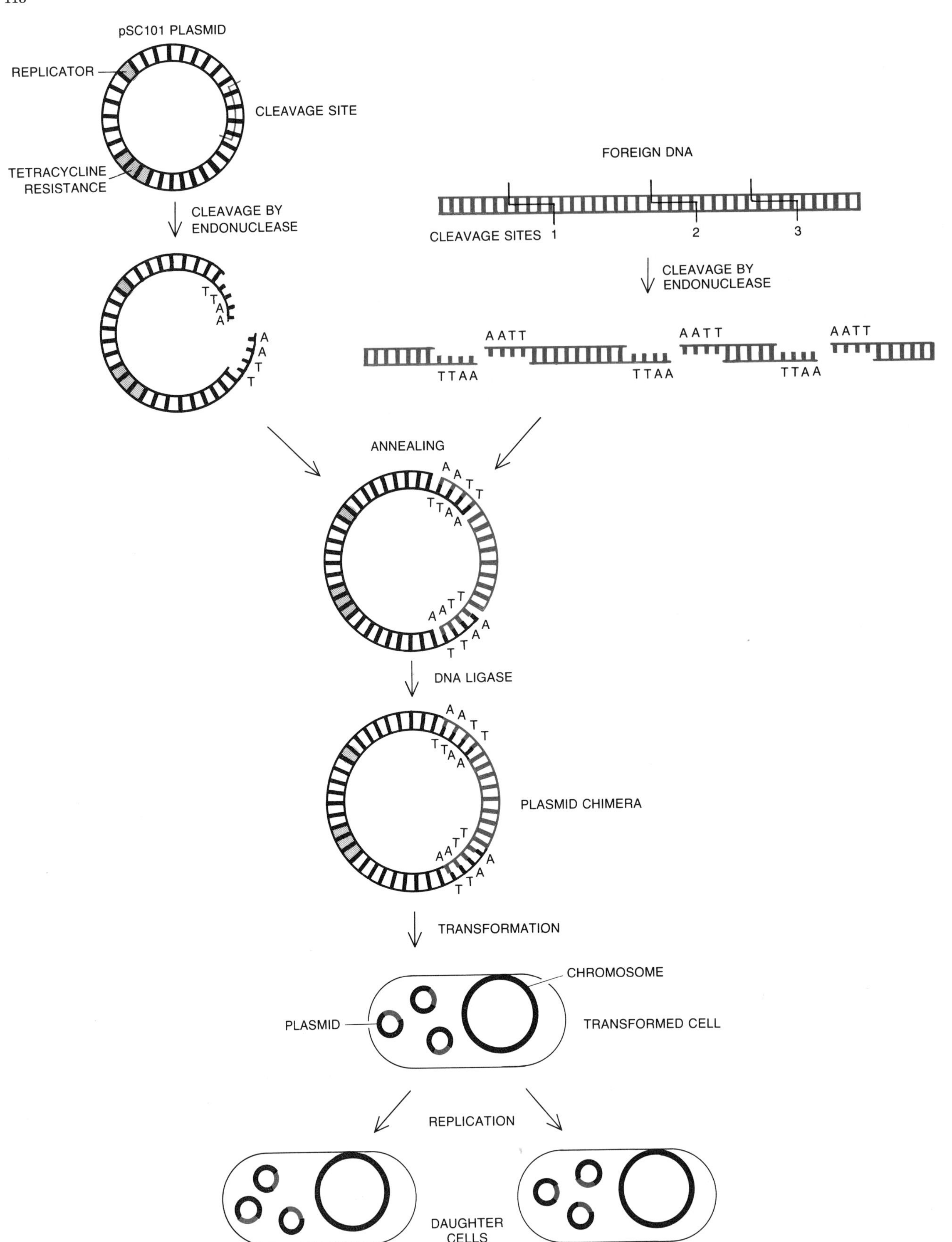

FOREIGN DNA is spliced into the *pSC*101 plasmid and introduced with the plasmid into the bacterium *Escherichia coli.* The plasmid is cleaved by the endonuclease *Eco RI* at a single site that does not interfere with the plasmid's genes for replication or for resistance to tetracycline (*top left*). The nucleotide sequence recognized by *Eco RI* is present also in other DNA, so that a foreign DNA exposed to the endonuclease is cleaved about once in every 4,000 to 16,000 nucleotide pairs on a random basis (*top right*). Fragments of cleaved foreign DNA are annealed to the plasmid DNA by hydrogen bonding of the complementary base pairs, and the new composite molecules are sealed by DNA ligase. The DNA chimeras, each consisting of the entire plasmid and a foreign DNA fragment, are introduced into *E. coli* by transformation, and the foreign DNA is replicated by virtue of the replication functions of the plasmid.

ant to both tetracycline and kanamycin. The plasmids isolated from such transformants contained the entire *pSC*101 DNA segment and also a second DNA fragment that carried the information for kanamycin resistance, although it lacked replication functions of its own. The results meant that the *pSC*101 could serve as a cloning vehicle for introducing at least a nonreplicating segment of a related DNA into *E. coli.* And the procedure was extraordinarily simple.

Could genes from other species be introduced into *E. coli* plasmids, however? There might be genetic signals on foreign DNA that would prevent its propagation or expression in *E. coli.* We decided to try to combine DNA from a plasmid of another bacterium, the *pI*258 plasmid of *Staphylococcus aureus,* with our original *E. coli* plasmid. The staphylococcal plasmid had already been studied in several laboratories; we had found that it was cleaved into four DNA fragments by *Eco RI.* Since *pI*258 was not native to *E. coli* or to related bacteria, it could not on its own propagate in an *E. coli* host. And it was known to carry a gene for resistance to still another antibiotic, penicillin, that would serve as a marker for selecting any transformed clones. (Penicillin resistance, like combined resistance to tetracycline and kanamycin, was already widespread among *E. coli* strains in nature. That was important; if genes from a bacterial species that cannot normally exchange genetic information with the colon bacillus were to be introduced into it, it was essential that they carry only antibiotic-resistance traits that were already prevalent in *E. coli.* Otherwise we would be extending the species' antibiotic-resistance capabilities.)

Chang and I repeated the experiment that had been successful with two kinds of *E. coli* plasmids, but this time we did it with a mixture of the *E. coli*'s *pSC*-101 and the staphylococcal *pI*258: we cleaved the mixed plasmids with *Eco RI* endonuclease, treated them with ligase and then transformed *E. coli.* Next we isolated transformed bacteria that expressed the penicillin resistance coded for by the *S. aureus* plasmid as well as the tetracycline resistance of the *E. coli* plasmid. These doubly resistant cells were found to contain a new DNA species that had the molecular characteristics of the staphylococcal plasmid DNA as well as the characteristics of *pSC*101.

The replication and expression in *E. coli* of genes derived from an organism ordinarily quite unable to exchange genes with *E. coli* represented a breach in the barriers that normally separate biological species. The bulk of the genetic information expressed in the transformed bacteria defined it as *E. coli,* but the transformed cells also carried replicating DNA molecules that had molecular and biological characteristics derived from an unrelated species, *S. aureus.* The fact that the foreign genes were on a plasmid meant that they would be easy to isolate and purify in large quantities for further study. Moreover, there was a possibility that one might introduce genes into the easy-to-grow *E. coli* that specify a wide variety of metabolic or synthesizing functions (such as photosynthesis or antibiotic production) and that are indigenous to other biological classes. Potentially the *pSC*101 plasmid and the molecular-cloning procedure could serve to introduce DNA molecules from complex higher organisms into bacterial hosts, making it possible to apply relatively simple bacterial genetic and biochemical techniques to the study of animal-cell genes.

Could animal-cell genes in fact be introduced into bacteria, and would they replicate there? Boyer, Chang, Helling and I, together with Morrow and Goodman, immediately undertook to find out. We picked certain genes that had been well studied and characterized and were available, purified, in quantity: the genes that code for a precursor of the ribosomes (the structure on which proteins are synthesized) in the toad *Xenopus laevis.* The genes had properties that would enable us to identify them if we succeeded in getting them to propagate in bacteria. The toad DNA was suitable for another reason: although we would be constructing a novel biological combination containing genes from both animal cells and bacteria, we and others expected that no hazard would result from transplanting the highly purified ribosomal genes of a toad.

Unlike the foreign DNA's of our earlier experiments, the toad genes did not express traits (such as antibiotic resistance) that could help us to select bacteria carrying plasmid chimeras. The tetracycline resistance conferred by *pSC*101 would make it possible to select transformed clones, however, and we could then proceed to examine the DNA isolated from such clones to see if any clones contained a foreign DNA having the molecular properties of toad ribosomal DNA. The endonuclease-generated fragments of toad ribosomal DNA have characteristic sizes and base compositions; DNA from the transformed cells could be tested for those characteristics. The genes propagated in bacteria could also be tested for nucleotide-sequence homology with DNA isolated directly from the toad.

When we did the experiment and analyzed the resulting transformed cells, we found that the animal-cell genes were indeed reproducing themselves in generation after generation of bacteria by means of the plasmid's replication functions. In addition, the nucleotide sequences of the toad DNA were being transcribed into an RNA product in the bacterial cells.

Within a very few months after the first DNA-cloning experiments the procedure was being used in a number of laboratories to clone bacterial and animal-cell DNA from a variety of sources. Soon two plasmids other than *pSC*101 were discovered that have a single *Eco RI* cleavage site at a location that does not interfere with essential genes. One of these plasmids is present in many copies in the bacterial cell, making it possible to "amplify," or multiply many times, any DNA fragments linked to it. Investigators at the University of Edinburgh and at Stanford went on to develop mutants of the virus lambda (which ordinarily infects *E. coli*) that made the virus too an effective cloning vehicle. Other restriction endonucleases were discovered that also make cohesive termini but that cleave DNA at different sites from the *Eco RI* enzymes, so that chromosomes can now be taken apart and put together in various ways.

The investigative possibilities of DNA cloning are already being explored intensively. Some workers have isolated from complex chromosomes certain regions that are implicated in particular functions such as replication. Others are making plasmids to order with specific properties that should clarify aspects of extrachromosomal-DNA biology that have been hard to study. The organization of complex chromosomes, such as those of the fruit fly *Drosophila,* is being studied by cloning the animal genes in bacteria. Within the past few months methods have been developed for selectively cloning specific genes of higher organisms through the use of radioactively labeled RNA probes: instead of purifying the genes to be studied before introducing them into bacteria, one can transform bacteria with a heterogeneous population of animal-cell DNA and then isolate those genes that produce a particular species of RNA. It is also possible to isolate groups of genes that are expressed concurrently at a particular stage in the animal's development.

The potential seems to be even broader. Gene manipulation opens the prospect of constructing bacterial cells, which can be grown easily and inexpensively, that will synthesize a variety of biologically produced substances such as antibiotics and hormones, or enzymes that can convert sunlight directly into food substances or usable energy. Perhaps it even provides an experimental basis for introducing new genetic information into plant or animal cells.

It has been clear from the beginning of experimentation in molecular cloning that the construction of some kinds of novel gene combinations may have a potential for biological hazard, and the scientific community has moved quickly to make certain that research in genetic manipulation would not endanger the public. For a time after our initial experiments the *pSC*101 plasmid was the only vehicle known to be suitable for cloning foreign DNA in *E. coli,* and our colleagues asked for supplies with which to pursue studies we knew were of major scientific and medical importance. Investigators normally facilitate the free exchange of bacteria and other experimental strains they have isolated or developed, but Chang and I were concerned that manipulation of certain genes could give rise to novel organisms whose infectious properties and ecological effects could not be predicted. In agreeing to provide the plasmid we therefore asked for assurance that our colleagues would neither introduce tumor viruses into bacteria nor create antibiotic-resistance combinations that were not already present in nature; we also asked the recipients not to send the plasmid on to other laboratories, so that we could keep track of its distribution.

When still other cloning vehicles were discovered, it became apparent that a more general mechanism for ensuring experimental safety in gene-manipulation research was advisable. The groundwork for such control had been established earlier: the National Academy of Sciences had been urged to consider the "possibility that potentially biohazardous consequences might result from widespread or injudicious use" of these techniques and had asked Paul Berg to form an advisory committee that would consider the issue. Berg too had been concerned about the potential hazards of certain kinds of experimentation for some years, and had himself decided to abandon plans to try to introduce genes from the tumor virus SV40 into bacteria because of the possible danger if the experiment were successful.

Berg brought together a number of investigators, including some who were then directly involved in molecular cloning, in the spring of 1974. In a report released in July and in a letter to leading professional journals the members of the committee expressed their "concern about the possible unfortunate consequences of indiscriminate application" of the techniques and formally asked all investigators to join them in voluntarily deferring two types of experiments (which had, as a matter of fact, been avoided by informal consensus up until that time). Experiments of Type I involved the construction of novel organisms containing combinations of toxin-producing capabilities or of antibiotic-resistance genes not found in nature. Type 2 experiments involved the introduction of DNA from tumor viruses or other animal viruses into bacteria; the committee noted that "such recombinant molecules might be more easily disseminated to bacterial populations in humans and other species, and might thus increase the incidence of cancer or other diseases."

The Academy committee was concerned largely because of our inability to assess the hazards of certain experiments accurately before the experiments were undertaken. Guidelines for safety had long been available in other areas of potentially hazardous research, such as studies involving known disease-causing bacteria and viruses, radioactive isotopes or toxic chemicals. Because of the newness of the microbial gene-manipulation methods, no such guidelines had yet been developed for work in this area, however; there was the possibility that potentially hazardous experiments might proceed before appropriate guidelines could be considered and implemented. We recognized that most work with the new methods did not and would not involve experiments of a hazardous nature but we recommended the deferral of Type I and Type II experiments until the hazards were more carefully assessed, until it was determined whether or not the work could be undertaken safely and until adequate safety precautions were available. The committee also proposed that an international meeting be held early in 1975 to consider the matter more fully.

Such a meeting was held in February at the Asilomar Conference Center near Pacific Grove, Calif. It brought together 86 American biologists and 53 investigators from 16 other countries, who spent three and a half days reviewing progress in the field of molecular cloning and formulating guidelines that would allow most types of new hereditary characteristics to be introduced into bacteria and

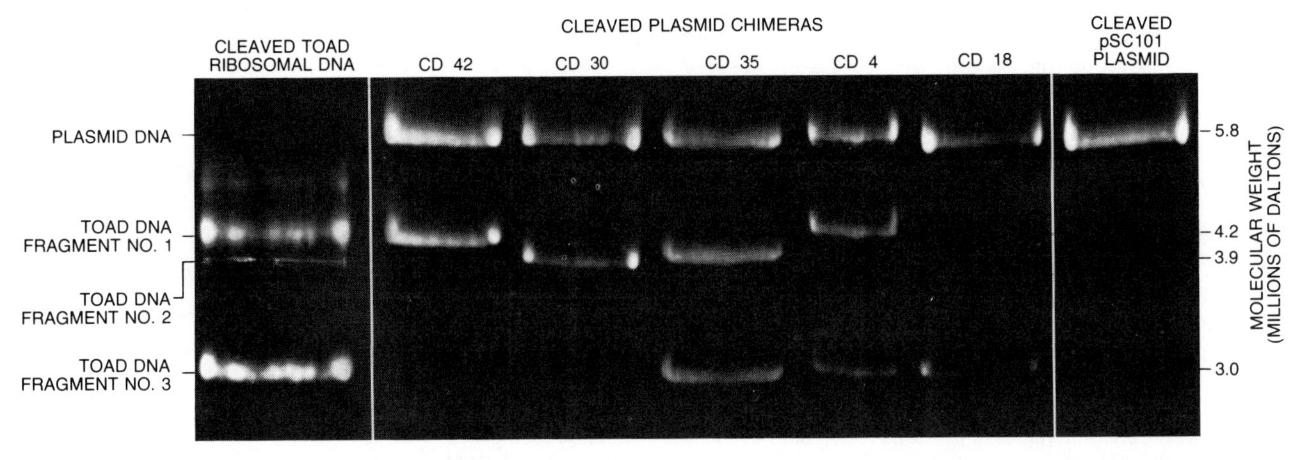

GEL ELECTROPHORESIS demonstrates the presence of toad DNA in chimeric plasmids. Fragments of DNA migrate through a gel at different rates under the influence of an electric current, depending on their size. Linear molecules of plasmid DNA (*right*) and the cleavage products of toad ribosomal DNA (*left*) therefore have characteristic sizes and migrate characteristic distances in a given time. The bands of DNA, visualized by a fluorescent dye, are photographed in ultraviolet. All five chimeric plasmids (*center*) contain a plasmid DNA molecule; in addition each chimera includes one or more fragments characteristic of original toad DNA.

viruses safely. Invited nonscientists from the fields of law and ethics participated in the discussions and decisions at Asilomar, along with representatives of agencies that provide Federal funds for scientific research; the meetings were open to the press and were fully reported. The issues were complex and there were wide differences of opinion on many of them, but there was consensus on three major points. First, the newly developed cloning methods offer the prospect of dealing with a wide variety of important scientific and medical problems as well as other problems that trouble society, such as environmental pollution and food and energy shortages. Second, the accidental dissemination of certain novel biological combinations may present varying degrees of potential risk. The construction of such combinations should proceed only under a graded series of precautions, principally biological and physical barriers, adequate to prevent the escape of any hazardous organisms; the extent of the actual risk should be explored by experiments conducted under strict containment conditions. Third, some experiments are potentially too hazardous to be carried out for the present, even with the most careful containment. Future research and experience may show that many of the potential hazards considered at the meeting are less serious and less probable than we now suspect. Nevertheless, it was agreed that standards of protection should be high at the beginning and that they can be modified later if the assessment of risk changes.

Physical containment barriers have long been used in the U.S. space-exploration program to minimize the possibility of contamination of the earth by extraterrestrial microbes. Containment procedures are also employed routinely to protect laboratory workers and the public from hazards associated with radioactive isotopes and toxic chemicals and in work with disease-causing bacteria and viruses. The Asilomar meeting formulated the additional concept of biological barriers, which involve fastidious cloning vehicles that are able to propagate only in specialized hosts and equally fastidious bacterial strains that are unable to live except under stringent laboratory conditions.

In the past the scientific community has commonly policed its own actions informally, responding to ethical concerns with self-imposed restraint. Usually, but not always, society at large has also considered the public well-being in determining how knowledge obtained by basic scientific research should be applied. Extensive public scrutiny and open discussion by scientists and nonscientists of the possible risks and benefits of a particular line of basic research has been rare, however, when (as in this case) the hazards in question are only potential and, for some experiments, even hypothetical. As this article is being written it is still too early to know what the long-range outcome of the public discussions initiated by scientists working in genetic manipulation will be. One can hope that the forthright approach and the rigorous standards that have been adopted for research in the cloning of recombinant DNA molecules will promote a sharper focus on other issues relevant to public and environmental safety.

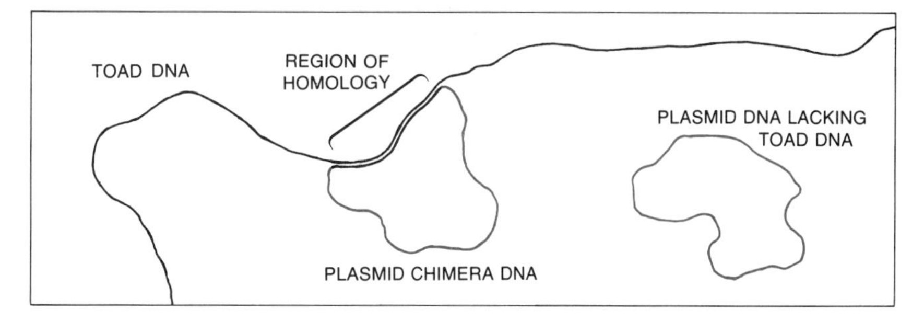

HETERODUPLEX ANALYSIS identifies regions of a toad DNA (*black*) that have been incorporated in a chimeric plasmid DNA molecule. DNA isolated from toad eggs and the DNA of the chimera are denatured, that is, each natural double-strand molecule is split into two single strands of DNA, by alkali treatment. The toad and the chimeric DNA's are mixed together, and any complementary sequences are allowed to find each other. The toad DNA incorporated in the chimeras has nucleotide sequences that are complementary to sequences in the DNA taken directly from the animal source. Those homologous sequences anneal to form heteroduplex double-strand DNA that can be identified in electron micrographs.

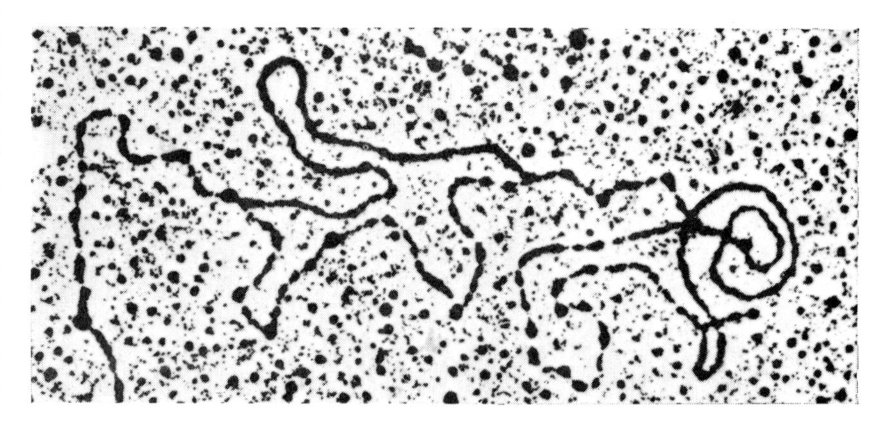

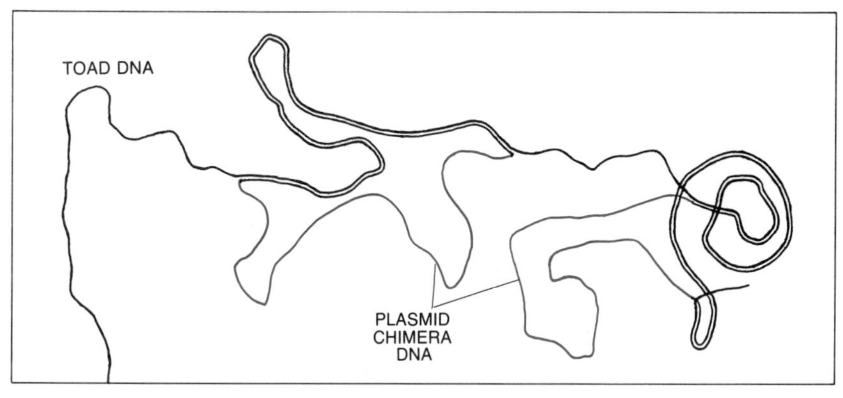

PRESENCE OF TOAD DNA in two separate chimeric plasmid molecules is demonstrated by an electron micrograph made by John F. Morrow at the Stanford University School of Medicine. As is indicated in the drawing (*bottom*), there are DNA strands from two plasmids and a strand of toad DNA. The micrograph shows thickened regions of DNA where nucleotide sequences are homologous and two single strands have been annealed. The toad DNA in the chimeras codes for ribosomes, and the space between the two heteroduplex regions is compatible with the spacing of multiple ribosomal genes in toad DNA.

III

THE CONTROVERSY

III THE CONTROVERSY

INTRODUCTION

No one doubts that genetic engineering can be beneficial. Opponents argue only that the benefits are not worth the potential risk. Furthermore few proponents would state unequivocally that there is *no* risk; only that the probability of a biological accident is so small that it can be ignored. For example, Robin Holliday of the National Institutes of Medical Research, London, roughly calculated the probabilities of various types of biological disasters, using pessimistic values, or high failure rates, for each possible contributing factor in such a disaster. He concluded, ". . . if 10 scientists in each of 100 laboratories carried out 100 experiments per year, the least serious accident would occur an average once in a million years."

But if we are to analyze the debate in any depth, we must first determine what we mean by "benefit." For example, some of the recombinant organisms have obvious practical value but, for most, the application seems limited to basic research. To a scientist this is not an important restriction, since history reveals that it is precisely fundamental research that has yielded the important advances in technology and medicine.

Let us now examine not only the kinds of benefits that have been discussed, but also some of the risks.

Practical Value of Recombinant Organisms. The practical benefits of recombinant DNA research are not hard to envision. For example, a bacterium that consumes oil has been created, and it may be useful in cleaning up oil spills. It is not yet as effective as natural bacteria but the technology for improving it exists. Then, "super" bacteria capable of rapid photosynthesis, efficient composting, or waste removal—though not yet in existence—are easily within the realm of possibility.

Recombinant organisms could also be designed to aid in the alleviation of human diseases. An *E. coli* strain capable of synthesizing insulin would provide a cheap source of the hormone to diabetics. More important though is the possibility of this bacterium producing *human* insulin rather than that of the animal species now used; the human type would have greater effectiveness and cause fewer problems. Human hormones such as the pituitary growth hormone, the world supply of which is totally inadequate for medical needs, might also be synthesized by an appropriately constructed bacterium. (In fact in mid-1977 Herbert Boyer of the University of California Medical Center created an *E. coli* capable of synthesizing somatostatin, a hormone secreted by the hypothalamus of the brain. When this hormone was first purified, 500,000 sheep brains were required to obtain 5 milligrams, which was purified for approximately two gallons of a culture of Boyer's new bacterium—costing about $2.50.) Expensive antibiotics could also be made cheaply from recombinant organisms. Clifford Grobstein gives other examples in his article.

Research Value of Recombinant Organisms. At present, recombinant organisms have hundreds of applications in basic research. For example, in much of the work in molecular biology and biochemistry, a specific protein must be purified in order that its enzymatic or physical properties can be examined. This protein might exist in such a small quantity (e.g., 10 molecules per cell) that thousands of liters of bacterial culture would be necessary to obtain a useful amount of the protein. However if the gene coding for the protein were carried on a λ phage, which can replicate and thus produce several hundred copies of the λ DNA per cell, the amount of protein per cell would be increased enormously. Alternatively the gene could be coupled to a region of a DNA molecule that contains a base sequence specifying initiation of synthesis of a specific RNA (a *promoter* sequence). The regulation of many promoters is so well understood that transcription (and thus synthesis of a particular protein molecule) can be turned off or on at the will of the experimenter. For instance, if the gene were coupled to the promoter for the *E. coli* gene responsible for metabolism of the sugar lactose (the Lac promoter), under appropriate growth conditions, as much as five percent of the total bacterial protein could be the specific protein needed.

Another application of this technology is in the study of the regulation of the synthesis of a mammalian messenger RNA (mRNA) molecule corresponding to a particular gene *X*. In such studies an mRNA molecule is detected by its ability to hybridize to a single-stranded DNA molecule having the complementary base sequence. However the portion of a DNA molecule corresponding to a single gene cannot, by ordinary methods, be isolated in sufficient quantity from a mammalian cell; thus the total cellular DNA is used. However, because mammalian cells contain an enormous number of genes, that fraction of the cellular DNA which is gene *X* is generally very small, often so small that the analysis is not possible. With the use of recombinant DNA techniques, the gene *X* can be joined to an *E. coli* plasmid. The plasmid DNA, which is easily isolated, thus becomes a source of a large amount of DNA, which may be as much as 30 percent gene *X*.

The arrangement of genes in mammalian cells is an important area of current research. Information is easily obtained by means of elegant techniques of electron microscopy, if a few micrograms (10^{-6} gram) of pure DNA are available. A hybrid plasmid such as that mentioned in the preceding paragraph provides an ample supply for this work.

The Potential Hazards and the Debate. Some critics maintain that the benefits of recombinant DNA research do not warrant the risk. They think its sole application is the simplification of tedious and expensive procedures, and that it will not make possible the impossible. One opposition group summarized this sentiment by asking if it makes a significant difference whether a particular problem takes 25 years or 30 years to solve, and if it would not be better to invest the five years in assessing the risk carefully. In short, are the potential risks justified merely because time and money are saved? An obvious ready answer is "no." But *is* the question so simple when in fact speed in fundamental research might produce, say, cures for human disease that much sooner? Furthermore because it is precisely the cost, in both money and time, that prevents the pursuit of many important studies, recombinant DNA techniques might indeed put the impossible within reach.

But all scientists share at least two major concerns. First, just as it is possible to create beneficial organisms, it is also possible to knowingly produce highly destructive and dangerous bacteria. For example, suppose a plasmid carrying a gene for penicillin resistance were introduced into a β-hemolytic *Streptococcus*, a bacterium that in nature never mutates to penicillin resistance. This bacterium causes strep throat, rheumatic fever, scarlet fever, and kidney disease. Penicillin is the most important and effective drug for combating this bacterium, but it would be useless against the recombinant resistant organism. Another dangerous organism that could easily be made is a strain of *E. coli* capable of synthesizing the toxin of diphtheria or of botulism. Thus harmless *E. coli,* which we all harbor in our large intestine, could be converted to a lethal pathogen. The construction of these bacteria would so obviously be disastrous that presumably no intelligent scientist would knowingly make such an organism. Nonetheless, to ensure that it does not happen even unintentionally, regulations banning such experimentation have been enacted. We will discuss some of them later.

Unfortunately some of the bacteria of practical value *might* also be dangerous. For example, if a human were infected with an *E. coli* strain that synthesizes a human hormone, the great outpouring of the hormone in the intestine, unregulated by body chemistry, could seriously upset the body's chemical balance and even cause death. A similar statement has frequently been made about the insulin-producing bacterium currently under construction. This is a bad example though because—as all diabetics who give themselves daily intramuscular insulin injections know—insulin cannot enter the body through the gastrointestinal route.

The second, much more controversial area of concern is that scientists do not really know what kind of organism will result when the DNA molecules from two species, which have been separated for millions of years, are combined. In particular, reservations (carefully examined in Grobstein's article) focus on bacteria in which DNA from animals, plants, or viruses has been introduced. The anxiety stems mostly from the fact that our understanding of gene expression in higher organisms is still primitive at best. Will, for example, a mammalian gene in *E. coli* produce, either directly or indirectly, a carcinogen? Will placement of a plant gene in *E. coli* yield a new bacterium causing plant disease? We know that certain viruses cause respiratory infections in some animals, and tumors in others, but we do not know why. What, then, might happen if a gene from a mouse virus were in a bacterium growing in a human? The answer to each of these questions is that we really do not know.

Some scientists raise less specific objections. For instance, they ask if there is not a reason for the great divergence—that has developed in the course of evolution—between prokaryotes and eukaryotes.* What might happen if these hypothetical ancient barriers were breached? In answer, proponents of recombinant DNA research cite the following points. (1) There is no real evidence for the existence of such barriers. (2) There has been ample possibility in nature for bacteria to take up eukaryote DNA. (3) The blocks to genetic mixing of species and interbreeding probably exist to prevent reproductive waste; that is, hybrid organisms, when they survive, are usually defective or, at the very least, nonfertile. Thus proponents argue that the probability of a biological monster being constructed is very low—especially if precautions are taken. The counterargument is simply that bacteria do grow—that if a dangerous bacterium escapes into nature, it will be there forever.

However, this point too is countered. In order for a bacterium to become a pathogen, its properties must include: (1) the ability to survive in the environment so that it can spread from animal to animal, (2) the ability to multiply within an infected host, (3) mechanisms for penetrating the skin or a mucosal surface, (4) the ability to spread systematically within the host, (5) resistance to the defense mechanisms of the host, and (6) the ability to produce a toxin or some mechanism to damage the host. Although the probability of creating a new organism with all of these properties seems small, this was considered in detail at the Workshop on Studies for Assessment of Potential Risks Associated with Recombinant DNA Experimentation held at Falmouth, Massachusetts

*A prokaryote is a cell that does not have its nucleus bounded by a membrane; eukaryote cells possess nuclear membranes.

in June of 1977. Participants (who included various members of the medical profession involved in infection disease and epidemiological research) heard reports of a large number of experiments in which animals and human volunteers were either exposed or purposely infected with large amounts of *E. coli* K12* under various conditions. Their unanimous conclusion was, ". . . *E. coli* K12 could not be inadvertently converted to an epidemic pathogen by inserted DNA molecules." Some participants went on to state, ". . . even deliberate attempts to produce an epidemic pathogen would require twenty years of full-time effort."

The Safeguards. If these dangers do in fact exist, how do the scientists themselves propose to avoid them? In 1973 at the annual Gordon Research Conference on Nucleic Acids in New Hampshire, the participants of the meeting raised the question of possible hazards in recombinant DNA research. In a letter to the editor of *Science*** they publicly affirmed their concern and stated their intention to hold a conference devoted entirely to this issue. Thus the Committee on Recombinant DNA Molecules was established and, after a preliminary meeting, called for the following voluntary restrictions:

1. DNA-containing information for synthesis of a toxin or for antibiotic resistance shall not be introduced into a bacterium not normally containing either of these.
2. The DNA of animals, animal viruses, and any tumor virus shall not be introduced into any bacterium.

These restrictions were to apply temporarily and a more formal international conference was planned. In the meantime serious debate, in the form of letters to the editors of *Science* (see Additional Readings), ensued. Erwin Chargaff of Columbia University eloquently argued that it is a mistake to use *E. coli* because it grows in humans and suggested seeking another host for recombinant DNA. Others replied that more is known about the genetics and physiology of *E. coli* than any other bacterium and that therefore any potential risk would be minimized by careful application of this knowledge. Furthermore laboratory strains of *E. coli* (in particular, *E. coli* K12) were so weakened by continual growth under laboratory conditions that they were no longer capable of colonization in the human intestine.

In February 1975 the Conference on Recombinant DNA Molecules was held at the Asilomar Conference Center, Pacific Grove, California. It was attended by 134 distinguished scientists from the United States and 18 foreign countries. The principal concern was to assess the safety factor and the various accepted methods for preventing the escape of recombinant organisms into nature. Also, the important new concept of *biological containment* was proposed: that is, host bacteria that, in order to grow, require highly unnatural conditions available only in the laboratory were to be used; thus if they escape to the environment, they have little possibility of growth. At this point the great value of *E. coli* K12 became evident because this strain is so well understood that it could easily be mutated and the appropriate variant selected.

The culmination of the Asilomar meeting was the subsequent development of the so-called NIH Guidelines, which define the advisable safeguards to observe in research with recombinant DNA. These Guidelines provide regulations for government-sponsored research but do not apply to industrial research. They are considered acceptable by the vast majority of biological scientists, but some opposition remains.

The NIH Guidelines establish three restrictions. First, they clearly specify those experiments for which the potential hazard may be so great that they ought not to be attempted at present, even though containment is possible.

E. coli K12 is the particular strain of *E. coli* used in recombinant DNA experiments.

**Volume 81, 1973, p. 114.

Thus the following types of experiments are not to be conducted: (1) the introduction of DNA from pathogenic organisms, from oncogenic viruses, or from cells containing oncogenic viruses into *E. coli* or other bacteria; (2) the formation of recombinant DNA molecules containing genes for the synthesis of toxins such as insect and snake venom or diphtheria and botulism toxin; (3) the creation of plant pathogens that might extend the range of infectivity to other plants; (4) the introduction of drug resistance into any microorganisms that are known not to acquire it in nature so that the effectiveness of the drug in treatment of disease in animals or plants might be reduced; (5) the deliberate introduction into the environment of any organism containing recombinant DNA.

The second restriction the Guidelines establish is the level of physical containment required to prevent or minimize release of bacteria containing recombinant DNA from the laboratory into the environment. The third is the level of biological containment: the Guidelines recommend the use of specific bacteria, plasmids, and phages that have been developed. For example, one permissible bacterium (named X1776 by Roy Curtiss, who isolated it) has nutritional requirements in combinations not found in nature, as well as membrane defects that cause its total dissolution unless certain chemicals are present. Furthermore, to avoid the possibility that a recombinant plasmid or phage DNA might be transferred to an *E. coli* lacking these defects, mutations are induced in the bacterium to reduce transmission of transmissible plasmids. The Guidelines also advise the use of nontransmissible plasmids and of mutant phages that are unable to reproduce.

The details of physical and biological containment and the criteria for selection of the particular level of containment are described in detail in the article by Grobstein.

The Cambridge Affair. An incident that brought the recombinant DNA issue most dramatically to public attention was the action in 1976 of the City Council of Cambridge, Massachusetts, which curtailed the research activities of scientists at Harvard University and the Massachusetts Institute of Technology (MIT), both in Cambridge. The Council declared a temporary moratorium on all recombinant DNA research in Cambridge for six months. The members' reasoning was that, as long as scientists were unable to agree on the safety issue, then they, as laymen unable to judge the validity of the arguments on either side, were obliged to be conservative; that it was better to determine first whether something was safe than to try to undo damage after an accident had occurred. The Harvard and MIT scientists reacted strongly to this decision, fearing that a precedent would be established for government regulation of the type of research that could be done. Thus ensued another series of debates (or confrontations), which has probably not yet ended, although the moratorium has been lifted and replaced by regulations more stringent than the NIH Guidelines. Many scientists argued that the NIH Guidelines are more than adequate, and they may in fact be. However one member of the City Council brought up the problem of self-regulation and, by way of analogy, asked how scientists would respond if industry were to be made responsible for establishing the standards for clean air and water.

To solve the problem, the City of Cambridge created a citizens' committee, the Cambridge Experimentation Review Board. Its responsibilities were to examine the Guidelines and review procedures for monitoring compliance with the safeguards and for handling an accident. The Board consisted of nine members in health care, engineering, urban affairs, business, and local politics. For four months they educated themselves on the biochemistry of recombinant DNA, meeting 20 times with various faculty members at Harvard, MIT, and other schools, with staff members of NIH in Washington and the Communicable Disease Center in Atlanta, and with vehement proponents and opponents of recombinant DNA research. In a final report submitted to the city manager of Cambridge in January 1977, the Board unanimously decided that recombinant DNA research could be carried out in Cambridge under the following conditions: (1) adherence to the NIH Guidelines, which were judged to provide "a sufficient number of safeguards . . . to protect the public against reasonable likelihood of a biohazard," (2) the use of NIH certified host-vector systems of EK-2 level when a P3 facility* is required, (3) continual screening of bacterial cultures for purity, (4) constant monitoring of the survival and escape of host organisms. Items (2) through (4) are more rigid than the guidelines, and were established to protect against "extremely unlikely possibilities." The report has largely resolved the issue of recombinant DNA research in Cambridge.

*See the article by Grobstein for a definition of these terms.

SUGGESTED ADDITIONAL READING FOR PART III

GUIDELINES FOR DNA HYBRID MOLECULES. M. Singer and D. Söll in *Science* Vol. 181, 1973, page 1114.

POTENTIAL BIOHAZARDS OF RECOMBINANT DNA MOLECULES. Committee on Recombinant DNA Molecules in *Science,* Vol. 185, 1974, page 303.

POTENTIAL BIOHAZARDS OF RECOMBINANT DNA MOLECULES. Committee on Recombinant DNA Molecules in *Proc. Nat. Acad. Sci. U.S.,* Vol. 71, 1974, pages 2593–2594.

SUMMARY STATEMENT OF THE ASILOMAR CONFERENCE ON RECOMBINANT DNA MOLECULES. P. Berg, D. Baltimore, S. Brenner, R. Roblin, and M. Singer in *Proc. Nat. Acad. Sci. U.S.,* Vol. 72, 1975, pages 1981–1984.

ASILOMAR CONFERENCE ON RECOMBINANT DNA MOLECULES. P. Berg. D. Baltimore, S. Brenner, R. Roblin, and M. Singer in *Science,* Vol. 188, 1975, pages 991–994.

RECOMBINANT DNA: NIH SETS STRICT RULES TO LAUNCH NEW TECHNOLOGY. N. Wade, in *Science,* Vol. 190, 1975, pages 1175–1179.

ASILOMAR CONFERENCE ON RECOMBINANT DNA MOLECULES. Editorial in *Nature,* Vol. 255, 1975, page 442.

RECOMBINANT DNA: GUIDELINES DEBATED AT PUBLIC HEARING. N. Wade in *Science,* Vol. 191, 1976, pages 834–836.

RECOMBINANT DNA: THE LAST WORD BEFORE THE LEAP. N. Wade in *Science,* Vol. 192, 1976, pages 236–238.

ON THE DANGERS OF GENETIC MEDDLING. E. Chargaff and F. Simring in *Science,* Vol. 192, 1976, pages 938–940.

GENETIC MANIPULATION OF MICRO-ORGANISMS: POTENTIAL BENEFITS AND HAZARDS. R. Curtiss in *Ann. Review of Microbiology,* Vol. 30, 1976, pages 507–533.

EVOLUTION, EPIDEMIOLOGY, AND RECOMBINANT DNA. B. Davis in *Science,* Vol. 193, 1976, page 442.

RECOMBINANT DNA RESEARCH: BEYOND THE NIH GUIDELINES. C. Grobstein in *Science,* Vol. 194, 1976, pages 1133–1135.

ON COUPLING INQUIRY AND WISDOM. R. L. Sinsheimer in *Federation Proceedings,* Vol. 35, 1976, pages 2540–2542.

SEEKING WISDOM IN RECOMBINANT DNA RESEARCH. P. Berg and M. Singer in *Federation Proceedings,* Vol. 35, 1976, pages 2542–2543.

RECOMBINANT DNA RESEARCH, VOLUME 1. DOCUMENTS RELATING TO THE NIH GUIDELINES FOR RESEARCH INVOLVING RECOMBINANT DNA MOLECULES. Department of Health, Education, and Welfare 1976. DHEW Publication No. (NIH)-76-1138. This includes many of the articles published by the Committee on Recombinant DNA Molecules, reports of unpublished meetings, and letters written to the Director of the National Institutes of Health. Available from the Superintendent of Documents, Washington, D.C. Stock #017-040-00398-6.

PLAYING GOD WITH DNA. A Lubow in *New Times,* April 1977, pages 48–53. This is a popular article containing information about the personabilities involved in the controversy and about the Cambridge debate.

GUIDELINES FOR THE USE OF RECOMBINANT DNA MOLECULE TECHNOLOGY IN THE CITY OF CAMBRIDGE. Cambridge Experimentation Review Board, 1977. Recommendations submitted to the City Manager. This remarkable document is available by writing to the office of the City Manager, City Hall, Cambridge, Massachusetts.

SHOULD GENETIC ENGINEERS BE CONTAINED? R. Holliday in *New Scientist*, Vol. 73, 1977, pages 399–401.

RECOMBINANT DNA RESEARCH. S. Wright, P. Reilly, and P. Scheie in *Science*, Vol. 195, 1977, pages 131–133.

RECOMBINANT DNA: FACT AND FACTION. S. Cohen in *Science*, Vol. 195, 1977, pages 654–657.

THE RECOMBINANT DNA DEBATE. Robert May in *Science*, Vol. 198, pages 1144–1145.

BIOHAZARD. Michael Rogers. Knopf, 1977.

PLAYING GOD. GENETIC ENGINEERING AND THE MANIPULATION OF LIFE. June Goodfield. Random House, 1977.

THE ULTIMATE EXPERIMENT. MAN-MADE EVOLUTION. Walker, 1977.

The Recombinant-DNA Debate

13

by Clifford Grobstein
July 1977

The four-year-old controversy over the potential biohazards presented by the gene-splicing method and the effectiveness of plans for their containment is viewed in a broader context

The guidelines for research involving recombinant-DNA molecules issued a year ago by the National Institutes of Health were the culmination of an extraordinary effort at self-regulation on the part of the scientific community. Yet the policy debate over recombinant-DNA research was clearly not laid to rest by the appearance of the NIH guidelines. Instead the debate has escalated in recent months both in intensity and in the range of public involvement. A watershed of sorts was reached in March at a public forum held by the National Academy of Sciences in Washington. The forum was in part a repeat performance by scientists arguing fixed positions that were established early in the debate. There were, however, new participants on the scene, and they presented a varied and rapidly shifting agenda. They made it clear that research with recombinant DNA had become a political issue. As one speaker remarked, the Academy forum may have been the last major public discussion of recombinant DNA arranged by the scientists involved in the research. Nonscientists at the forum, by word and deed, reiterated the theme that science has become too consequential either to be left to the self-regulation of scientists or to be allowed to wear a veil of political chastity.

Science of course is crucially consequential to society, precisely because it is an intensifying source of both benefits and risks. Research with recombinant DNA may provide major new social benefits of uncertain magnitude: more effective and cheaper pharmaceutical products; better understanding of the causes of cancer; more abundant food crops; even new approaches to the energy problem. These and other possible outcomes are envisioned in "best-case scenarios" for the future application of recombinant-DNA technology. "Worst-case scenarios" can also be conceived: worldwide epidemics caused by newly created pathogens; the triggering of catastrophic ecological imbalances; new tools for militarists and terrorists; the power to dominate and control the human spirit.

Both the best-case and worst-case scenarios are largely speculative; the gap between them symbolizes the large degree of uncertainty that surrounds this major step forward in molecular genetics. The material basis of biological heredity has been broken into in the past two decades, and it seems as though each of the fragments has acquired a life of its own. In this resulting period of instability fear threatens to override wonder as the implications of the research diffuse more widely. The fear is not so much of any clear and present danger as it is of imagined future hazards. The classic response to such fears is rigid containment: the Great Wall, the Maginot Line, the cold war. All are manifestations of the effort to provide absolute security against unpredictable risks, and yet each generates its own risk. The escalation of the recombinant-DNA debate has a component of this kind of behavior, but there is a more rational component as well.

The first round of the fateful debate began in 1974, when investigators at the leading edge of work in this field declared a voluntary moratorium on several types of experiment judged to be conceivably risky. A set of techniques had been developed that made it possible to cut the long, threadlike molecules of DNA into pieces with the aid of certain enzymes, to recombine the resulting segments of DNA with the DNA of a suitable vector, or carrier, and to reinsert the recombinant into an appropriate host cell to propagate and possibly to function.

The significance of the new develop-

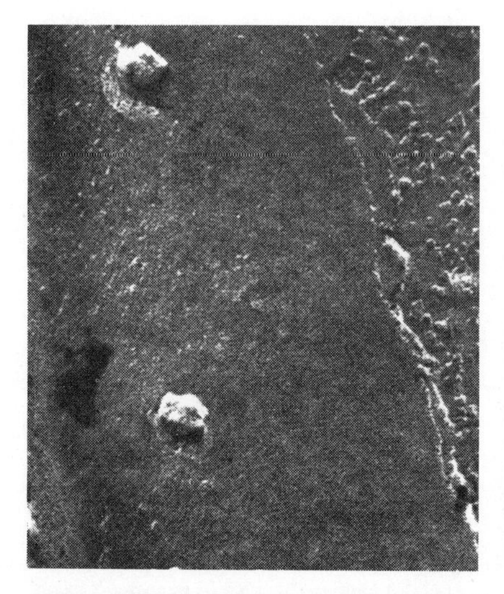

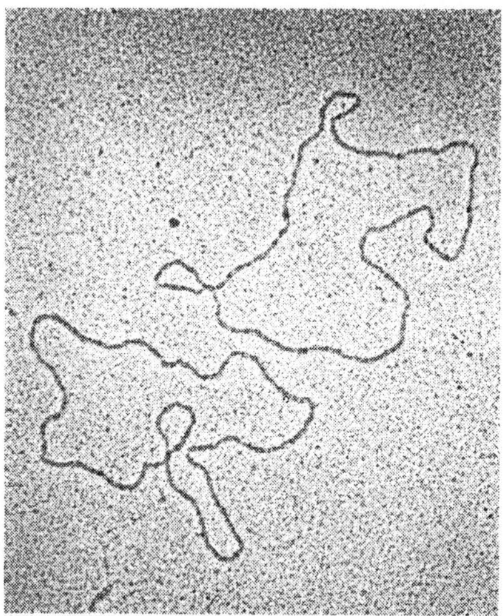

BACTERIOPHAGE ΦX174 AND ITS DNA are portrayed in this pair of electron micrographs. The virus infects the common intestinal bacterium *Escherichia coli*. In the micrograph at left, made by Jack D. Griffith of the Stanford University School of Medicine and Andrew Staehelin of the University of Colorado, two φX174 particles are seen attached to surface of an *E. coli* cell. In micrograph at right, made by Griffith, the DNA molecules of two φX174 viruses are seen in their double-strand form; each molecule is about 18,000 angstroms long.

ments is rooted in the central biological role of DNA as the transmitter of genetic information between generations. The transmission of the encoded genetic message depends on the ability of a cell to generate exact replicas of the parental DNA and to allocate the replicas among the offspring. In addition the success of genetic transmission depends on the ability of the offspring to "express" the encoded information properly by referring to it to control essential life processes. The mechanism of genetic expression in higher organisms is at present only dimly understood, and the discovery of the new recombinant-DNA techniques seemed immediately to open a broad new avenue to increased knowledge in this field.

The detailed mechanisms of genetic replication and expression are enormously complex. The essence of the matter, however, is found in the famous "double helix" structure of DNA. Both of the two long, interwound and complementary strands of the DNA molecule are made up of four kinds of nucleotides, cytosine, guanine, adenine and thymine (abbreviated *C, G, A* and *T*), which are linked end to end like a train of boxcars. The genetic message of each strand is embodied in the particular sequence of nucleotides, any one of which may follow any other. For example, the sequence *CATTACTAG* contains five identifiable English words: *CAT, AT, TACT, ACT* and *TAG*. The genetic message, however, is "written" in triplets: *CAT, TAC* and *TAG*. In general each triplet "codon" determines, through a series of intermediate steps, the position of a specific amino acid in a protein molecule.

Proteins, like nucleic acids, can be visualized as long trains of boxcars coupled end to end; here, however, the subunits are amino acids rather than nucleotides. The sequence of nucleotides in a given DNA molecule determines the sequence of amino acids in a particular protein, with each triple-nucleotide codon placing one of 20 possible amino acids at each successive position in the protein chain. The sequence of amino acids in turn specifically establishes both the structure and the function of the protein. Thus the nucleotide sequence of DNA precisely specifies the protein-building properties of the organism. Moreover, virtually every property of the organism, from enzymatic action to eye color, depends on protein structure in one way or another.

The transmission of the essential genetic information between generations depends on the precise replication of the nucleotide sequences of DNA. The mechanism for replication stems from the complementary relation between the two strands of the DNA molecule. A sequence on one strand (for example *CATTACTAG*) lies immediately opposite a complementary sequence (*GTAATGATC*) on the other strand. The strands are complementary because *C* and *G* are always opposite on the intercoiled strands, as are *A* and *T*. Complementarity depends on the special chemical affinity, or binding, between *C* and *G* on the one hand and *A* and *T* on the other. The sum of these bonds, repeating along the length of the strands, is what holds the strands together in the double helix. Under appropriate conditions affinity is reduced and the two strands can unwind and separate. The single strands can again pair and rewind when conditions for high affinity are restored.

Double-strand DNA replicates by means of an extension of these properties. The unwinding and separation of the strands begins at a localized site along the DNA molecule. In the presence of suitable enzymes and free nucleotides a new chain is formed next to the exposed portion of each unpaired older chain. Each nucleotide lines up next to its opposite number (*C* next to *G, A* next to *T*). The complementary sequence thus established is then linked end to end by an enzyme that closes the nucleotide couplings. When the replication process has traveled along the entire length of the original double helix, two new helixes identical with the first one have been formed. The replication of DNA is the most fundamental chemical reaction in the living world. It fully accounts for the classical first principle of heredity: like begets like.

If DNA replication always worked without error, life would be far more homogeneous than it is. Here, however, a second classical principle of heredity intervenes: the principle of mutational variation, or the appearance in the offspring of new hereditary characteristics not present in the progenitors. Mutations arise through error, at least partly in the replication process. For example, the substitution of one nucleotide by another changes the triplet codon and puts a different amino acid in the corresponding position in the resulting protein. Single-nucleotide errors lead to single-amino-acid errors. Thus, a single-nucleotide error is responsible for the human disease sickle-cell anemia. Most mutations are not such simple, single-nucleotide exchanges; nevertheless, they correlate directly with altered, transposed or deleted nucleotide sequences in DNA. When these changes appear in a gene (that is, a segment of DNA that codes the amino acid sequence of a particular protein), a change in the protein and hence in the hereditary properties it controls is the result.

Therein lies the crux of recombinant-DNA technology. It makes possible for the first time the direct manipulation of nucleotide sequences. Changes in nucleotide sequence that are produced by "natural" errors are random, even when their overall frequency is artificially increased. In natural populations Darwinian selection "chooses" among the random errors, increasing the representation in breeding populations of those errors that lead to more offspring in particular environments. Artificial selection, practiced by human beings for millenniums, favors errors that meet human needs (agricultural breeding) or whims (exotic-pet breeding). The success of both natural and artificial selection, however, is dependent on the random occurrence of desirable mutations. There was no way to direct genetic change itself until recombinant-DNA techniques came along. The new techniques enable one to deliberately introduce known and successful nucleotide sequences from one strain or species into another, thereby conferring a desired property.

The recombinant-DNA approach involves experimental ingenuity and detailed knowledge of the DNA molecule. It begins with an attack on DNA by the proteins called restriction enzymes, which are isolated from bacteria. The enzyme attack breaks the double chain

GENETIC CODE of an extremely small bacterial virus, the bacteriophage designated φX174, is given by the sequence of letters on the opposite page. The letters stand for the four nucleotides cytosine, guanine, adenine and thymine, which are linked end to end to make up each strand of the normally double-strand DNA molecule. The genetic message embodied in each strand of DNA is represented by the particular sequence of nucleotides, any one of which may follow any other. In the φX174 virus the DNA molecule, which has only a single circular strand for part of its life cycle, consists of approximately 5,375 nucleotides; the nucleotides are grouped into nine known genes, which are responsible in turn for coding the amino acid sequences of nine different proteins. For example, the dark-color segment of the molecule, called gene *J*, codes for a small protein that is part of the virus; this segment also happens to be the shortest gene in the φX174 genome. The complete nucleotide sequence for the DNA in φX174 was worked out recently by Frederick Sanger and his colleagues at the British Medical Research Council Laboratory of Molecular Biology in Cambridge. About 2,000 pages of this type would be required to show the nucleotide sequence for the DNA in the chromosome of a typical single-cell bacterium; roughly a million pages would be needed to similarly display the genetic code embodied in DNA molecules that make up chromosomes of a mammalian cell.

GAGTTTTATCGCTTCCATGACGCAGAAGTTAACACTTTCGGATATTTCTGATGAGTCGAAAAATTATCTTGATAAACGAGGAATTACTACTGCTTGTTTA
TCAACTACCGCTTTCCAGCGTTTCATTCTCGAAGAGCTCGACGCGTTCCTATCCAGCTTAAAAGAGTAAAAGGCGGTCGTCAGGTGAAGCTAAATTAAGC
AACGATTCTGTCAAAAACTGACGCGTTGGATGAGGAGAAGTGGCTTAATATGCTTGGCACGTTCGTCAAGGACTGGTTTAGATATGAGTCACATTTTGTT
TGAACTGAGTACTAAAGAATGGATAATCACCAACTTGTCGTAGCCTGAGTCTATCATTAGGTGCGAGAAAATTTTACAGTTGTTCTCTTAGAGATGGTAC
TACTGAACAATCCGTACGTTTCCAGACCGCTTTGGCCTCTATTAAGCTCATTCAGGCTTCTGCCGTTTTGGATTTAACCGAAGATGATTTCGATTTTCTG
TCGCTCCCATAGGATGTTTCAGGTCGCATGGTATTTGCGTTCGGAGTTGCGTCGCTGCTCGTGCTCTCGCCAGTCATCGTTAGGTTTGAAACAATGAGCA
TTCCTGCTCCTGTTGAGTTTATTGCTGCCGTCATTGCTTATTATGTTCATCCCGTCAACATTCAAACGGCCTGTCTCATCATGGAAGGCGCTGAATTTAC
CAGTCATTCTTGCAGTCACAAAGGACGCGCATGTGCGTTCCATTTGCGCTTGTTAAGTCGCCGAAATTGGCCTGCGAGCTGCGGTAATTATTACAAAAGG
GCAGAAGAAAACGTGCGTCAAAAATTACGTGCGGAAGCAGTGATGTAATGTCTAAAGGTAAAAAACGTTCTGGCGCTCGCCCTGGTCGTCCGCAGCCGTT
TTAAATAGGAGTTCATTCCCCGGCTTCGGGGACGTTAATTTTAACAACTGGTGGATGTATGGTTTCTGCTCGCGGAAATGCGAACGGAAATCATGGAGCG
ATGTCTAATATTCAAACTGGCGCCGAGCGTATGCCGCATGACCTTTCCCATCTTGGCTTCCTTGCTGGTCAGATTGGTCGTCTTATTACCATTTCAACTA
TACAGATGTCATCTCAGTTATCGTTCCGGTGCTGCGTTACCTCTTTCTGCCTCTCGCGGTTGCCGCAGGTAGAGCTTCCTCAGCGGTCGCTATTGGCCTC
TTTTACTTTTTATGTCCCTCATCGTCACGTTTATGGTGAACAGTGGATTAAGTTCATGAAGGATGGTGTTAATGCCACTCCTCTCCCGACTGTTAACCAA
GATATCTATAGTTTATTGGGACTTTGTTTACGAATCCCTAAAATAACCATAGTCCCAATTAGCACGGTTCTTTTCGCCGTACCAGTTATATTGGTCATCA
CGTATTTTAAAGCGCCGTGG--ATGCCTGACCGTACCGAGGCTAACCCTAATGAGCTTAATCAAGATGATGCTCGTTATGGTTTCCGTTGCTGCCATCT
TCGTCGAACGTCTGGGTATTACAGTTATCTACACCATCTTCAGCAGTAAACCGCTCTTTCGAGTCAGAGTCCTCCTTCGCCTCGTCAGGTTTACAAAAAC
TATGCTAATTTGCATACTGACCAAGAACGTGATTACTTCATGCAGCGTTACCATGA-GTTATTTCTTCATTTGGAGGTAAAACCTCATATGACGCTGACA
ACTTGTGCTGGTCTTTTGACCGGATTGCTGCAAACCAGTCAAGGTAGTTGTAGTATCGGTCTACGGGTCTCTAATCTCGCGTACTGTTCATTTCCTGCCA
ACAGACCTATAAACATTCTGTGCCGCGTTTCTTTGTTCCTGAGCATGGCACTATGTTTACTCTTGCGCTGGTTCGTTTTCCGCCTACTGCGACTAAAGAG
GGAAGTATCTTTAAAGTGCGCCGCCGTTCAACGGTATGTTTTGTCCCAGCGGTCGTTATAGCCATATTCAGTTTCGTGGAAATCGCAATTCCATGACTTA
ATGTTTTCCGTTCTGGTGATTCGTCTAAGAAGTTTAAGATTGCTGAGGGTCAGTGGTATCGTTATGCGCCTTCGTATGTTTCTCCTGCTTATCACCTTCT
TTGACTTGCTGACTTTGTGACCAGTATTAGTACCAACGCTTATTCATGCGCAAGAACGTTTAGTGGTCTTCCGCCAAGGACTTACTTACCCTTCGGAAGT
GTTGCAGTGGATAGTCTTACCTCATGTGACGTTTATCGCAATCTGCCGACCACTCGCGATTCAATCATGACTTCGTGATAAAAGATTGAGTGTGAGGTTA
TTTCAGACTTTGTACTAATTTGAGGATTCGTCTTTTGGATGGCGCGAAGCGAACCAGTTGGGGAGTCGCCGTTTTTAATTTTAAAAATGGCGAAGCCAAT
TATTTCTCGCCACAATTCAAACTTTTTTTCTGATAAGCTGGTTCTCACTTCTGTTACTCCAGCTTCTTCGGCACCTGTTTTACAGACACCTAAAGCTACA
GGACTAATCGCCGCAACTGTCTACATAGGTAGACTTACGTTACTTCTTTTGGTGGTAATGGTCGTAATTGGCAGTTTGATAGTTTATATTGCAACTGCT
TTGTTTCAGTTGGTGCTGATATTGCTTTTGATGCCGACCCTAAATTTTTTGCCTGTTTGGTTCGCTTTGAGTCTTCTTCGGTTCCGACTACCCTCCCGAC
TAACGGCCCGCATGCCCCTTCCTGCAGTTATCAGTGTGTCAGGAACTGCCATATTATTGGTGGTAGTACCGCTGGTAGGTTTCCTATTTGTAGTATCCGT
AACGTCTACGTTGGTTTCATGGTTTGGTCTAACTTTACCGCTACTAAATGCCGCGGATTGGTTTCGCTGAATCAGGTTATTAAAGAGATTATTTGTCTCC
AACTGGCGGAGGTTTGTTAAATCTGTACCGCGGTGGTCGTTCTCGTCTTCGTTATGGCGGTCGTTATCGTGGTTTGTATTTAGTGGAGTGAATTCACCGA
AAAGCCGCCTCCGGTGGCATTCAAGGTGATGTGCTTGCTACCGATAACAATACTGTAGGCATGGGTGATGCTGGTATTAAATCTGCCATTCAAGGCTCTA
CTTCACGGTCGGACGTTGCATGGAAGTTCTTCAGGAAATGGTCGAAATCGTTATCGTGTGCTTTGTTTTGATCCCCGCCGGAGTAGTCCCAATCCTTGTA
TGCCGTTTCTGATAAGTTGCTTGATTTGGTTGGACTTGGTGGCAAGTCTGCCGCTGATAAAGGAAAGGATACTCGTGATTATCTTGCTGCTGCATTTCCT
GGTCAACGTAAAATCATTCGAGAAAAACTAAGAGTTTAGGCCGCAGTTGGTATGGTCGTCTCCTTCGTAGTCGTGGTCGTGCGAGGGTTCGTAATTCGAG
ACAATCAGAAAGAGATTGCCGAGATGCAAAATGAGACTCAAAAAGAGATTGCTGGCATTCAGTCGGCGACTTCACGCCAGAATACGAAAGACCAGGTATA
CCTTTGGACGACAACGAACCTTTCTAACCACAAAAGGTATTATCTGCGTTGCGTCGTCATCTGAGGAAGACA-CTTATTCGTTCGTAGAGTAAAACACGT
GAGATTATGCGCCAAATGCTTACTCAAGCTCAAACGGCTGGTCAGTATTTTACCAATGACCAAATCAAAGAAATGACTCGCAAGGTTAGTGCTGAGGTTG
GGTCTTCGTCGTAGTCACTGCTGTAATCTTTATAGGAAACGTCATCGCGGTTATACTCTTCTCGGTATGGCGACTAAGACGCAAACGACTACTTGATTCA
TGTGGTTGATATTTTTCATGGTATTGATAAAGCTGTTGCCGATACTTGGAACAATTTCTGGAAAGACGGTAAAGCTGATGGTATTGGCTCTAATTTGTCT
TAAGTCTTCCCATTATTCTTGCTTGGTATTTTTTCGGAGGTTCTAAACCTCCGTACTTTTGTATGTTAACCCTCCCACAGTTAGGACTGCCAATAAAGGA
GTCACGCTGATTATTTTGACTTTGAGCGTATCGAGGCTCTTAAACCTGCTATTGAGGCTTGTGGCATTTCTACTCTTTCTCAATCCCCAATGCTTGGCTT
CGGCAGTTGTATGTATAGTGGTAATAGCTTGAGTTGCGGGACGTATGCTTTTCTGTCTTAGAGAAGGTTCTCGAACTACGCCAATAGGTAGACGAATACC
CATAAGGCTGCTTCTGACGTTCGTGATGAGTTTGTATCTGTTACTGAGAAGTTAATGGATGAATTGGCACAATGCTACAATGTGCTCCCCCAACTTGATA
CCCGCAAGTCGTCGGTCGAACGTTTTGACGCATTGGCAGAAGAGCAAGAGATTTTTGGTAAAAAGCAGGGGAAGCCCCGCCACCAGATATCACAATAATT
TCTTAAGGATATTCGCGATGAGTATAATTACCCCAAAAAGAAAGGTATTAAGGATGAGTGTTCAAGATTGCTGGAGGCCTCCACTAAGATATCGCGTAGA
GATTAGCCAGCAGTCGGTTGCACTCTCACAGTTTTTGCTATTTGGTTGGTAGTCGTACTCGGACAGCGTAACGTAAGTAGTTTGCGACTTATCGTTTCGG
AGGCGTTTTATGATAATCCCAATGCTTTGCGTGACTATTTTCGTGATATTGGTCGTATGGTTCTTGCTGCCGAGGGTCGCAAGGCTAATGATTCACACGC
CGATGGACATCCTTCACAGGCGTATTTCACGTGGCGTACCTTTACTTCTGCCGGTAATCGACATGGTATGAGTCCGTGTGTTTTTATGACTATCGTCAGC
GTTGACCCTAATTTTGGTCGTCGGGTACGCAATCGCCGCCAGTTAAATAGCTTGCAAAATACGTGGCCTTATGGTTACAGTATGCCCATCGCAGTTCGCT
TGTATCTTTGGTTGTCGGTATATTGACCATCGAAATTCGCCGAGTGGAAATCGTAGTTGTCCGGTGTTGGTTGGTCTTGCACTTTTTCGCAGGACGCACA
GGCTAAATACGTTAACAAAAAGTCAGATATGGACCTTGCTGCTAAAGGTCTAGGAGCTAAAGAATGGAACAACTCACTAAAAACCAAGCTGTCGCTACTT
CGAACCATTCAACCTAATTCGTGAGGCACCTGTCTAAACAGTAACACTCGTAAAAGTAGGGCTTCAACGCCGAGTAAGACTAAGACTTGTCGAAGAACCC
TGGGTTACGACGCGACGCCGTTCAACCAGATATTGAAGCAGAACGCAAAAAGAGAGATGAGATTGAGGCTGGGAAAAGTTACTGTAGCCGACGTTTTGGC
ACGTCCAACCTATGCGGTTAGTAAAAATAGCTTCGCGCGTATTTAAACTCGTCTAAACAGCAGTGTCCAACGCGG

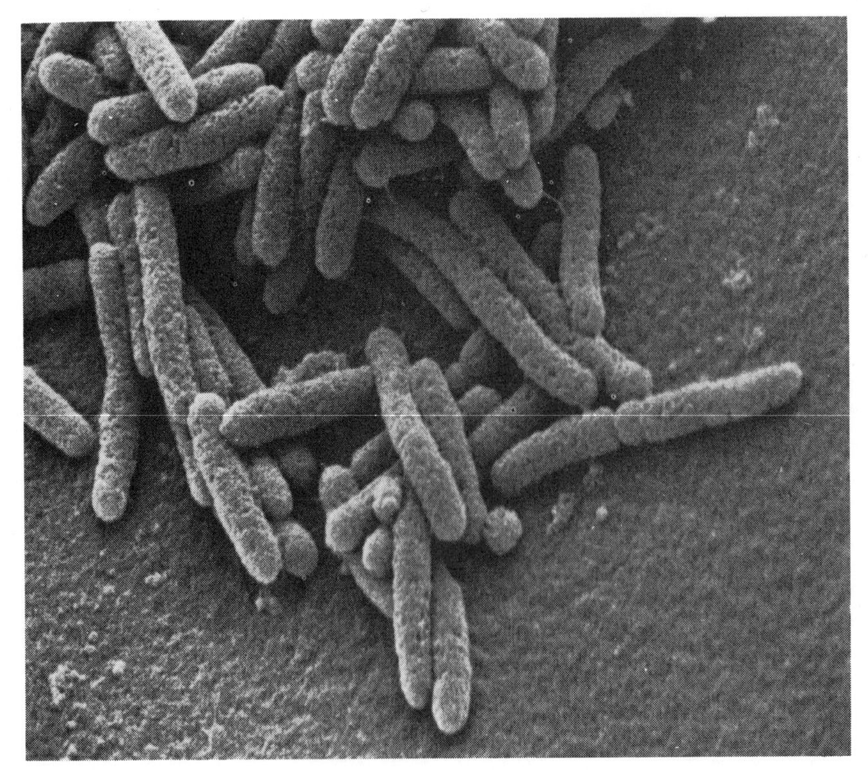

PILE OF E. COLI CELLS appears in this scanning electron micrograph made by David Scharf. Some of the cells have been caught in the act of asexual reproduction (cell division); a few appear to be transferring their DNA by means of the threadlike connection characteristic of the process known as conjugation. *E. coli* bacteria are considered by most investigators to be most suitable host cells for recombinant-DNA experiments. Magnification is 11,000 diameters.

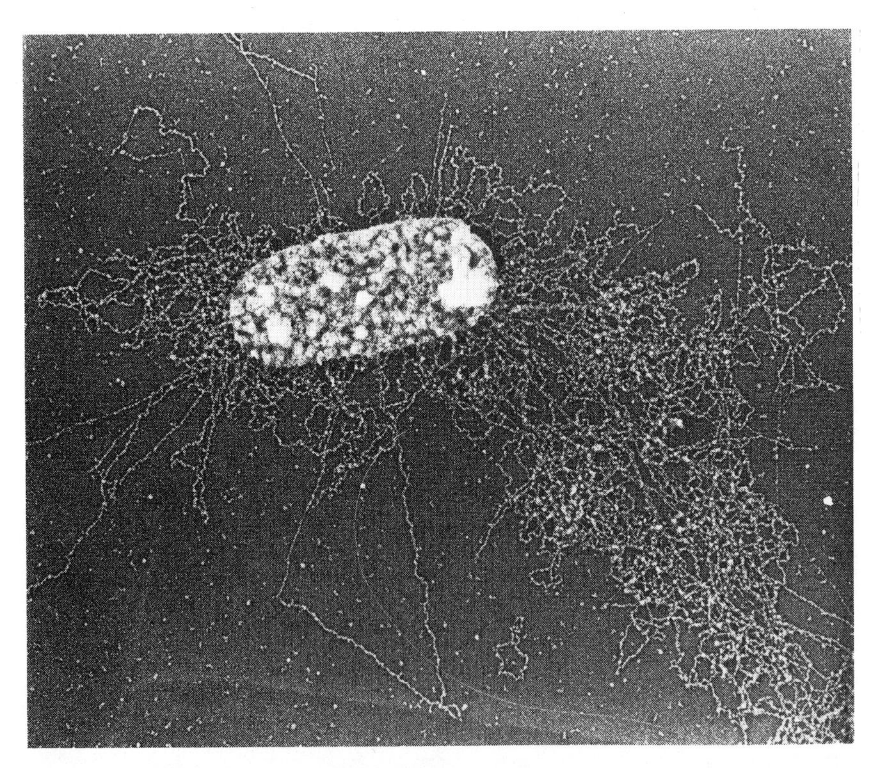

E. COLI SPEWS OUT DNA through its chemically disrupted cell wall in this electron micrograph by Griffith. Most of the DNA is in the form of a single large molecule of double-strand DNA, which constitutes the chromosome of this simple prokaryotic organism. In addition the *E. coli* bacterium may have one or more of the independently replicating loops of DNA known as plasmids; one of these smaller extrachromosomal DNA molecules can be seen near the bottom. Plasmids derived from *E. coli* cells play an important role in recombinant-DNA research, since they form one class of vectors, or carriers, into which segments of "foreign" DNA can be spliced prior to their being reinserted into an appropriate host cell to propagate, thereby duplicating not only their own native nucleotide sequence but also the foreign sequence.

of DNA at particular sequences, say at the sequence *CATTAC*, which is opposite the complementary sequence *GTAATG*. The break does not always occur at the same point on the two strands. It may, for example, be between the two *T*'s in the first strand but just to the right of the *ATG* in the second strand. On separation one piece therefore ends in *TAC*, whereas the other ends in *ATG*. Since the single-strand ends are complementary, they will under suitable conditions stick side by side, and they can then be coupled together end to end. If the same restriction enzyme is used on the DNA from two different sources, both of which have the appropriate target sequence, then sequences with the same "sticky" ends will result. By taking advantage of this stickiness two sequences from any source can be recombined into a single DNA molecule.

The only further step necessary is to put the recombinant DNA into a suitable host organism. The recombinant must have the ability to penetrate the host and become part of its genetic system. An effective way to accomplish this has been developed for the common intestinal bacterium *Escherichia coli*. In addition to its single large circular chromosome the *E. coli* bacterium may have one or more independently replicating, smaller loops of DNA known as plasmids. The plasmids can be isolated from the bacteria, broken open by restriction enzymes and used as one component of a recombinant. After linking up the plasmid DNA with the "foreign" DNA the circular form of the plasmid can be restored and the structure returned to a whole cell. There it can resume replication, duplicating not only its own native sequence but also the foreign one. A strain of bacteria is thus obtained that will yield an indefinite number of copies of the inserted nucleotide sequence from the foreign source.

Standing alone, none of this appears to be particularly momentous or threatening; it is only a new and intriguing kind of chemistry applied to living organisms. Given the complexity of living organisms and the still more complex world of social phenomena, however, this new chemistry quickly builds into varied new potentials, both speculative and real. Suppose, for example, one were to isolate the nucleotide sequence necessary to produce a potent toxin and to transfer it to *E. coli*, usually a harmless inhabitant of every human intestinal tract. Would a dangerous new pathogen be created? Would the transformed *E. coli* release a toxin in the human gut? Might such a new pathogen escape from control and induce epidemics? Questions of this kind have answers, but they take time to find. To gain some time for reflection investigators in 1974 called for a partial and temporary moratorium on those experiments thought to be potentially the riskiest. The separation of the certainly safe experiments from the less certainly safe ones became the chief function of the guidelines released by the NIH in June, 1976. The guidelines, which replaced the temporary moratorium, were derived from worst-case analyses of various kinds of experiments; the object was to evaluate the possible range of hazards and to prescribe appropriate matching safeguards in order to minimize the unknown risks. The guidelines assigned heavy responsibility to individual investigators, and they buttressed this responsibility with special monitoring committees in the sponsoring institutions and in the funding agency.

If such regulations have been adopted, why is debate continuing? Briefly, it is because the matching of estimated risk and prescribed containment adopted by the guidelines is regarded by critics as being inadequate in dealing with potential biohazards and incomplete in failing to address other important issues. The most vocal critics have presented their own worst-case analyses in the scientific and general press. These accounts have led to widespread alarm and to public-policy deliberations at the level of local communities, states and the Federal Government. The expressed concerns of the critics have generated a revised agenda for what is now emerging as a broadened second round of policymaking.

Potential biohazards and estimated degrees of risk continue to dominate the debate. The NIH guidelines balance the estimated risk of a given experiment and recommend specific measures for containing the risks. (Risk, it must be remembered, means possible danger, not demonstrated danger.) Those experiments judged to present an excessive risk are entirely proscribed. At the other end of the spectrum experiments judged to present an insignificant risk require only the safeguards of good laboratory practice. Between these extremes the guidelines establish various levels of estimated risk and prescribe combinations of suitably increasing physical and biological containment. The release into the environment of any recombinant organisms is forbidden.

Unfortunately, given the growing but still limited state of knowledge, wide disagreement is possible, both as to estimated degrees of risk and as to the efficacy of the proposed containment. Some critics project fragmentary information into the inevitable spread of dangerous, newly created organisms, threatening both the public health and the environment. Some defenders project the same fragmentary information to the conclusion that the NIH guidelines are already overly cautious. They believe the actual hazard under existing precautions will turn out to be no greater than that routinely faced in the use of automobiles, jet aircraft and other accepted technologies. The wide range of estimates is possible because of the multiplicity of conceivable experiments and because experience and critical data are inadequate for certainty on many points. One fact that is certain is that no known untoward event has yet resulted from recombinant-DNA research.

What emerges on the new policy agenda, then, is the need for effective policy-oriented research to reduce the current uncertainty as to the risk of particular kinds of experiments. For example, there is dispute over the use of *E. coli* as a host for recombinant DNA. One side argues that scientists must be mad to pick a normal human inhabitant (and a sometime human pathogen) to serve as a host for recombinant DNA. This view, in extreme form, demands the suspension of all recombinant-DNA research until an organism safer than *E. coli* can be found. The other side argues (1) that the vast amount of information available on *E. coli* makes it invaluable, (2) that the *K*-12 strain of *E. coli* actually used in laboratory research has been so modified genetically in adapting to laboratory conditions that it survives only with difficulty in the human intestine and (3) that new strains of *K*-12 have been developed with additional genetic deficiencies that will make survival outside of laboratory conditions essentially impossible. The use of such genetically deficient strains is what is meant by the term "biological containment." The concept is supported by proponents of the research as an efficacious new approach to safety and derided by critics as likely to be circumvented by natural recombination.

Such differences of opinion are normally reduced by scientists to experimental questions. For example, the suitability of the *K*-12 strain of *E. coli* as an experimental organism can be judged only from the effect of recombinant genes on the ecological relations of *E. coli* within the human intestine, including the degree of success of recombinant strains in competing with other strains of *E. coli* and with other organisms. Information on these matters is growing. Such questions, however, are not normally subjects of profound scientific interest. They have recently become matters of priority only because they may provide information that would be useful in arriving at a policy decision. Research on policy-oriented questions has never had a very high status among scientists engaged in basic research or even among those engaged in applied research. Therefore policy-oriented research must be encouraged through special funding mechanisms and through suitable new institutional arrangements. A regulatory agency for recombinant-DNA research and other conceivably hazardous kinds of research is urgently needed outside the NIH, and it should include a research component. The Center for Disease Control and its National Institute for Occupational Safety and Health come

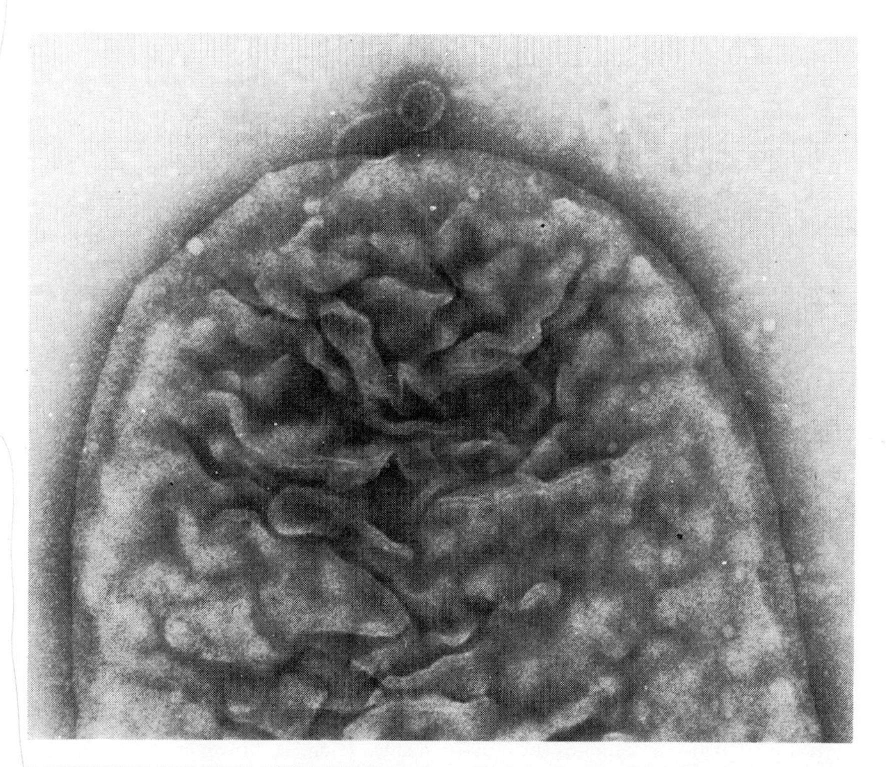

BACTERIAL VIRUS IS ATTACHED to the wall of an *E. coli* cell in this electron micrograph made by Maria Schnoss of the Stanford School of Medicine. This particular virus, named bacteriophage lambda, normally infects the bacterium by injecting its DNA into the host cell through a long taillike appendage. The magnification is approximately 140,000 diameters.

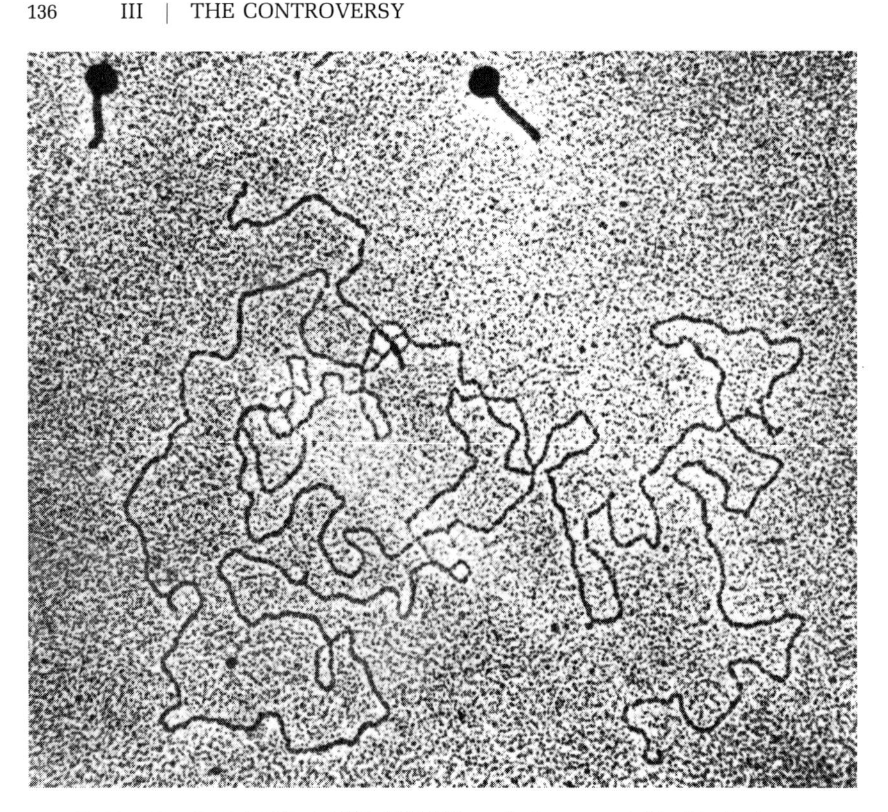

BACTERIOPHAGE LAMBDA AND ITS DNA are both represented in this electron micrograph provided by Griffith. Two complete lambda viruses are at the top; the long double-strand DNA molecule of a disrupted lambda is below them. DNA from bacteriophage lambda can also serve as a vector for recombinant-DNA experiments involving *E. coli* host cells.

to mind as possible models for such a dual-purpose agency.

Also related to the question of biohazards is a controversy over the desirability of centralizing recombinant-DNA research facilities. Some of those who fear severe dangers from recombinant organisms have urged that the potentially more hazardous research be concentrated in remote places with extremely stringent containment procedures. Those who minimize the hazard are opposed to the concentration concept because it would tend to separate the research from the intellectual mainstream and would be unnecessarily expensive in facilities. The argument has been particularly strenuous with respect to experiments requiring *P*3 facilities, which are defined as those necessary to contain "moderate risk" experiments. *P*4 facilities for "high risk" experimentation are expected to be fewer in number because of their high cost; generally speaking they are likely also to be comparatively isolated. The current NIH guidelines provide little direction in these matters. A decision on a firmer policy belongs on the discussion agenda. Particularly urgent is careful consideration of such intermediate possibilities as the use of centralized, high-risk facilities for making particular recombinations for the first time. These activities, together with preliminary testing of new recombinants for possible hazards, might also be carried out by the proposed new regulatory agency.

A special case that emphasizes the advantages of initial testing in a central facility is provided by what are called "shotgun" experiments. These experiments, which offer special advantages to the investigator, may also present special hazards. Shotgun experiments involve exposing the total DNA of a given organism to restriction enzymes in order to obtain many DNA fragments. The fragments are then each recombined with DNA from a suitable vector and the recombinants are randomly reinserted into *E. coli* host cells. The next step is to spread the *E. coli* cells on a nutrient substrate so that each recipient cell, containing a particular inserted foreign sequence, grows into a colony. If the experiment is successful, the yield is a "library" of all the nucleotide sequences of a particular organism, each sequence growing in a separate strain and accessible to manipulation and cross-combination at will.

This experimental approach is laborious but far less so than anything else available for the exploration of the complex genetic systems of higher organisms. There is, however, a risk of unknown magnitude that portions of the DNA with unknown or repressed functions might duplicate and create unanticipated hazards. The result might be particularly unfortunate if the original DNA preparation were to contain genetic material from parasites or from viruses associated with the species under study. Under the NIH guidelines, therefore, shotgun experiments are regarded as being more dangerous than those involving purified and characterized DNA. Experiments in this category are treated as being increasingly more dangerous as the test organism under study is biologically more like the human organism. Thus experiments with primate DNA are considered to be more dangerous than experiments with mouse DNA. This approach appears to represent a reasonable precaution with respect to human health hazards, but it is less reasonable with respect to potential ecological effects. For example, shotgun recombinants involving DNA from plant sources could conceivably lead to ecologically dangerous effects if they were to escape into the environment. Shotgun procedures might therefore be best conducted first in special centralized facilities that could also act as storage and distribution centers for the recombinant products once they had been tested for safety.

These examples suggest several advantages for the creation of a Center for Genetic Resources. The center might not only carry out DNA recombinations suspected to be hazardous but also function to preserve genetic information contained in threatened natural species and in special strains of cells or organisms developed for research and other purposes. Stored genetic information can be expected to be increasingly important in the future. For example, new genetic infusions into domesticated stocks of plants and animals from their wild progenitors have long been used to strengthen the response of the domesticated stocks to changing conditions of husbandry. The sources of wild progenitors are threatened by the reduction of wild habitats all over the world.

The possibility of a biohazard need not arise only as a by-product of basic research. The practical applications of recombinant-DNA techniques, together with the applied research and development leading to them, are at least equally likely sources. For example, recombinant techniques may enormously expand the use of bacteria (and other microorganisms) for the production of certain proteins and other pharmacological products. Microorganisms have long played an essential role in the food, beverage, pharmaceutical and chemical industries, and more precise genetic control of their characteristics has already yielded large benefits. The recombinant-DNA techniques not only offer advances on current practice but also suggest a new realm of "bacterifacture" in which the rapid, controlled growth of microorganisms is coupled to the production of specific products normally made only by higher organisms. Included among the possibilities are the production of insulin, blood-clotting factors and immunological agents. The probability of those possibilities ever being realized is no more easily assessed than the risks, but success in realizing them clearly could provide substantial economic and social benefits. Accordingly entrepreneurial interests have been aroused.

The NIH guidelines are silent on the matter of commercial applications other than stipulating that large-scale experiments (beyond production batches of 10 liters) with recombinants "known to make harmful products" be prohibited unless specially sanctioned. The guidelines also require detailed reporting of proposed recombinant-DNA experiments, a provision that runs counter to the protection of proprietary interest. There have been discussions of these matters between the NIH and representatives of industry. In addition industry spokesmen have testified at Congressional hearings. It is known that some industrial research already is under way and that representatives of industry generally endorse the precautionary approach of the NIH guidelines, but they are resistant to limitations on proprietary rights and on the size of batch production. Moreover, patent policy has come up as an issue and there has been some uncertainty in the Department of Commerce as to how it should be handled. Indeed, the possible commercial applications of recombinant-DNA techniques have yet to be publicly evaluated as a serious policy question, and they must be high on the agenda of the next round of discussions.

The problems of commercial applications lead from immediate issues to broader ones and to a larger time frame. Recombinant-DNA techniques have revived the debate over "genetic engineering" and have once again raised questions about the applications of fundamental biomedical research to technology, to the quality of life and to the future of society. Recombinant DNA has now joined nuclear fission, overpopulation, famine and resource shortages in the doomsday scenarios of "creative pessimism." These issues are even more difficult to deal with objectively than those related to potential biohazards, but they are plainly apparent in the general public discussion and in the public statements of respected scientists.

For example, Robert L. Sinsheimer of the California Institute of Technology has persistently raised issues that are in part practical and in part philosophical. Along with George Wald of Harvard University and Erwin H. Chargaff of the Columbia University College of Physicians and Surgeons, he suggests that the entire recombinant-DNA approach to gaining an understanding of the complexities of higher genetic systems is misbegotten. The argument is not that the approach may not work but that its alleged huge risks are unnecessary because less risky, although slower, means are available. Sinsheimer emphasizes the fundamental difference between simple prokaryotic organisms such as bacteria and complex eukaryotic organisms, including human beings. Prokaryotes, typically one-cell organisms, have a single, comparatively simple chromosome floating freely within the cell body, whereas eukaryotic cells have a nucleus that is bounded by a membrane and contains a number of far more complex chromosomes. The paleontological record suggests that prokaryotes existed on the earth for a billion or more years before the more complex eukaryotes arrived on the scene. Sinsheimer proposes that throughout the evolution of the eukaryotes there has been a genetic barrier between them and the prokaryotes, behind which eukaryotes have developed their more complex mechanisms of genetic control. To transfer these mechanisms, which are possibly the key to the evolutionary success and enormous diversity of eukaryotes, to prokaryotes may introduce, he says, incalculable evolutionary dangers. The prokaryotes may be made far more effective, both as competitors and as parasites, negating an ancient evolutionary strategy.

Sinsheimer's argument has won only a few vocal adherents among biologists, and he himself concedes that it is speculative. Nevertheless, his argument

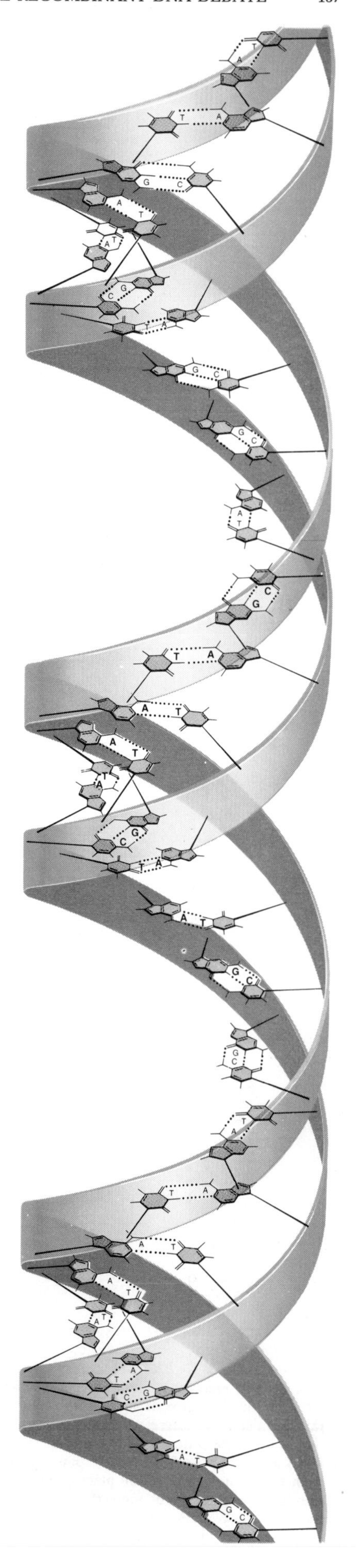

DOUBLE-HELIX STRUCTURE OF DNA is evident in this simplified diagram of a short segment of the deoxyribonucleic acid (DNA) molecule. The sugar and phosphate groups that are linked end to end to form the outer structural "backbones" of the double-strand molecule are represented schematically here by the two helical colored bands. The inner portion of each polynucleotide chain, drawn in somewhat greater detail, consists of a variable sequence of four kinds of bases: two purines (adenine and guanine, or *A* and *G*) and two pyrimidines (thymine and cytosine, or *T* and *C*). The two chains, which run in opposite directions, are held together by hydrogen bonds (*dotted black lines*) between pairs of bases. Adenine is always paired with thymine, and guanine is always paired with cytosine. The planes of the bases are perpendicular to the common axis of the two helixes. The diameter of the double helix is 20 angstroms. Adjacent bases are separated by 3.4 angstroms along the axis and are displaced successively around the axis by an angle of 36 degrees. The structure therefore repeats after 10 bases on each chain (360 degrees), or at intervals of 34 angstroms. The genetic information is stored in the sequence of bases along each chain. In this case the sequence *CATTACTAG* on one strand is identified in boldface type opposite complementary sequence *GTAATGATC* on other strand.

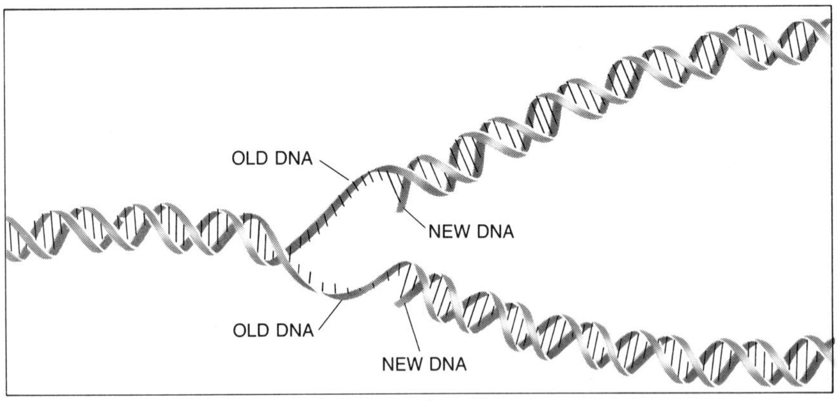

REPLICATION OF DNA depends on the complementary relation between the nucleotide sequences on the two strands of the DNA molecule. Under appropriate chemical conditions the hydrogen bonds between the bases are weakened and the two strands can unwind and separate. In the presence of suitable enzymes and free nucleotides a new chain can be formed next to the exposed portion of each unpaired older chain. The complementary sequence that is formed by each nucleotide lining up next to its opposite is then linked end to end by an enzyme that "zips up" the nucleotide couplings. In this way two new helixes identical with the first can be formed.

has attracted significant public attention, and it is widely cited to support opposition to continued recombinant-DNA research. Bernard D. Davis, a Harvard Medical School microbiologist, has provided a rebuttal, particularly with respect to the concept of a genetic barrier between prokaryotes and eukaryotes. He believes there has been an ample and continuous opportunity for the exchange of DNA between the two groups. He points out that bacteria can take up naked DNA from their immediate environment and that *E. coli* would be exposed to such DNA arising from dead human cells in the human intestine. Microorganisms might similarly take up DNA in the process of decomposing dead animals. Therefore, Davis argues, most recombinants probably have already been tried in the natural evolutionary arena and have been found wanting. Reasoning on analogy with extensive information on pathogenic bacteria, Davis concludes that under the existing NIH guidelines the probability for survival in nature of laboratory-produced prokaryote-eukaryote recombinants is vanishingly small.

This clash of opinion on a major biological issue illustrates the difficulty of assessment of even comparatively value-free questions when critical information is fragmentary. The controversy over the risk-benefit ratio becomes even more intense when issues involve substantial value judgments as well. Here again a concern of skeptics and opponents of recombinant techniques is sharply articulated by Sinsheimer. He asks: "Do we want to assume the basic responsibility for life on this planet? To develop new living forms for our own purposes? Shall we take into our own hands our own future evolution?" Since the questions include such concepts as responsibility, purpose and control of the future, they clearly involve considerations beyond science alone.

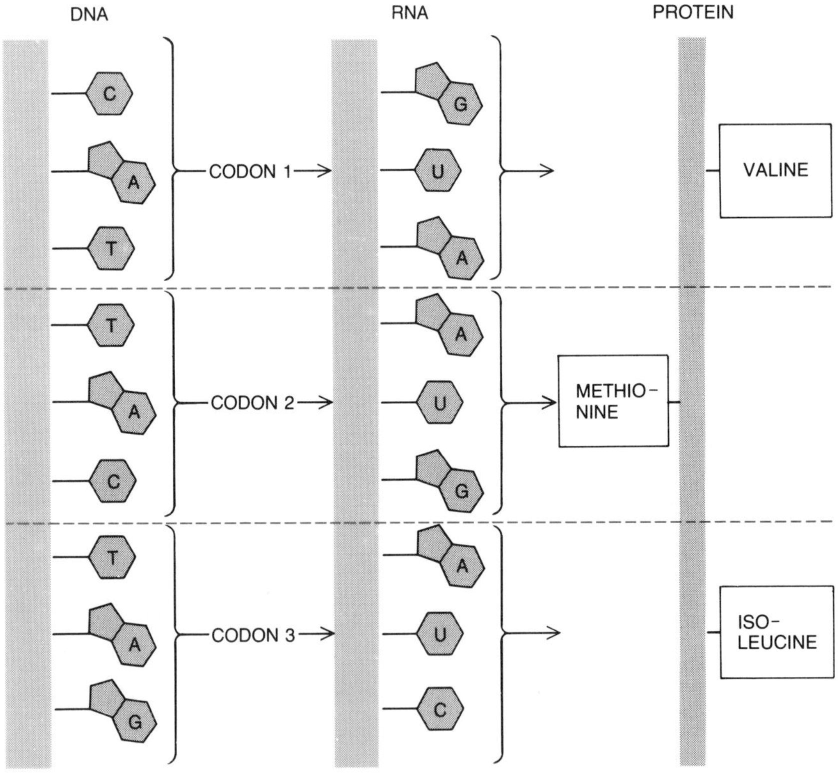

ROLE OF DNA IN PROTEIN SYNTHESIS is suggested by this highly schematic diagram. The genetic message contained in the nucleotide sequence *CATTACTAG*, for example, is "written" in the form of the triplet "codons" *CAT*, *TAC* and *TAG*. Each codon determines, through a series of intermediate steps involving a molecule of ribonucleic acid (RNA), the position of a specific amino acid in a protein molecule. Thus the sequence of nucleotides in a given DNA molecule specifies the corresponding sequence of amino acids in a particular protein, with each triple-nucleotide codon placing one of 20 possible amino acids at each successive position in the protein chain. Since the sequence of amino acids in turn establishes both the structure and the function of the protein, the nucleotide sequence of DNA determines virtually every property of organism. Letter *U* stands for the pyrimidine uracil, a constituent of RNA.

The human species has, of course, been altering life on this planet from the beginnings of human culture. When hunting and gathering gave rise to animal husbandry and agriculture, human choice and purpose began to influence the evolution of selected species. Unconscious human selection was replaced by deliberate plant and animal breeding, and the further development of human culture is now clearly altering the entire ecosystem. Moreover, the biocultural progression of the human species, based partly on human purpose, is undoubtedly altering the human gene pool and will slowly modify the species in unpredictable ways. Nevertheless, the advent of recombinant-DNA techniques has obviously enhanced the prospects for genetic engineering and has restressed the need to assess its implications. Can it be assumed that success in introducing recombinant DNA's into *E. coli* means that there will be similar success in introducing them into the human species? If it can, what is the probable time frame for applying the technique to the human species? Is it accurate and responsible to suggest that we have almost in hand control of "our own future evolution"? These certainly are questions for scientific assessment, and they should have a prominent place on the new policy agenda.

Sinsheimer has gone into still another controversial area, not only for the scientific community but also for the entire society. Arguing that time may be needed to "pace" new genetic knowledge to human capacities for putting nature to intelligent use, he wonders whether "there are certain matters best left unknown, at least for a time." This is high heresy in the scientific community, whose fundamental premise is that the growth of knowledge is the driver and not the captive of other values. The rejection of the concept of "forbidden knowledge" was part of the heroic period at the beginning of modern science, when it included willingness to face the Inquisition and the stake. Having been seared by the nuclear flame and now confronting the more subtle implications of the innermost language of life, 20th-century scientists fear not the stake but the judgment of history. Chargaff, a pioneer in the investigations that led to the decipherment of the genetic language, says: "My generation, or perhaps the one preceding mine, has been the first to engage, under the leadership of the exact sciences, in a destructive colonial warfare against nature. The future will curse us for it."

Sinsheimer and Chargaff, along with a number of philosophers, historians and sociologists of science, are clearly suggesting that the possible consequences of knowing must be consciously included in decisions about the directions of the search for knowledge itself. No issue cuts more deeply to the core of modern science. The self-doubt expressed by some scientists reflects a general questioning in the U.S. of the net benefits of science and technology. Cost-benefit analysis is a current preoccupation, and it is being increasingly applied to the generation of knowledge itself. It is hard enough to assess what we may gain or lose from particular new knowledge; it is even harder to assess the costs of not having it. This problem is epitomized by the recombinant-DNA controversy. The rise of molecular genetics in the U.S. is the direct product of a series of decisions made after World War II that provided funds for biomedical research. The objective was the conquest of the "killer" diseases: cancer, heart disease and stroke. Those diseases are still much with us, although they are better understood and cared for. Meanwhile, out of Federally supported research also came the impetus that led to the discovery of the double helix, the genetic code, the structure of proteins and recombinant DNA. In a classic "double take" the public is now asking whether it has been buying health and well-being or chimeric monsters. Is molecular genetics and all biomedical technology a sorcerer's apprentice? Are we increasing rather than lessening our burden of pain and anxiety?

The last question leads to yet another issue. Biohazard and ecohazard may arise inadvertently, but "sociohazard" may be the product of deliberate malev-

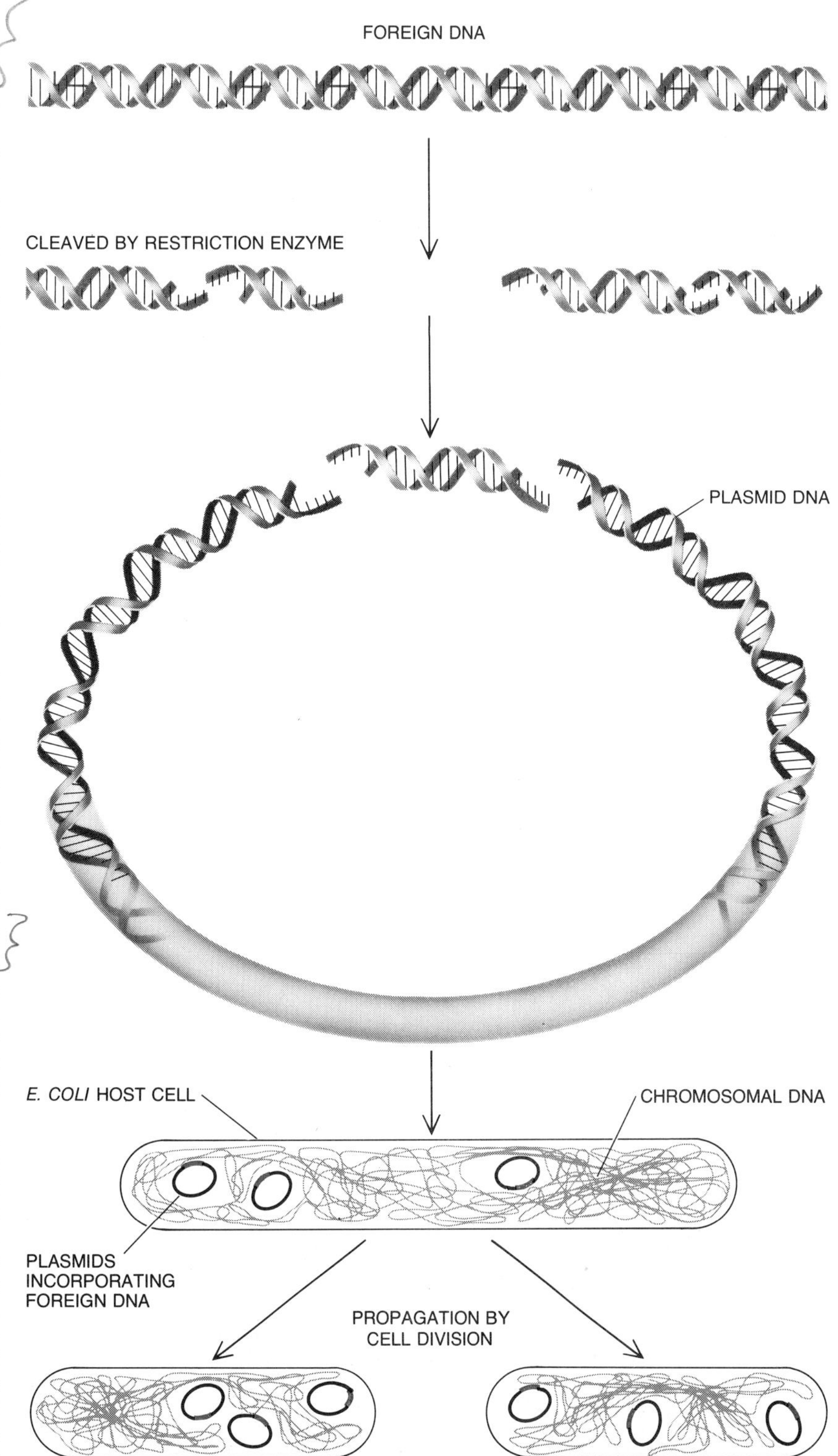

RECOMBINANT-DNA TECHNIQUE makes it possible for the first time to deliberately introduce nucleotide sequences from the DNA of one strain or species of organism into the DNA of another. The DNA of the "foreign" organism is first treated with restriction enzymes, which cleave the double-strand molecule at particular nucleotide sequences (typically thousands of base pairs apart) on a random basis. The same enzyme is then used to cleave the DNA of a suitable vector, in this case a plasmid isolated from *E. coli* bacteria. Since the break caused by the enzyme does not occur at the same point on both strands, the chemical treatment results in a mixture of DNA segments that have complementary single-strand ends. Under suitable conditions the "sticky" ends of two different sequences can be coupled to form a single DNA molecule. For example, after recombining the foreign DNA with the plasmid DNA the circular form of the plasmid can be restored and the structure can be inserted into a suitable host cell (in this case *E. coli*), where the plasmid can resume replication, thereby propagating an indefinite number of "cloned" copies of the inserted nucleotide sequence from the foreign source.

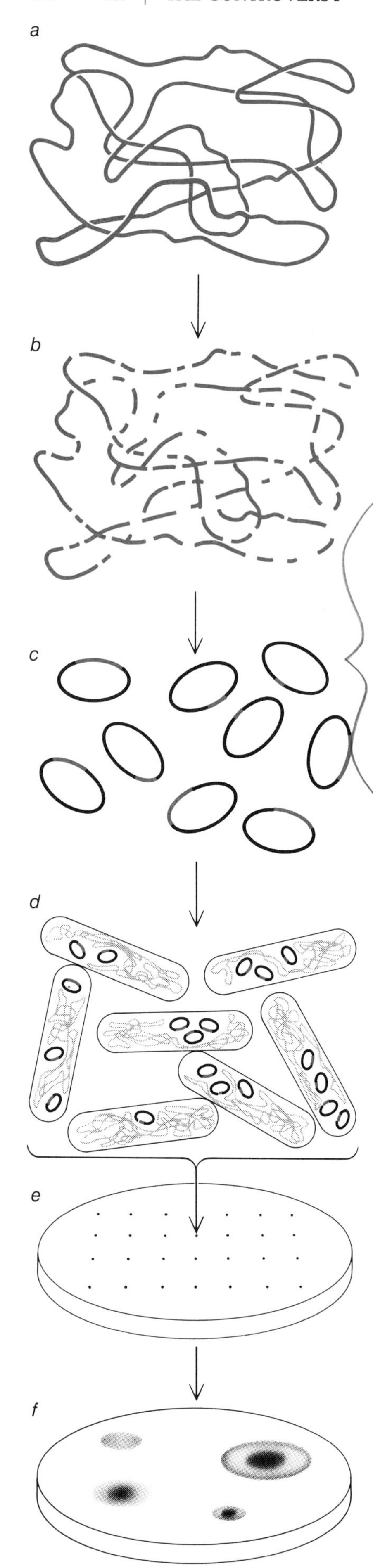

"SHOTGUN" EXPERIMENT is a type of recombinant-DNA experiment in which the total DNA of an organism (*a*) is exposed to restriction enzymes in order to yield many fragments (*b*), which are then recombined with the DNA from a suitable vector (*c*) and randomly reinserted with the vector into the host cells (*d*). The *E. coli* hosts are next spread on a nutrient substrate (*e*) so that each recipient cell, containing a particular inserted foreign nucleotide sequence, can grow into a colony (*f*). The result, if the experiment is successful, is a "library" of all the nucleotide sequences of the organism. Under the guidelines issued by the National Institutes of Health last year shotgun experiments are regarded as being potentially more hazardous than those involving purified and characterized DNA, since it is not known whether portions of the DNA with unknown or repressed functions might cause unexpected problems.

olence. The U.S. is a signatory to an international legal convention that has renounced biological warfare, including research to produce the necessary agents. Not all countries have taken this step, and public renunciation without adequate inspection cannot ensure that covert activities do not exist. Opponents of recombinant-DNA research see its techniques as being ideally suited to serve malevolent purposes, either as agents of organized warfare or of sabotage and terrorism. The techniques do not require large installations or highly sophisticated instrumentation. Contrary views have not denied this but have noted that recombinant-DNA techniques would not be the first technology to have potential malevolent applications. Explosives have such applications, but society does not completely ban them; it takes prudent precautions against their misuse.

Nevertheless, the issue of the possible misuse of recombinant-DNA technology deserves a place on the policy agenda, because it emphasizes the need for international discussion of the implications and management of recombinant-DNA research and recombinant-DNA applications. It can be argued that the U.S. is not ready for such discussion until its own policies are in better order. It is not too early, however, to begin the internal consideration of how best to approach the international arena.

These are the chief issues that have emerged from the policy debate so far. It is a not inconsiderable list. The debate has not been raging on every street corner, but it became strenuous enough in Cambridge last summer to have repercussions across the continent. For example, an evaluation presented by a panel of nonscientists to the Cambridge City Council was not too different in content from one produced by a task force of the Quality of Life Board of the City of San Diego, where I live. Both groups accepted within their community the continuance of recombinant-DNA research requiring *P*3 facilities but sought somewhat greater assurances of safety than those provided by the NIH guidelines. Meanwhile the Attorney General of the State of New York held a public hearing, and a joint hearing was conducted by two committees of the Assembly of the State of California Legislature. Legislation regulating DNA research was later introduced in the California Assembly, and it is still under consideration. Congress has also held several hearings and various items of regulatory legislation have been introduced in both the Senate and the House of Representatives. These local, state and Federal initiatives emphasize the necessity to get on with the policy agenda.

The agenda should be viewed in at least two time frames: immediate and longer range. A consensus has been growing that there is an immediate need to give the quasi regulation represented by the NIH guidelines a statutory base. In particular, regulation must be extended to activities not supported by Federal agencies, especially in the industrial sector. However this is to be done, it is important to maintain flexibility, since the problems to be dealt with will change as greater knowledge and experience are acquired.

Moreover, given the complexity of the longer-term issues, immediate legislation probably should be provisional and limited. A mechanism should be included, however, that actively leads toward a more definitive future policy. This requires provision for a new, comprehensive assessment of all the issues raised by recombinant-DNA research, including the probable effectiveness of the regulatory devices put in place under the NIH guidelines.

The need for such a new national assessment is demonstrated by the nature of the critical challenge to the product of the earlier assessment. First, it has been alleged that the 1975 Asilomar conference establishing the pattern for the NIH guidelines was dominated by scientists involved in the research, and therefore it could not yield a broad enough perspective. Second, it is argued that the earlier assessment was devoted primarily to the question of potential biohazards and did not address in any depth other gravely important questions. The passage of time has added several more points: that circumstances already have changed as research has progressed, that experience has grown and that a wider range of opinion has come to bear on the issue. Whatever format is adopted for the reappraisal of the recombinant-DNA issue, the public must be assured that the process is a comprehensive and objective one.

Whoever undertakes this new national review should first carefully examine the current situation, including the actual effectiveness of the regulatory mecha-

PHYSICAL CONTAINMENT	BIOLOGICAL CONTAINMENT (FOR *E. COLI* HOST SYSTEMS ONLY)		
	*EK*1	*EK*2	*EK*3
*P*1	DNA from nonpathogenic prokaryotes that naturally exchange genes with *E. coli* Plasmid or bacteriophage DNA from host cells that naturally exchange genes with *E. coli*. (If plasmid or bacteriophage genome contains harmful genes or if DNA segment is less than 99 percent pure and characterized, higher levels of containment are required.)		
*P*2	DNA from embryonic or germ-line cells of cold-blooded vertebrates DNA from other cold-blooded animals and lower eukaryotes (except insects maintained in the laboratory for fewer than 10 generations) DNA from plants (except plants containing known pathogens or producing known toxins) DNA from low-risk pathogenic prokaryotes that naturally exchange genes with *E. coli* Organelle DNA from nonprimate eukaryotes. (For organelle DNA that is less than 99 percent pure higher levels of containment are required.)	DNA from nonembryonic cold-blooded vertebrates DNA from moderate-risk pathogenic prokaryotes that naturally exchange genes with *E. coli* DNA from nonpathogenic prokaryotes that do not naturally exchange genes with *E. coli* DNA from plant viruses Organelle DNA from primates. (For organelle DNA that is less than 99 percent pure higher levels of containment are required.) Plasmid or bacteriophage DNA from host cells that do not naturally exchange genes with *E. coli*. (If there is a risk that recombinant will increase pathogenicity or ecological potential of host, higher levels of containment are required.)	
*P*3	DNA from nonpathogenic prokaryotes that do not naturally exchange genes with *E. coli* DNA from plant viruses Plasmid or bacteriophage DNA from host cells that do not naturally exchange genes with *E. coli*. (If there is a risk that recombinant will increase pathogenicity or ecological potential of host, higher levels of containment are required.)	DNA from embryonic primate-tissue or germ-line cells DNA from other mammalian cells DNA from birds DNA from embryonic, nonembryonic or germ-line vertebrate cells (if vertebrate produces a toxin) DNA from moderate-risk pathogenic prokaryotes that do not naturally exchange genes with *E. coli* DNA from animal viruses (if cloned DNA does not contain harmful genes)	DNA from nonembryonic primate tissue DNA from animal viruses (if cloned DNA contains harmful genes)
*P*4		DNA from nonembryonic primate tissue DNA from animal viruses (if cloned DNA contains harmful genes)	

"SHOTGUN" EXPERIMENTS USING *E. COLI K*-12 OR ITS DERIVATIVES AS THE HOST CELL AND PLASMIDS, BACTERIOPHAGES OR OTHER VIRUSES AS THE CLONING VECTORS

EXPERIMENTS IN WHICH PURE, CHARACTERIZED "FOREIGN" GENES CARRIED BY PLASMIDS, BACTERIOPHAGES OR OTHER VIRUSES ARE CLONED IN *E. COLI K*-12 OR ITS DERIVATIVES

SOME EXAMPLES of the physical and biological containment requirements set forth in the NIH guidelines for research involving recombinant-DNA molecules, issued in June, 1976, are given in this table. The guidelines, which replaced the partial moratorium that limited such research for the preceding two years, are based on "worst case" estimates of the potential risks associated with various classes of recombinant-DNA experiments. Certain experiments are banned, such as those involving DNA from known high-risk pathogens; other experiments, such as those involving DNA from organisms that are known to exchange genes with *E. coli* in nature, require only the safeguards of good laboratory practice (physical-containment level *P*1) and the use of the standard *K*-12 laboratory strain of *E. coli* (biological-containment level *EK*1). Between these extremes the NIH guidelines prescribe appropriate combinations of increasing physical and biological containment for increasing levels of estimated risk. (In this table containment increases from upper left to lower right.) Thus physical-containment levels *P*2, *P*3 and *P*4 correspond respectively to minimum isolation, moderate isolation and maximum isolation. Biological-containment level *EK*2 refers to the use of new "crippled" strains of *K*-12 incorporating various genetic defects designed to make the cells' survival outside of laboratory conditions essentially impossible. Level *EK*3 is reserved for an *EK*2-level host-vector system that has successfully passed additional field-testing. Because of the very limited availability of *P*4 facilities and because no bacterial host-vector system has yet been certified by the NIH as satisfying the *EK*3 criteria, the recombinant-DNA experiments now in progress in the U.S. with *E. coli* host systems are with a few exceptions limited to those in the unshaded boxes. Experiments with animal-virus host systems (currently only the polyoma and SV40 viruses) require either the *P*3 or the *P*4 level of physical containment. Experiments with plant-virus host systems have special physical-containment requirements that are analogous to the *P*1-to-*P*4 system.

nisms provided by the NIH guidelines. Particular attention needs to be paid to the local institutional biohazards committees mandated by the NIH guidelines. Beyond the responsibility assigned to the principal investigator these committees are the only source of local surveillance and standard-setting. Their composition and charge are unique, yet their authority and procedures are stipulated only generally in the NIH guidelines. They may well need the stimulus and support of external interests to carry out their important task. Moreover, no provision has been made for budgeting what may turn out to be their considerable cost for technical surveillance, personnel training and medical monitoring. Like all insurance, security against biohazard must be bought. The cost should be borne as an additional expense of the research, not as a competitor for existing funds.

Similarly, the actual performance of the NIH study sections, which are mandated by the guidelines to be independent evaluators of biohazards and containment, needs to be examined. Study sections are already heavily overloaded with the job of evaluating scientific quality. Yet these part-time peer groups are asked to assume another difficult function. If the responsibility is to be taken seriously, it too will entail additional costs.

Of special importance for early attention is an effective monitoring system for following the actual directions of recombinant-DNA research. The techniques involved are so rich in possibilities, whether for fundamental research or applications, for benefit or risk, that "early warning" is essential. Systematic following of the directions of investigators' interests, from applications for support through informal communication to formal publication, is essential to the early detection and assessment of either risks or opportunities. Needless to say, monitoring is particularly difficult in industrial research. It might therefore be desirable to limit or postpone certain development efforts pending closer study and greater knowledge of the underlying problems.

Equally urgent is a determined effort toward a more effective assessment of risks and their limitation. The specific assignment of responsibility for this kind of policy-oriented research should be an early recommendation of the body undertaking the reassessment. Given the differing perspectives required by regulation and the NIH mission to promote health-related research, the regulatory function probably belongs elsewhere in the long run. On the other hand, given the need for careful study of the implications of relying on existing agencies or of establishing a new one, the temporary continued assignment of this responsibility to the NIH may be desirable. This interim solution, if it is adopted, must be accompanied by additional funding to carry it out effectively.

Considerations of biohazards and physical and biological containment have necessarily had a high priority in this early phase of recombinant-DNA research. Many informed observers believe, however, that these concerns will decline in importance as research continues and experience grows. Therefore although the current furor makes a rational approach to the biohazards question an essential part of any successful recombinant-DNA policy, this approach does not exhaust the longer-term requirements and may even distort them. More crucial in the long run may be several other issues that have been raised directly or indirectly.

For example, in investing in fundamental genetic research that can profit from recombinant-DNA techniques, what relative priorities should be assigned to potential applications? In the past the national strategy in biomedical research has been to invest directly in basic research, without declared objectives, while also investing in specific objectives, allowing some of the latter support to "trickle down" to basic research. Thus an investigator of the interaction of viruses and cells, say, might be alternatively or simultaneously supported by funds for fundamental investigation and by funds intended for promoting the development of an effective therapy for cancer. What should be the priorities among possible practical applications of molecular genetics? Competing lines of inquiry include the microbiological synthesis of drugs, specific human gene therapies, the improved efficiency of photosynthesis, nitrogen fixation by food crops, enhanced agricultural production and so on. There are quite different potential risks and benefits in each of these directions, and all are unlikely to be maximally supported at once. In the new areas that are opening up is a new research strategy called for? If it is, by what procedures should it be formulated and how should it be implemented?

It is widely recognized that there is a logical continuum running from basic research through applied research and development to technological application. It is also recognized that movement along this continuum is neither smooth nor fully predictable and that varying motivations and institutional arrangements operate along its length. Recombinant-DNA techniques are the product of fundamental investigation, supported almost entirely by the partnership of the Federal Government and the universities. For the moment, at least, the techniques are likely to remain useful primarily in that area. The techniques may also be useful for various industrial purposes, however. Given the nature of the original investment as well as the complex issues raised, should technological uses, at least for a time, be kept under Federal control? Should some of the return from successful applications be employed to recycle the original investment of Federal resources? Should this promising new technology be a prototype for establishing a revolving capital fund to support a more stably financed basic-research effort?

The possibilities of genetic engineering and evolutionary control illustrate the fundamental dilemmas raised by the new capabilities conferred by scientific knowledge. Society has entered an age of intervention, in which the automatic operation of natural processes is increasingly, through informed intervention, brought consciously into the orbit of human purpose. Many events that humanity formerly could regard only as a boon or a scourge—an act of God or of nature—are now the partial product of human decision and intervention. If human beings do not have the capability today to invent new organisms or to initiate life itself, they may soon have that capability. If they cannot today consciously and fully control the behavior of large ecosystems, that power is not far beyond what has already been achieved. The humility of individuals understandably shrinks from awesome powers that were earlier assigned to divine will. It was not, however, the humility of individuals that conferred these emerging capabilities or is called on to control them today. It was the social interaction of individuals, operating through social institutions, that brought us to the present fateful decision making. Imperfect though they are, our social institutions built the platform for the age of intervention.

The policy challenge we face, refracted in the exquisite structure and potential of the double helix, is whether we can create institutions able to transform the fruits of an age of reason into the achievements of an age of intervention. There are voices today urging us not only to eschew conscious intervention but also to distrust and limit the uses and consequences of reason itself. Perhaps it needs to be restated that it was, after all, natural selection that evoked the double helix and all it conveys. Included among the products are human knowledge and judgment, to which has now passed the duty of designing social processes and structures that can cope with the manipulability of the double helix itself.

The concept and control of the double helix signal a new frontier of biocultural progression. A stereoscopic vision that includes both "creative pessimism" and "creative optimism" is now required. Neither alone can do justice to the profound revelations human beings have recently experienced. A single eye is particularly limited in yielding depth and perspective. For the age of intervention at least two are needed.

EPILOGUE

We have seen that precautions have been taken to ensure that a potentially hazardous organism cannot escape from the laboratory and become established in the environment. Most scientists believe these safeguards to be more than adequate (and some consider them quite bothersome); yet, some critics still ask: Can our planet afford even a single mistake?

More detailed information on the arguments is contained in the references for the introduction to Section III. Even if one finds it impossible to reach an unequivocal decision, it is hoped that this volume will facilitate the forming of a thoughtful opinion. As eloquently stated by the Cambridge Experimentation Review Board in their report, ". . . . we wish also to express our sincere belief that a predominantly lay citizen group can face a technical scientific matter of general and deep public concern, educate itself appropriately to the task, and reach a fair decision."

BIBLIOGRAPHIES

1 BASIC MOLECULAR BIOLOGY

1. Nucleic Acids

THE STRUCTURE OF THE HEREDITARY MATERIAL. F. H. C. Crick in *Scientific American,* Vol. 191, No. 4; October, 1954.

COLD SPRING HARBOR SYMPOSIA ON QUANTITATIVE BIOLOGY, VOL. XXI: GENETIC MECHANISMS: STRUCTURE AND FUNCTION. The Biological Laboratory, Cold Spring Harbor, L.I., New York, 1956.

THE ORGANIZATION AND DUPLICATION OF CHROMOSOMES AS REVEALED BY AUTORADIOGRAPHIC STUDIES USING TRITIUM-LABELED THYMIDINE. J. Herbert Taylor, Philip S. Woods and Walter L. Hughes in *Proceedings of the National Academy of Sciences,* Vol. 43, No. 1, pages 122–128; January, 1957.

A SYMPOSIUM ON THE CHEMICAL BASIS OF HEREDITY. William D. McElroy and Bentley Glass. Johns Hopkins Press, 1957.

2. How Cells Make Molecules

ENZYMATIC SYNTHESIS OF DEOXYRIBONUCLEIC ACID. Arthur Kornberg in *The Harvey Lectures, 1957–1958,* Series 53, pages 83–112; 1959.

HISTORICAL AND CURRENT ASPECTS OF THE PROBLEM OF PROTEIN SYNTHESIS. Paul C. Zamecnik in *The Harvey Lectures, 1958–1959,* Series 54, pages 256–281; 1960.

THE INTERPHASE NUCLEUS. Alfred E. Mirsky and Syozo Osawa in *The Cell,* Vol. II, pages 677–770. Academic Press Inc., 1961.

THE ISOLATION OF SUBCELLULAR COMPONENTS. Vincent C. Allfrey in *The Cell,* Vol. I, pages 193–290. Academic Press Inc., 1959.

3. The Genetic Code: III

THE GENETIC CODE, VOL. XXXI: 1966 COLD SPRING HARBOR SYMPOSIA ON QUANTITATIVE BIOLOGY. Cold Spring Harbor Laboratory of Quantitative Biology, in press.

MOLECULAR BIOLOGY OF THE GENE. James D. Watson. W. A. Benjamin, Inc., 1965.

RNA CODEWORDS AND PROTEIN SYNTHESIS, VII: ON THE GENERAL NATURE OF THE RNA CODE. M. Nirenberg, P. Leder, M. Bernfield, R. Brimacombe, J. Trupin, F. Rottman and C. O'Neal in *Proceedings of the National Academy of Sciences,* Vol. 53, No. 5, pages 1161–1168; May, 1965.

STUDIES ON POLYNUCLEOTIDES, LVI: FURTHER SYNTHESES, IN VITRO, OF COPOLYPEPTIDES CONTAINING TWO AMINO ACIDS IN ALTERNATING SEQUENCE DEPENDENT UPON DNA-LIKE POLYMERS CONTAINING TWO NUCLEOTIDES IN ALTERNATING SEQUENCE. D. S. Jones, S. Nishimura and H. G. Khorana in *Journal of Molecular Biology,* Vol. 16, No. 2, pages 454–472; April, 1966.

4. Ribosomes

RECONSTITUTION OF FUNCTIONALLY ACTIVE RIBOSOMES FROM INACTIVE SUBPARTICLES AND PROTEINS. Keiichi Hosokawa, Robert K. Fujimura and Masayasu Nomura in *Proceedings of the National Academy of Sciences,* Vol. 55, No. 1, pages 198–204; January, 1966.

STRUCTURE AND FUNCTION OF E. COLI RIBOSOMES, V: RECONSTITUTION OF FUNCTIONALLY ACTIVE 30S RIBOSOMAL PARTICLES FROM RNA AND PROTEINS. P. Traub and M. Nomura in *Proceedings of the National Academy of Sciences,* Vol. 59, No. 3, pages 777–784; March, 1968.

STRUCTURE AND FUNCTION OF *Escherichia coli* RIBOSOMES; VI: MECHANISM OF ASSEMBLY OF 30 S RIBOSOMES

5. The Chemical Structure of Proteins

Automatic Recording Apparatus for Use in the Chromatography of Amino Acids. Stanford Moore, Daniel H. Spackman and William H. Stein in *Federation Proceedings*, Vol. 17, No. 4, pages 1107–1115; December, 1958.

The Disulfide Bonds of Ribonuclease. D. H. Spackman, William H. Stein and Stanford Moore in *The Journal of Biological Chemistry*, Vol. 235, No. 3, pages 633–647; March, 1960.

General Properties of Proteins. Joseph S. Fruton and Sofia Simmonds in *General Biochemistry*, pages 14–44. John Wiley & Sons, Inc., 1958.

Molecular Complementarity and Antidotes for Alkylphosphate Poisoning. Irwin B. Wilson in *Federation Proceedings*, Vol. 18, No. 2, Part I, pages 752–758; July, 1959.

The Sequence of the Amino Acid Residues in Performic Acid Oxidized Ribonuclease. C. H. W. Hirs, Stanford Moore and William H. Stein in *The Journal of Biological Chemistry*, Vol. 235, No. 3, pages 648–659; March, 1960.

II THE BIOLOGICAL BASIS OF GENETIC MANIPULATION

6. The Recognition of DNA in Bacteria

Host-induced Modification of T-Even Phages Due to Defective Glucosylation of Their DNA. Stanley Hattman and Toshio Fukasawa in *Proceedings of the National Academy of Sciences of the United States of America*, Vol. 50, No. 2, pages 297–300; August 15, 1963.

General Virology. S. E. Luria and James E. Darnell, Jr. John Wiley & Sons, Inc., 1967.

Restriction of Nonglucosylated T-Even Bacteriophage: Properties of Permissive Mutants of *Escherichia coli* B and K12. Helen R. Revel in *Virology*, Vol. 31, No. 4, pages 688–701; April, 1967.

DNA Modification and Restriction. Werner Arber and Stuart Linn in *Annual Review of Biochemistry*, Vol. 38, pages 467–500; 1969.

Restriction of Nonglucosylated T-Even Bacteriophages by Prophage P1. Helen R. Revel and C. P. Georgopoulos in *Virology*, Vol. 39, No. 1, pages 1–17; September, 1969.

7. The Molecule of Infectious Drug Resistance

Extrachromosomal Inheritance in Bacteria. Richard P. Novick in *Bacteriological Reviews*, Vol. 33, No. 2, pages 210–235; June, 1969.

The Problems of Drug-resistant Pathogenic Bacteria. Conference of the New York Academy of Sciences, edited by Eugene L. Dulaney and Allen I. Laskin in *Annals of the New York Academy of Sciences*, Vol. 182; June 11, 1971.

Molecular Structure of Bacterial Plasmids. Royston C. Clowes in *Bacteriological Reviews*, Vol. 36, No. 3, pages 361–405; September, 1972.

Drug-Resistance Factors and Other Bacterial Plasmids. G. G. Meynell. The MIT Press, 1973.

8. How Viruses Insert Their DNA into the DNA of the Host Cell

The Linear Insertion of a Prophage into the Chromosome of *E. coli* shown by Deletion Mapping. Naomi C. Franklin, William F. Dove and Charles Yanofsky in *Biochemical and Biophysical Research Communications*, Vol. 18, No. 5–6, pages 910–923; March 12, 1965.

Episomes. Allan M. Campbell. Harper & Row, Publishers, 1969.

The Bacteriophage Lambda. Edited by A. D. Hershey. Cold Spring Harbor Laboratory, 1971.

Note on the Structure of Prophage λ. Phillip A. Sharp, Ming-Ta Hsu and Norman Davidson in *Journal of Molecular Biology*, Vol. 71, No. 2, pages 499–501; November 14, 1972.

Integrative Recombination of Bacteriophage Lambda DNA *in vitro*. Howard A. Nash in *Proceedings of the National Academy of Sciences of the United States of America*, Vol. 72, No. 3, pages 1072–1076; March, 1975.

DNA Insertion Elements, Plasmids and Episomes. Edited by A. I. Bukhari, J. Shapiro and S. Adhya. Cold Spring Harbor Laboratory, in press.

9. The Isolation of Genes

Ribosomal Cistrons and the Nucleolar Organizer. H. Wallace and M. L. Birnstiel in *Biochimica et*

Biophysica Acta, Vol. 114, No. 2, pages 296–310; February 21, 1966.

SPECIFIC GENE AMPLIFICATION IN OOCYTES. Donald D. Brown and Igor B. Dawid in *Science*, Vol. 160, No. 3825, pages 272–280; April 19, 1968.

PURIFICATION AND SOME CHARACTERISTICS OF 5S DNA FROM XENOPUS LAEVIS. Donald D. Brown, Pieter C. Wensink and Eddie Jordan in *Proceedings of the National Academy of Sciences of the United States of America*, Vol. 68, No. 12, pages 3175–3179; December, 1971.

A COMPARISON OF THE RIBOSOMAL DNA'S OF XENOPUS LAEVIS AND XENOPUS MULLERI: THE EVOLUTION OF TANDEM GENES. Donald D. Brown, Pieter C. Wensink and Eddie Jordan in *Journal of Molecular Biology*, Vol. 63, No. 1, pages 57–73; January 14, 1972.

10. The Synthesis of DNA

ENZYMATIC SYNTHESIS OF DNA. Arthur Kornberg. John Wiley & Sons, Inc., 1961.

ENZYMATIC SYNTHESIS OF DEOXYRIBONUCLEIC ACID, VIII: FREQUENCIES OF NEAREST NEIGHBOR BASE SEQUENCES IN DEOXYRIBONUCLEIC ACID. John Josse, A. D. Kaiser and Arthur Kornberg in *The Journal of Biological Chemistry*, Vol. 236, No. 3, pages 864–875; March, 1961.

ENZYMATIC SYNTHESIS OF DNA, XXIV: SYNTHESIS OF INFECTIOUS PHAGE φX174 DNA. Mehran Goulian, Arthur Kornberg and Robert L. Sinsheimer in *Proceedings of the National Academy of Sciences*, Vol. 58, No. 6, pages 2321–2328; December 15, 1967.

11. RNA-Directed DNA Synthesis

ONCOGENIC RIBOVIRUSES. Frank Fenner in *The Biology of Animal Viruses: Vol. II*. Academic Press, 1968.

CENTRAL DOGMA OF MOLECULAR BIOLOGY. Francis Crick in *Nature*, Vol. 227, No. 5258, pages 561–563; August 8, 1970.

MECHANISM OF CELL TRANSFORMATION BY RNA TUMOR VIRUSES. Howard M. Temin in *Annual Review of Microbiology: Vol. XXV*. Annual Reviews, Inc., 1971.

THE PROTOVIRUS HYPOTHESIS: SPECULATIONS ON THE SIGNIFICANCE OF RNA-DIRECTED DNA SYNTHESIS FOR NORMAL DEVELOPMENT AND FOR CARCINOGENESIS. Howard M. Temin in *Journal of the National Cancer Institute*, Vol. 46, No. 2, pages III–VII; February, 1971.

12. The Manipulation of Genes

CONSTRUCTION OF BIOLOGICALLY FUNCTIONAL BACTERIAL PLASMIDS *IN VITRO*. Stanley N. Cohen, Annie C. Y. Chang, Herbert W. Boyer and Robert B. Helling in *Proceedings of the National Academy of Sciences of the United States of America*, Vol. 70, No. 11, pages 3240–3244; November, 1973.

REPLICATION AND TRANSCRIPTION OF EUKARYOTIC DNA IN *ESCHERICHIA COLI*. John F. Morrow, Stanley N. Cohen, Annie C. Y. Chang, Herbert W. Boyer, Howard M. Goodman and Robert B. Helling in *Proceedings of the National Academy of Sciences of the United States of America*, Vol. 71, No. 5, pages 1743–1747; May, 1974.

POTENTIAL BIOHAZARDS OF RECOMBINANT DNA MOLECULES. P. Berg et al. in *Proceedings of the National Academy of Sciences of the United States of America*, Vol. 71, No. 7, pages 2595–2599; July, 1974.

FIRST ASM CONFERENCE ON EXTRACHROMOSOMAL ELEMENTS IN BACTERIA in *Microbiology—1974*. American Society for Microbiology, 1975.

REPORT OF THE WORKING PARTY ON THE EXPERIMENTAL MANIPULATION OF MICROORGANISMS. Her Majesty's Stationery Office, London, 1975.

III THE CONTROVERSY

13. The Recombinant-DNA Debate

THE MANIPULATION OF GENES. Stanley N. Cohen in *Scientific American*, Vol. 233, No. 1, pages 24–33; July, 1975.

MOLECULAR BIOLOGY OF THE GENE. J. D. Watson. Addison-Wesley Publishing Co., Inc., 1977.

NATIONAL INSTITUTES OF HEALTH GUIDELINES FOR RESEARCH INVOLVING RECOMBINANT DNA MOLECULES. Public Health Service, U.S. Department of Health, Education, and Welfare, June 23, 1976.

RECOMBINANT DNA RESEARCH: BEYOND THE NIH GUIDELINES. Clifford Grobstein in *Science*, Vol. 194, No. 4270, pages 1133–1135; December 10, 1976.

NOTES OF A BIOLOGY-WATCHER: THE HAZARDS OF SCIENCE. Lewis Thomas in *The New England Journal of Medicine*, Vol. 296, No. 6, pages 324–328; February 10, 1977.

RECOMBINANT DNA: FACT AND FICTION. Stanley N. Cohen in *Science*, Vol. 195, No. 4279, pages 654–657; February 18, 1977.

INDEX